Practical ESM Analysis

For a listing of recent titles in the
Artech House Electronic Warfare Library,
turn to the back of this book.

Practical ESM Analysis

Sue Robertson

ARTECH
HOUSE

BOSTON | LONDON
artechhouse.com

Library of Congress Cataloging-in-Publication Data
A catalog record for this book is available from the U.S. Library of Congress.

British Library Cataloguing in Publication Data
A catalog record for this book is available from the British Library.

ISBN-13: 978-1-63081-528-8

Cover design by John Gomes

685 Canton Street
Norwood, MA 02062

10 9 8 7 6 5 4 3 2 1

Contents

Preface

Electronic warfare (EW) is now extremely relevant worldwide with recognition of the importance of the electromagnetic spectrum in current war zones and for future conflicts. The increasing military interest in EW in almost all countries and the emergence of EW industries across the globe means that there is a need for EW knowledge transfer. Electronic support measures (ESM) are one of the core disciplines of EW, and this book aims to set out the information needed to understand why ESM systems see the world in the way that they do.

Before starting work on ESM systems it is important to know what the radar frequency (RF) environment looks like to an ESM system. It should be noted that radars do not always appear to ESM systems as they are specified by their manufacturers and that ESM systems rarely work as efficiently as their developers would claim.

It is not the aim of this book to describe the hardware and software of ESM systems, but rather to explain why an ESM system sees the RF environment in the way that it does. Some common misconceptions about this subject are explained and ideas for improving the performance of ESM systems are explored. The overriding aim is to provide the ESM analyst with the tools to diagnose system issues, to devise effective tests and trials for ESM equipment, and to manage expectations about the performance of these systems.

This book has been written following many years of data analysis experience on different types of ESM systems mounted on platforms in all physical domains. EW is rightly also considered by modern military strategists to be a domain that is pertinent to all the traditional domains of land, air, sea, subsurface, and space.

In writing this book, I hope to provide ESM analysts with the book that I wish had been available at the start of my EW career.

CHAPTER 1

Introduction to Electronic Support Measures Systems

The domination of the electromagnetic spectrum begins with effective surveillance of both the communications and radar domains, and it is with the radar domain that this book is concerned. Electronic surveillance or electronic support measures (ESM) systems detect, locate, and identify radars. This knowledge can then be used by electronic countermeasures (ECM) systems to disrupt or degrade enemy radars by jamming their transmissions and denying them use of the radar frequency (RF) spectrum for communications and situational awareness. Electronic attack (EA) systems use electromagnetic energy to further disrupt enemy electronic systems.

Before starting to consider ESM systems it is important to address a popular misconception about the density of radar pulses in the RF environment. ESM systems are typically designed to cope with the detection of millions of pulses per second on a routine basis. However, in reality, even in very dense RF environments, there are not more than a few thousand pulses per second available for detection by the ESM system (unless the ESM platform is the subject of a target tracking radar). Most radars are scanning radars and are only detectable by the ESM when they are pointing towards the ESM receiver.

This situation is analogous to a lighthouse illuminating the sea to provide guidance for shipping. A ship can only see the light when the light sweeps past it and does not see it when the light is pointing in any other direction. The ship can see the light from a great distance, but the lighthouse keeper cannot see the ship unless it is at close range. This fact provides another analogy with ESM and radars, as the ESM system can "see" the radar at a far greater range than the radar can detect the ESM platform The radar signal only has to travel in one direction for detection by the ESM system. For the radar to receive its own signal reflected from the ESM platform, the distance traveled by the signal is twice as far. The radar range equation shows that the maximum range from the radar to the ESM system is proportional to the square root of the radar power, whereas the maximum detection range of the radar is proportional to the fourth root of the radar power.

ESM systems all exhibit the same issues that influence their performance, notably direction of arrival (DOA) errors, uncertainty in radar location estimation, multitracking, and ambiguous radar identification. This book addresses these issues and offers solutions for improving ESM system performance. The separation of ESM antennas, often mounted at the extremities of a platform, plays a significant role in the generation of DOA errors. This effect is described in detail in this book along with the accompanying multipath effects. ESM processes such as pulse de-interleaving and radar location estimation each have their own chapter. Conventional wisdom about RF scan strategies for narrowband ESM receivers is challenged and new methods for developing RF scan strategies are proposed.

The unambiguous identification of radar signals is a challenge for even the most sophisticated RF library matching process. The reasons for this difficulty are explored and suggestions for improving the generation of radar databases and methods for successful library matching are given. The visualization of ESM data is important in the analysis of ESM performance. Graphical methods for depicting test and trials data are described together with simulated examples of ESM testing or trial data. The processes to be carried out to prepare for trials are discussed, together with the techniques needed to simulate the RF environment.

A particular problem of ESM systems is multitracking (i.e., where more than one track is created for a single radar). It is almost unheard of for an ESM system to produce one track, and only one track, for each radar. There are many causes of this phenomenon, explained in this book, together with methods for reducing multitracking.

One of the most important factors affecting ESM performance is multipath. Three chapters are devoted to multipath beginning with a classification of reflection type, of which multipath is the most significant. The reasons why multipath affects DOA measurement in all types of ESM systems are explored. Issues such as radar beamwidth, ESM antenna configuration, and the range from the ESM to radar, all factors that exacerbate multipath effects, are discussed.

This book does not contain many equations, relying instead on text and diagrams to explain the principles underlying the effects described in each section. There are many examples of simulated data to back up explanations. Appendixes contain more detailed calculations and include a method of modeling a radar beamwidth and an explanation of reflection coefficients, important in the understanding of multipath effects.

1.1 ESM Systems: Painting a Picture of the RF Environment

An ESM system detects radar pulses and produces tracks on the ESM operator's screen, ideally one for each radar in the RF environment, as shown in Figure 1.1. Some ESM systems produce location fixes for the radars and plot them on a map together with error ellipses to show the confidence with which the location has been calculated.

The inputs to the ESM system are radar pulses in an interleaved stream from all the radars in the RF environment of the ESM platform. Figure 1.2 shows pulses arriving from the individual radars. There are 6 radars in this example, all at different DOAs. Each symbol on the graph represents a group of pulses (an intercept) from a radar. Pulses are seen from scanning radars when the peak of the radar beam sweeps past the ESM system. The symbols on Figure 1.2 for each radar are separated by the radar scan period. The duration for which pulses are seen for each scan depends on radar parameters such as beamwidth and pulse repetition interval (PRI) and on the range from the radar to the ESM platform.

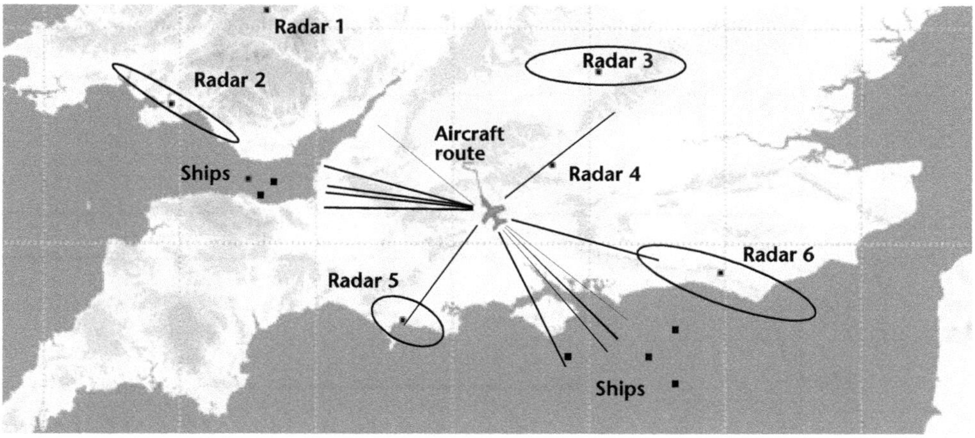

Figure 1.1 The picture of the RF environment created by an ESM system.

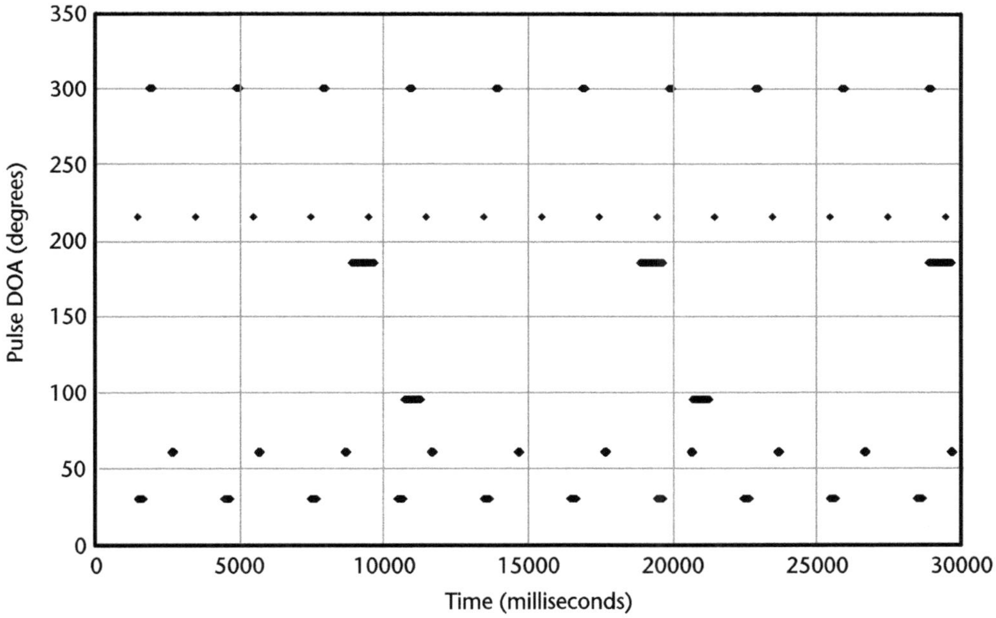

Figure 1.2 ESM pulse data.

The first task of the ESM processor is to deinterleave the pulse data from individual radars and to form intercepts, one for each radar. Intercepts are then correlated into an existing track for each radar held in the ESM track table, or a new track is created if there are no suitable candidate tracks for correlation. Intercept to track correlation takes place using parameters that are measured for the intercept (i.e., RF and DOA and the PRI that is calculated as part of the de-interleaving process). The resultant track data can be plotted as shown in Figure 1.3 where the DOA profiles of the track updates provide a convenient way of looking at the ESM environment.

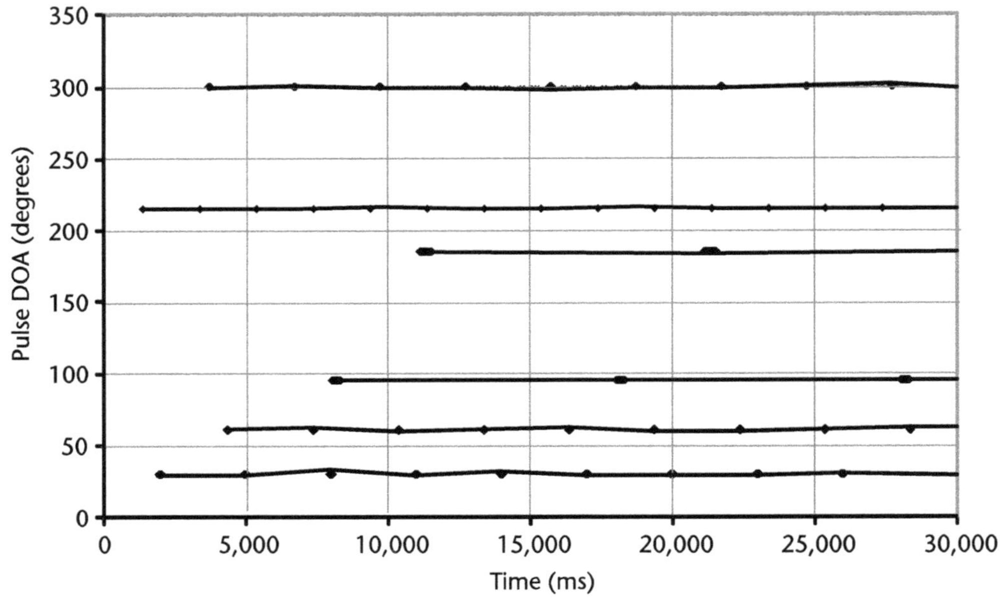

Figure 1.3 ESM track data.

Identification of the radars is a very important task so that threat radars can be found and appropriate countermeasures can be invoked. There are thousands of different types of radar so to uniquely specify each radar in the ESM radar library is not a reasonable objective. Most systems keep several possible identities for each ESM track, and it is the job of experienced ESM operators to make sense of the information that is presented to them by the ESM system.

1.2 ESM as an ELINT System

Electronic surveillance systems were traditionally divided into two categories, those for tactical use called electronic surveillance or ESM systems, and those for gathering data for the population of strategically important databases, known as electronic intelligence (ELINT) systems [1].

The operator of the ESM system relies on algorithms within the system to produce an initial picture of the RF environment and then applies expert interpretation to use the information for mission situation awareness.

The operator of the ELINT system puts more emphasis on collecting data from radars in the RF environment so that accurate characteristics of the radars and their locations can be derived during post-mission analysis. The recording of the received radar pulses is therefore an important feature of an ELINT system.

In recent years the boundary between ESM and ELINT systems has become blurred, with many ESM systems that have pulse data recording systems also being capable of acting as ELINT systems. The techniques for analysing ESM data presented in this book apply equally to ESM and ELINT systems.

1.3 ESM as a Radar Warning Receiver

The ESM system can be used for strategic ELINT purposes and it can also be used in a tactical role as a radar warning receiver (RWR). For an ESM system to be useful as an RWR, its radar library must be programmed with the parameters of specific threat radars. Weapon systems often have separate target acquisition, target tracking, and weapon guidance radars. The ESM system operating as an RWR should be capable of recognizing when one of these radars is present in the RF environment and change its RF scan strategy, if necessary, to look for the other related radars. The ESM operator should be alerted so that appropriate countermeasures can be implemented in a timely manner.

1.4 ESM Operators and Platforms

It has been more than 100 years since the first use of radio receiving equipment to pinpoint the location of German transmitters during World War I. Captain H. J. Round, working for Marconi, began carrying out experiments with direction-finding radio equipment for the army in France in 1915.

After World War II, radio direction finding (RDF), originally applied only in the communications domain, was broadened to encompass systems operating in radar bands. This led to the development of a family of NATO ESM systems, such as the shipboard U.S. AN/WLR-1-AN/WLR-6 equipment and comparable airborne systems.

Modern ESM systems operate in all physical domains and are mounted on all types of vehicles. Man-portable ESM systems are currently under development so that individual war-fighters can have access to information about the RF environment. In conflict situations cooperating platforms and unmanned vehicles share data to allow an accurate picture of the RF environment to be developed.

1.5 Topics Unique to This Book

There are several ESM issues that are unique to this book. Methods of modeling the radar azimuth beamshape and determining optimum RF scan strategies are described. Ideas for the mitigation of multipath effects, multitracking, and radar library pattern matching are all topics that are not covered elsewhere in ESM literature.

1.5.1 Radar Beamshape Calculation

It is essential to be able to model the radar beamshape, especially for scanning radars having narrow beamwidths. For these radars, the emitted pulse amplitude varies greatly from the peak of the radar beam to the troughs between the main beam lobe and the sidelobes. The effect of the radar beamshape is seen on the DOA profile of pulses received by amplitude comparison ESM systems. Chapter 5 describes this effect and the method of simulating the radar beamshape is described in Appendix A.

1.5.2 RF Scan Strategy Generation

ESM systems usually cover the RF range from 2 to 18 GHz as a minimum. Many of them also have the capability to detect signals below 2 GHz and some can detect high-frequency signals of the order of 30 GHz. If the ESM system is not wide open (i.e., it cannot detect all frequencies simultaneously), a strategy for sampling each RF band in turn must be devised. An effective RF scan strategy allows adequate probability of intercept in all RF bands deemed important in a particular mission. A method for implementing an effective RF scan strategy is given in Chapter 4.

1.5.3 Multipath Considerations

Multipath is the most important cause of errors in DOA measurement for all types of ESM system. A detailed description of how multipath affects the pulse amplitude and phase appears in Chapter 15. Chapter 16 gives explanations of how factors such as the radar beamshape, range to the radar, and reflection geometry affect multipath and shows that the effects of multipath are different for short and long ranges to the radar. Chapter 17 discusses measures that can be taken to reduce multipath effects for all three types of DOA measurement: amplitude comparison, time difference of arrival (TDOA), and phase difference systems.

1.5.4 Multitracking Mitigation

There are many causes of multitracking (where more than one track is created for a single radar). DOA errors are a major cause of extra tracks but errors in the calculation of PRI due to pulses being missed during the detection of the pulse sequence also account for multitracking. Chapter 14 offers some techniques for reducing the effects of multitracking.

1.5.5 Radar Identification Improvements

Misidentification of a radar is one of the most serious limitations of ESM systems. The small capacity of current radar libraries in terms of the number of possible mode lines for the description of radars means that they are completely inadequate. Traditional library matching techniques are challenged and methods for improving this important ESM task are discussed in Chapter 10.

Reference

[1] Wiley, R. G., *ELINT: The Interception and Analysis of Radar Signals*, Norwood, MA: Artech House, 2006.

CHAPTER 2

Radar Parameters and How They Affect ESM Systems

There are thousands of diverse radars operating worldwide. Of these, approximately 800 are weapons-related and the rest are used for purposes such as air traffic control, weather determination, shipborne navigation, situation awareness, and border and harbor surveillance.

A knowledge of the likely parameter ranges for the radars and the types of agility that may be exhibited is needed in the design of the deinterleaving, track creation, and identification processes to be carried out by the ESM system. Such information is also useful in the design of a signal that can defeat an ESM. This is because there are some parameter ranges that are more populous than others and some parameter agility levels that are difficult for current ESM systems to determine and associate with a single radar.

2.1 Pulsed and Continuous-Wave Radars

A pulsed radar [1] transmits high-power, high-frequency pulses and waits for the echo of the transmitted pulse before it transmits the next pulse, as shown in Figure 2.1. The choice of pulse repetition frequency determines the range and resolution of the radar. Target ranges and bearings are calculated from the radar antenna position and time of arrival (TOA) of the reflected signal.

Pulsed radars are used for applications such as air traffic control, ship navigation, and border surveillance as well as for threat functions including target acquisition and tracking and missile guidance. Pulsed radars can also be used to measure target velocities. This type of radar is known as a Doppler radar. It uses high a high pulse repetition frequencey (PRF) to avoid velocity ambiguities, but can have numerous range ambiguities.

Continuous-wave (CW) radars continuously transmit a high-frequency signal. The reflected energy is also received and processed continuously. An unmodulated CW radar transmits at constant amplitude and frequency. A CW radar transmitting an unmodulated signal can measure speed by using the Doppler effect. However, it cannot measure range and it cannot differentiate between two reflecting objects.

To measure range, a CW signal must be modulated using frequency shifting. A frequency-modulated CW (FMCW) radar that constantly changes in frequency around a fixed reference is used to detect stationary objects. If the target is moving,

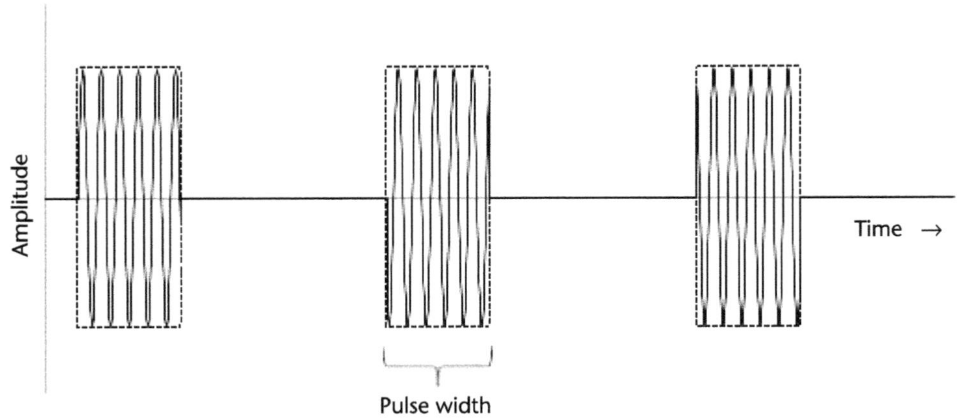

Figure 2.1 Pulsed radar.

the Doppler frequency shift can be used to find out if target is approaching or receding.

2.2 Pulse Descriptor Words

The process of extraction of separate intercepts for each radar from a set of interleaved pulses from several radars and the determination of the identities of the radars relies on the measurement of fundamental parameters of the pulses. RF, pulse width (PW), and amplitude are measured by the ESM for each pulse, together with the pulse TOA and its DOA.

The parameters of each pulse are recorded in pulse descriptor words (PDW), examples of which are given in Table 2.1.

Following the deinterleaving process (described in Chapter 8), PRI, a parameter needed for assigning radar identity is calculated from the TOA difference between pulses received from a single radar.

The recording of PDW data is extremely useful in analyzing the performance of the ESM system and also for populating electronic intelligence (ELINT) databases where information about radars at the pulse data level is essential for the construction of effective radar libraries for ESM systems and radar warning receivers.

Table 2.1 Example of PDW Data

Pulse No.	*TOA*	*RF (MHz)*	*PW (μs)*	*DOA (°)*	*Amplitude (dBmi)*
1	10:01:01.345	3,045	0.7	50	–34.5
2	10:01:01.367	9,340	0.3	120	–38.5
3	10:01:01.389	2,984	1.5	253	–23.5

2.3 Radar Frequency

The frequency of the signal is one of the most important parameters in the identification of a radar. Most radars of interest to the ESM system operate at between 0.4 and 18 GHz. A few higher frequency radars may also operate in the environment of the ESM. When referring to radars, it is usual to talk about different RF bands. There are two main designations of radar bands, IEEE and NATO [2], as shown in Tables 2.2 and 2.3. Both systems are commonly in use, so are defined here, but in this book, the NATO designation is used.

The most populous radar bands are E/F-band and I-band, as shown in Figure 2.2, a histogram of the percentage of radars likely to be operating in each RF band up to 18 GHz.

From Figure 2.2, it can be seen that by far the most populous RF band is I-band, between 9 GHz and 10 GHz. Most of the radars in this RF range have frequencies between 9.25 and 9.45 GHz and are shipborne radars. Radars carrying out such diverse functions as ship navigation, coastal surveillance, and airborne intercept typically operate at between 9.2 and 9.5 GHz. When taking into account all types of radars, over 30% of radar types operate in the range 9 to 10 GHz. Another RF band where ship radars operate is 3.03 to 3.08 GHz. These two bands are populous because ships over 300 tons and passenger ships are mandated to have an I-band (~9-GHz) radar and ships over 3,000 tons must also have an E/F band (~3-GHz) radar.

At about 1 GHz, there are many airborne radars such as Identification Friend or Foe (IFF) and traffic collision avoidance systems (TCAS). Air Traffic Control (ATC) radars operate at E-band with typical RF at 2.7 to 2.9 GHz. Many of these types of radars use dual RFs. In G-band, there are weather radars operating at 5 to 6 GHz. Radar altimeters are also in this band, usually at 4.2 to 4.4 GHz.

The RF range from 30 to 300 GHz has wavelengths in millimeters. Compared to lower bands, radio waves in this band have high atmospheric attenuation (i.e., they are absorbed by the gases in the atmosphere). Therefore, they have a short range and attenuation by rain [3] is a serious problem even over short distances. Millimeter-wave radar is used for short-range fire-control radar mounted on tanks and aircraft and for automated guns on ships to shoot down incoming missiles.

Table 2.2 IEEE RF Bands

RF Band	*RF Range*	*Band Wavelength*	*RF Band*	*RF Range*	*Band Wavelength*
I	0–200 MHz	Up to 1.5m	X	8,000–12,000 MHz	3.75–2.5 cm
G	200–250 MHz	1.5–1.2m	Ku	12,000–18,000 MHz	2.5–1.67 cm
P	250–500 MHz	1.2m–60 cm	K	18,000–26,000 MHz	1.67–1.15 cm
L	500–1,500 MHz	60–20 cm	Ka	26,000–40,000 MHz	11.54–7.5 mm
S	2,000–4,000 MHz	15–7.5 cm	V	40,000–75,000 MHz	7.5–4 mm
C	4,000–8,000 MHz	7.5–3.75 cm	W	75,000–111,000 MHz	4–2.7 mm

Table 2.3 NATO RF Bands

RF Band	*RF Range*	*Band Wavelength*	*RF Band*	*RF Range*	*Band Wavelength*
A	0–250 MHz	Up to 1.2m	H	6–8 GHz	5–3.75 cm
B	250–500 MHz	1.2–0.6m	I	8–10 GHz	3.75–3 cm
C	0.5–1 GHz	60–30 cm	J	10–20 GHz	3–1.5 cm
D	1–2 GHz	30–15 cm	K	20–40 GHz	15–7.5 mm
E	2–3 GHz	15–10 cm	L	40–60 GHz	7.5–5 mm
F	3–4 GHz	10–7.5 cm	M	60–100 GHz	5–3 mm
G	4–6 GHz	7.5–5 cm			

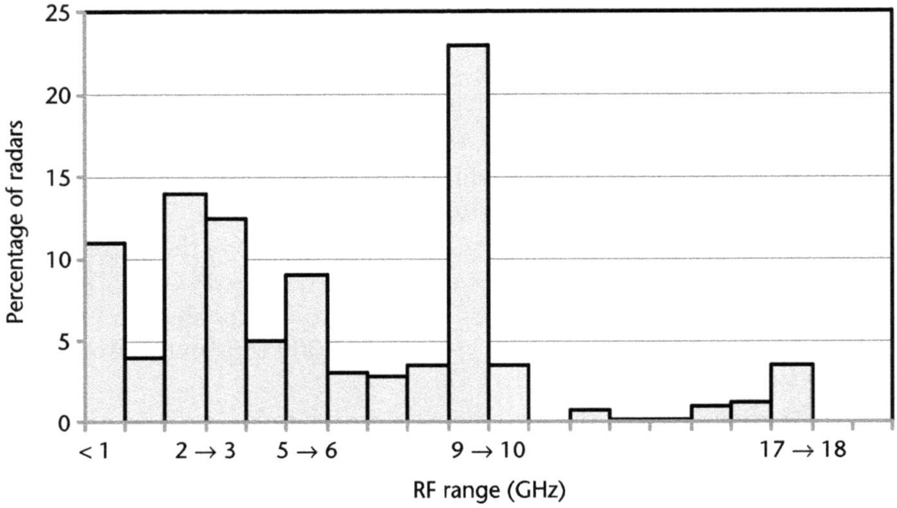

Figure 2.2 Typical histogram of RF of all radars up to 18 GHz.

The small wavelength of millimeter waves allows these radars to track the stream of outgoing bullets as well as the target, allowing the computer fire control system to change the aim of the gun to bring them together.

Threat radars may be found in any RF band, from low RF target acquisition radars such as Flat Face, operating in C-band, to Rapier fire control radars operating in J-band. More details on the types of threat radar and their typical parameters are given in Chapter 3.

2.3.1 RF Agility

The type of RF emitted can be fixed, agile from pulse to pulse, or agile from batch to batch of pulses. Some agile signals follow a specific pattern of changing RF from pulse to pulse or from batch to batch, whereas others change RF on a pseudorandom basis. The extent of RF agility varies from just a few megahertz difference between the RF levels to hundreds of megahertz difference between maximum and minimum values for active electronically scanned array (AESA) radars. Some scanning radars have up to 20 different RF levels available, with tens of megahertz difference between levels. In general, radars with differences of more than 20 MHz between RF levels are batch-to-batch agile, while those with smaller RF differences between levels are pulse-to-pulse agile. However, AESA radars carrying out several different functions simultaneously exhibit pulse-to-pulse RF agility with vastly different RFs between consecutive pulses.

Some radars operate in several different agility modes with both pulse-to-pulse and batch-to-batch modes available. Table 2.4 illustrates the two RF agility modes of a typical current radar. The difference in RF between levels is 15 MHz and the difference in RF varies in a set sequence from pulse to pulse. In the batch-to-batch case, there are four pulses emitted at each RF.

There are several reasons for the radar to use RF agility apart from the desire to cause disruption to an ESM system. For example, AESA radars use different RFs to carry out different functions simultaneously. These radars can have RF agility of hundreds of megahertz as shown in Figure 2.3, a simulation of the frequency agility that might be seen from pulse to pulse for an AESA radar.

2.4 Pulse Repetition Interval

The PRI of a signal is an important parameter both in the deinterleaving and identification stages of ESM processing. The PRI is defined as the time between the leading edges of two successive pulses, as shown in Figure 2.4.

Table 2.4 Example of a Radar with Pulse-to-Pulse and Batch-to-Batch RF Agile Modes

Pulse-to-Pulse RF Agility					*Batch-to-Batch RF Agility*				
Pulse No.	*RF No.*	*RF (MHz)*	*PRI (μs)*	*PW (μs)*	*Pulse No.*	*RF Burst*	*RF (MHz)*	*PRI (μs)*	*PW (μs)*
1	1	2,740	670	3	1	A	2,740	670	3
2	3	2,765	632	3	2	A	2,740	632	3
3	4	2,795	632	3	3	A	2,740	632	3
4	5	2,780	632	3	4	A	2,740	632	3
5	2	2,725	652	3	5	B	2,765	652	3
6	1	2,740	580	3	6	B	2,765	580	3
7	3	2,765	672	3	7	B	2,765	672	3
8	4	2,795	505	3	8	B	2,765	505	3
9	5	2,780	599	3	9	C	2,725	599	3
10	2	2,725	599	3	10	C	2,725	599	3
11	1	2,740	599	3	11	C	2,725	599	3
12	3	2,765	619	3	12	C	2,725	619	3
13	4	2,795	600	3	13	D	2,780	600	3
14	5	2,780	672	3	14	D	2,780	672	3
15	2	2,725	565	3	15	D	2,780	665	3
16	1	2,740	605	3	16	D	2,780	605	3
17	3	2,765	632	3	17	E	2,795	632	3
18	4	2,795	632	3	18	E	2,795	632	3
19	5	2,780	632	3	19	E	2,795	632	3
20	2	2,725	650	3	20	E	2,795	650	3
21	1	2,740	652	3	21	A	2,740	652	3
22	3	2,765	652	3	22	A	2,740	652	3
23	4	2,795	652	3	23	A	2,740	652	3

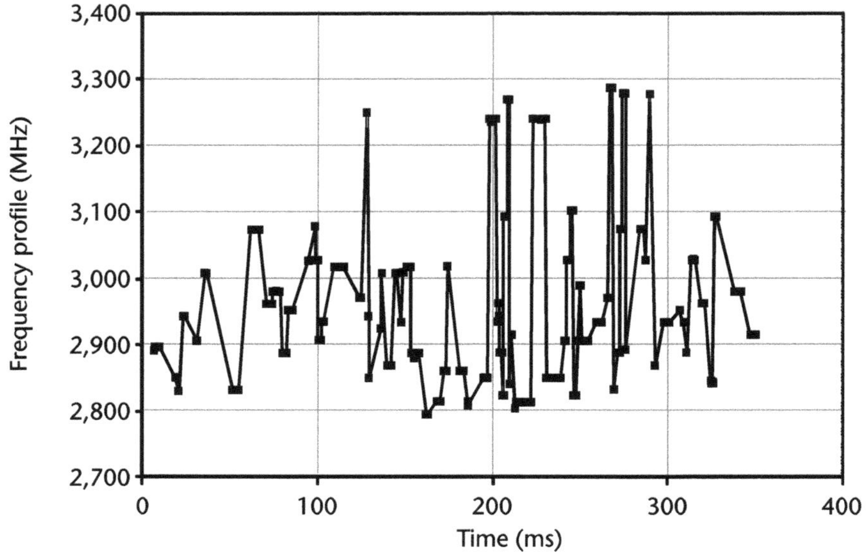

Figure 2.3 Example of radar with 400-MHz RF agility.

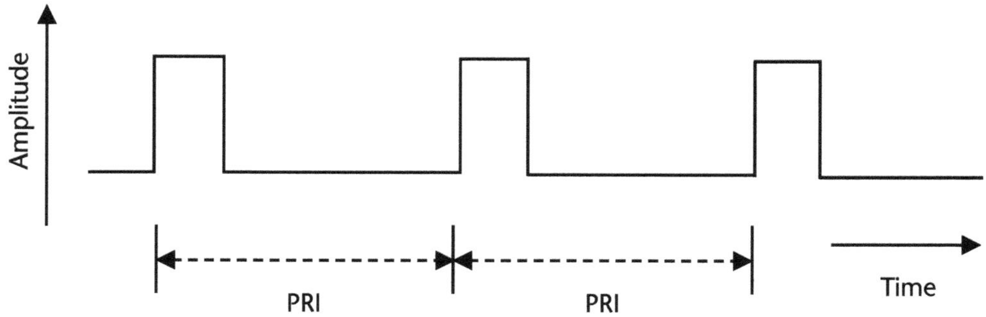

Figure 2.4 Definition of PRI.

Pulse repetition frequency (PRF) is the inverse of PRI:

$$PRF = 1 / PRI \tag{2.1}$$

A long PRI corresponds to a low PRF, for example a PRI of 2 ms gives a PRF of 1/0.002 or 500 Hz, that is, 500 pulses per second (pps), whereas a PRI of 500 μs has a PRF of 2,000 pps.

PRI types are classified as fixed, staggered, jittered, or switch and dwell. Pulse groups (doublets or triplets) are used for some applications such as missile guidance.

2.4.1 Fixed PRI

Many radars operate with a simple fixed PRI (i.e., there is only one PRI element in the pulse sequence). The range of PRI values for different radar types varies from a few tens of microseconds to several milliseconds. Types of radar that typically operate with fixed PRI are ship navigation and weather radars. For example, weather radars operate at RF about 5 GHz with fixed PRI of about 3 ms as shown in Table 2.5.

2.4.2 Staggered PRI

Most radars operate a staggered PRI sequence. The most common number of elements in a staggered PRI sequence is three, although complex pulse sequences may have several tens of elements. The sequence in which the PRI elements are emitted adds significantly to the deinterleaving problem. For example, the radar shown in Table 2.6 has 12 different elements, but has a PRI pattern that repeats only every 30 pulses.

2.4.3 Jittered PRI

In this type of PRI, a single PRI element is varied by a few percent from a mean value on the emission of each pulse. This type of PRI may disrupt the operation of predictive gate processors (see Chapter 8). In predictive gate systems, the limits that are set up to find the next pulse from the radar must be wide enough to cope with the potential PRI jitter, but narrow enough to prevent pulses from other radars from being included in the pulse chain. In a dense RF environment, where pulses are being received every few tens of microseconds, this will cause problems for the PRI calculation process.

For example, a PRI of 800 μs that is jittered by ±5% may change its PRI by up to 80 μs from pulse to pulse. Table 2.7 shows an example of such a pulse sequence. Many shipborne radars operating at I-band have jittered PRI sequences, and as many are pulses detected within a small RF band, the correct operation of the deinterleaver is compromised.

2.4.4 Switch and Dwell PRI

In this PRI type, a set of pulses is emitted at one PRI, and then the PRI is changed for the emission of the next set of pulses. Two examples are shown in Tables 2.8 and 2.9. The first has 303 pulses at a PRI of 14 μs, followed by 305 pulses at 12 μs, and then a set of pulses with PRI 13 μs. In the second example, the dwell is only for five pulses and the PRI range is between 800 μs and 1.2 μs. The radar in

Table 2.5 Example of a Radar with a Fixed PRI

Pulse No.	*RF (MHz)*	*PRI (μs)*	*PW (μs)*
1	5,345	3,320	3
2	5,346	3,320	3
3	5,345	3,320	3
4	5,346	3,320	3
5	5,345	3,320	3
6	5,345	3,320	3
7	5,345	3,320	3
8	5,346	3,320	3

Table 2.6 Example of a Radar with a Staggered PRI with 12 Elements and a 30-Pulse PRI Sequence

Pulse No.	*RF (MHz).*	*PRI (μs)*	*PW (μs)*	*Pulse No.*	*RF (MHz)*	*PRI (μs)*	*PW (μs)*
1	2,780	504	6	26	2,780	530	6
2	2,780	532	6	27	2,781	530	6
3	2,780	532	6	28	2,780	619	6
4	2,780	532	6	29	2,780	600	6
5	2,780	552	6	30	2,780	600	6
6	2,779	500	6	31	2,780	504	6
7	2,780	572	6	32	2,780	532	6
8	2,780	505	6	33	2,779	532	6
9	2,780	599	6	34	2,780	532	6
10	2,781	599	6	35	2,780	552	6
11	2,780	599	6	36	2,780	500	6
12	2,780	619	6	37	2,780	572	6
13	2,780	600	6	38	2,780	505	6
14	2,781	572	6	39	2,780	599	6
15	278	565	6	40	2,779	599	6
16	2,780	505	6	41	2,780	599	6
17	2,780	530	6	42	2,780	619	6
18	2,780	530	6	43	2,780	600	6
19	2,780	500	6	44	2,780	572	6
20	2,779	599	6	45	2,780	565	6
21	2,780	532	6	46	2,780	505	6
22	2,780	532	6	47	2,781	530	6
23	2,780	599	6	48	2,780	530	6
24	2,780	572	6	49	2,780	500	6
25	2,780	504	6	50	2,780	599	6

Table 2.7 Example of Pulses with a Jittered PRI

Pulse No.	*RF (MHz)*	*PRI (μs)*	*PW (μs)*
1	9,393.0	802	0.35
2	9,393.0	830	0.35
3	9,393.1	783	0.34
4	9,393.2	795	0.35
5	9,393.0	832	0.35
6	9,393.0	780	0.33
7	9,393.2	805	0.34
8	9,393.2	784	0.35
9	9,393.2	815	0.33
10	9,393.1	803	0.33

Table 2.8 causes problems for the track correlation process as the large number of pulses at each PRI may deceive the ESM into believing that separate radars exist for each PRI.

Table 2.8 Example of a Switch and Dwell PRI with Short PRI and Long Dwell

Pulse No.	*RF (MHz)*	*PRI (μs)*	*PW (μs)*
1	9,757	14	0.32
2	9,755	14	0.32
3	9,757	14	0.28
4	9,757	14	0.32
5	9,758	14	0.32
6	9,758	14	0.28
297 more pulses at 14 μs PRI			
304	9,757	12	0.28
305	9,758	12	0.28
306	9,757	12	0.28
307	9,758	12	0.28
308	9,757	12	0.28
309	9,757	12	0.32
299 more pulses at 12 μs PRI			
609	9,757	12	0.32
610	9,757	13	0.32
611	9,757	13	0.32
612	9,757	12	0.32

Table 2.9 Example of a Switch and Dwell PRI with Long PRI and Short Dwell

Pulse No.	*RF (MHz)*	*PRI (μs)*	*PW (μs)*
1	9,757	1,071	16.8
2	9,757	1,071	16.8
3	9,757	1,072	16.8
4	9,756	1,071	16.8
5	9,757	1,071	16.8
6	9,757	1,021	16.8
7	9,757	1,021	16.8
8	9,757	1,021	16.8
9	9,757	1,021	16.8
10	9,757	1,021	16.8
11	9,756	981	16.8
12	9,757	980	16.7
13	9,756	981	16.8
14	9,758	981	16.8
15	9,757	981	16.7
16	9,757	981	16.7

An example of a type of radar using switch and dwell PRI is a pulse Doppler radar operating with dwell sequence of three PRIs at only 6.0/6.5/7.0 μs with a dwell of 10 ms (or an integer multiple thereof) on each PRI. PWs are in the range 1.0 to 2.0 μs. PW changes occur in conjunction with PRI in an attempt to maintain a constant duty cycle. The small difference in the PRI levels combined with the coincident PRI and PW changes cause difficulties for ESM deinterleaving algorithms and it is likely that multiple tracks will be created for this type of radar. More details on multitracking are given in Chapter 14.

2.4.5 Pulse Group Repetition Interval

The pulse group repetition interval (PGRI) is the sum of all PRI elements before a repeat in the PRI sequence occurs. This is a useful parameter as some ESM pulse deinterleaving algorithms are able to determine the PGRI but cannot identify the individual elements. The PGRI may encompass as few as three or four PRI elements for a radar with a simple stagger, as depicted in Figure 2.5, or as many as several tens of elements for a radar with a more complex PRI sequence.

There is one category of radars where PGRI has a rather different meaning. Missile guidance radars that send commands to missiles using pulse position modulation (PPM) typically have several pulses separated by very short PRIs, followed by a larger gap in the pulses.

The position of the doublet or triplet of pulses within the pulse group sends specific commands to the missile. Figure 2.6 shows an example of the type of pulse groups that are typical of missile guidance radars using PPM.

Two missiles are being guided by the signal in Figure 2.6. Each missile has a separate subframe marker pulse triplet and the next triplet relays the guidance command to the missile. The spacing of the command triplet indicates the type of command (e.g., heading or pitch change).

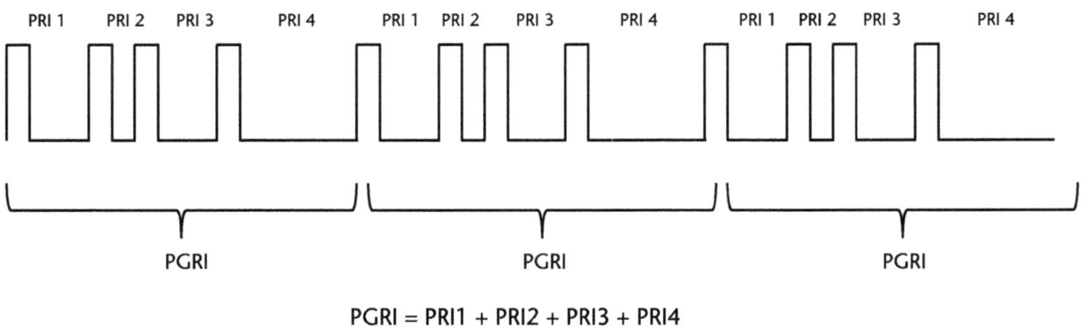

Figure 2.5 Example of a pulse sequence showing the PGRI.

Frame	1 (Duration ~ 6 ms)											
Sub-frame	A (Duration ~ 3ms)						B (Duration ~ 3ms)					
PRI (μs)	9	4	1420	6	15	1546	5	2	1573	3	12	1405
Function	Sub-frame marker for missile A		Missile command for Missile A				Sub-frame marker for missile B		Missile command for Missile B			

Figure 2.6 Example of a pulse sequence for a missile guidance radar using PPM.

The position of the triplet within the command portion of the subframe shows the magnitude of the course change needed.

2.5 Pulse Width

The radar PW is the duration of the pulse, typically measured in microseconds. Most pulses are not rectangular in shape and have a profile with sloping rise and fall edges. The PW [4] is measured from the point at which the pulse has achieved half its peak amplitude until the pulse amplitude has dropped to the same point on the falling edge of the pulse, as shown in Figure 2.7.

PW is the least agile of the fundamental radar parameters with some radars using two PWs, but not many using more values than this. Figure 2.8 shows a histogram of PWs for current radars. Over 30% of the radars have PWs of 1 microsecond or less.

2.5.1 Pulse Rise Times

The pulse rise time is defined as the time taken for the pulse amplitude to increase from 10% to 90% of its peak amplitude. More than 50% of current radars have rise times of 50 ns or less, with about 15% of these radars having rise times of 45 to 50 ns. Another 25% of radars have rise times that are greater than 50 ns, but less than 100 ns. Figure 2.9 shows histograms giving an indication of the percentage of radars with various pulse rise times.

2.6 Pulse Modulation

The most common form of intentional pulse modulation is frequency modulation (FMOP), called chirp [5]. The transmitted waveform consists of a rectangular pulse of constant amplitude with the frequency increasing linearly from f_1 to f_2 over the duration of the pulse. Figure 2.10 shows an example of a pulse with a linear chirp

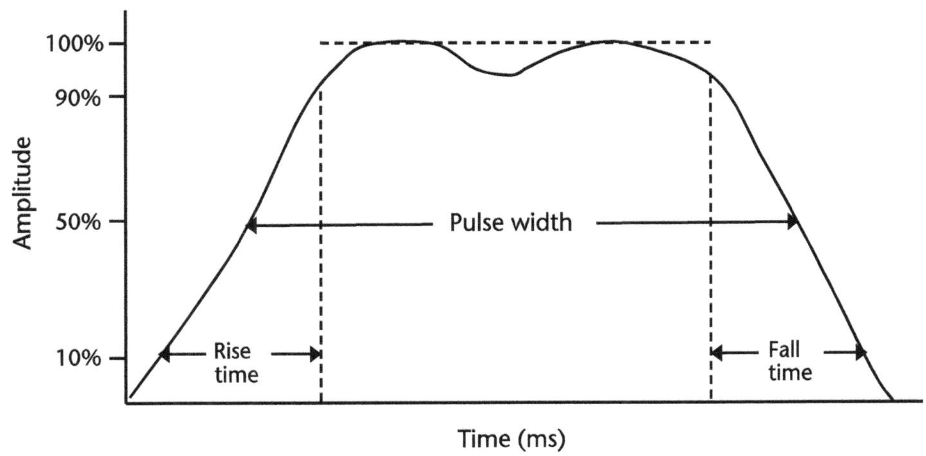

Figure 2.7 PW definition.

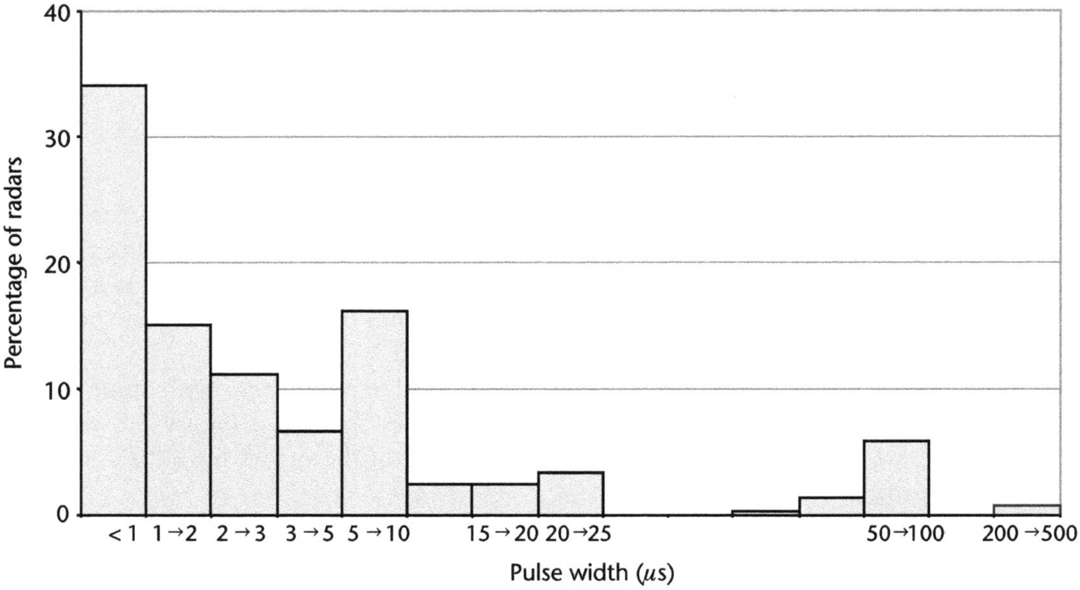

Figure 2.8 Histogram of radar PWs.

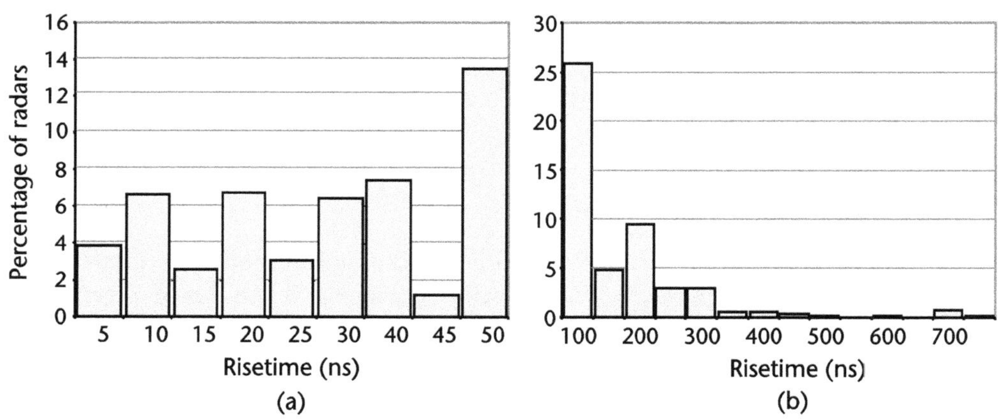

Figure 2.9 Pulse rise times.

of 20 MHz. Current radars may have chirps of up to 125 MHz over a 1 μs pulse, although the most usual values are much less than this, typically 5 to 10 MHz. A chirp of 5 MHz should not cause problems in the measurement of RF for ESM systems.

Some radars use nonlinear FM waveforms, as shown in Figure 2.11. The pulse in Figure 2.11 has a V-chirp, a linear chirp of 10 MHz for the first half of the pulse and then a linear chirp of –10 MHz for the second half of the pulse. The RF profile may be recognized by the ESM receiver, but it is likely to fail to correctly measure the center RF of the pulse and will find it difficult to determine the chirp rate for the pulse.

Phase modulation is another common form of pulse modulation, where a long pulse of duration T is divided into N subpulses of width τ. The phase of each

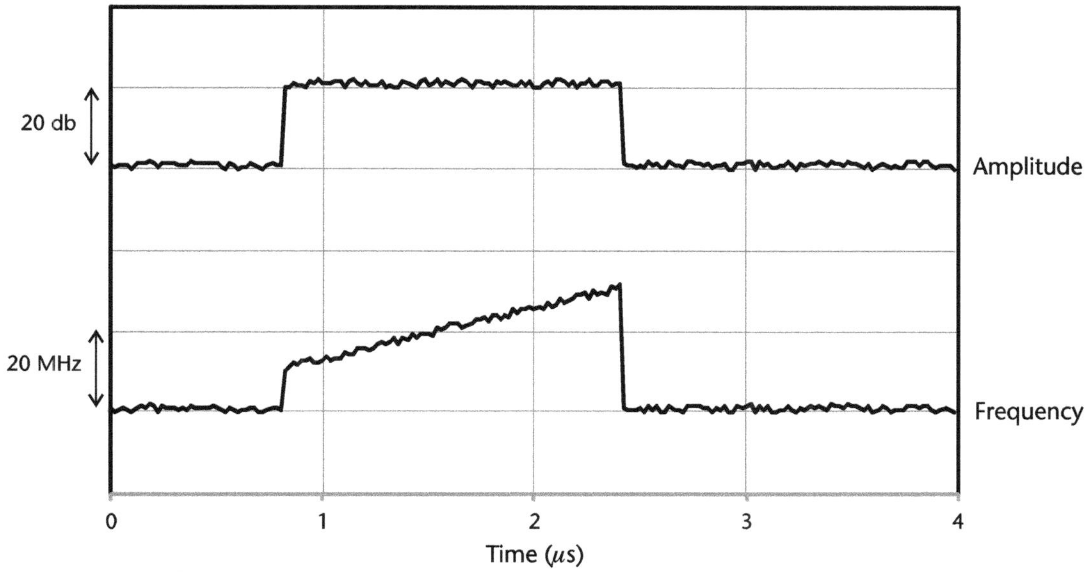

Figure 2.10 Intrapulse amplitude and frequency profiles for a pulse with 20-MHz linear chirp.

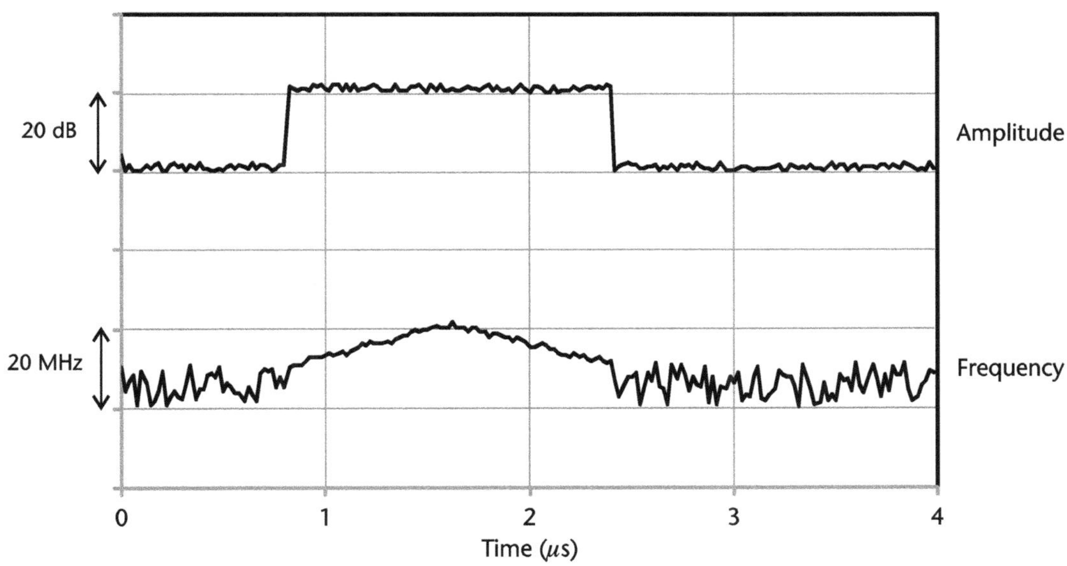

Figure 2.11 Intrapulse amplitude and frequency profiles for a pulse with a 10-MHz V-chirp.

subpulse is either 0 or π radians. The output of a matched filter is a spike of width t with an amplitude N times greater than that of the long pulse. The pulse compression ratio is $N = T/\tau = BT$ where $B = 1/\tau$ is the bandwidth.

The output waveform extends a distance T to either side of the central spike. The portions of the waveform lying outside the main spike are called time sidelobes. It is desirable to select a phase-coded waveform such that its autocorrelation function should have equal time sidelobes.

Binary phase-coded sequences of 0 and π that result in equal time sidelobes after passing through a matched filter are called Barker codes [6]. Figure 2.12 shows an example of a Barker code of length 13. The (+) indicates 0 phase and (–) indicates π radians phase.

Figure 2.13 shows the amplitude and frequency profiles for a 13-bit Barker coded pulse.

Although there are spikes seen in the amplitude and frequency profiles for the Barker-coded pulse there should be no significant disruption to either the frequency or PW measurement due to the use of this type of coding.

2.7 Radar Beamshapes

Radar beams are generally of two types those with small azimuth beamwidths coupled with large elevation beamwidths, sometimes called fan beams, or beams having small beamwidths in both azimuth and elevation, called pencil beams.

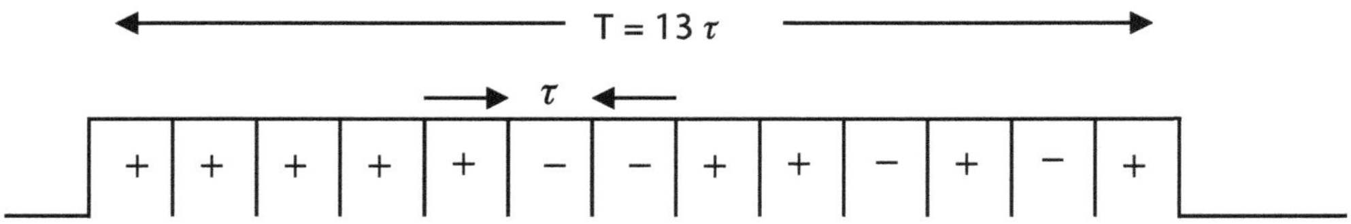

Figure 2.12 A Barker phase-coded pulse with 13 equal subdivisions of either 0° or 180° phase.

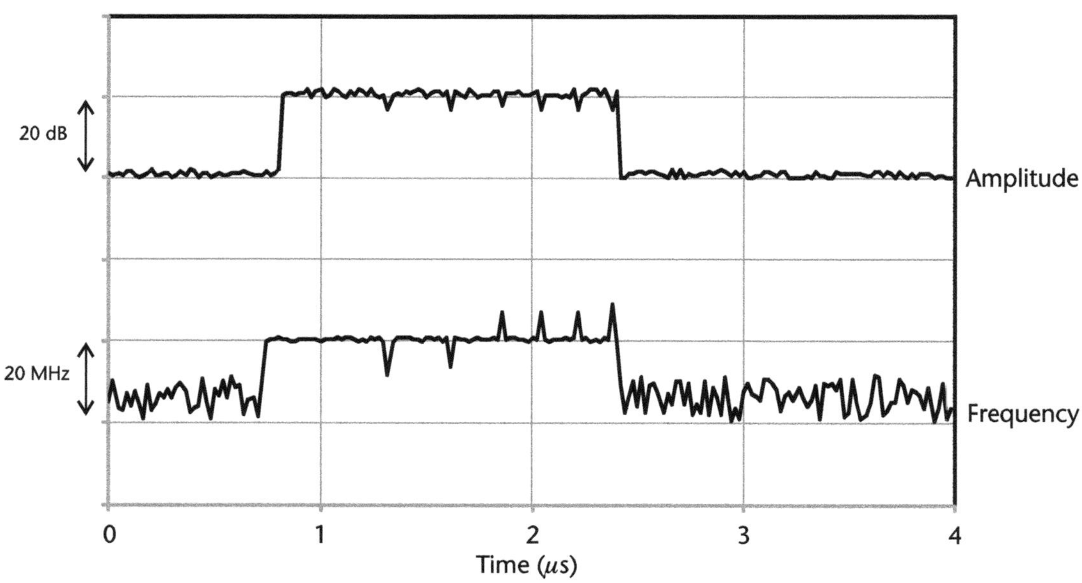

Figure 2.13 Intrapulse amplitude and frequency profiles for a 13-bit Barker-coded pulse.

The most common radar antenna beam shape is one with a $(\sin(x)/x)^2$ distribution in azimuth and a cosec^2 beam shape in elevation [7]. Figure 2.14 gives an indication of such antenna beam profiles in azimuth and elevation.

The way that the azimuth beamshape appears to the ESM system is very important for the analysis of ESM systems, so the derivation of the beamshape is considered here, with the detailed calculations needed for generating an exact simulation of radar azimuth beamshape given in Appendix A.

The most important parameter in defining the beamshape is the 3-dB beamwidth, defined as the level at which the amplitude of the beam has dropped to half its peak amplitude. The 3-dB beamwidth is taken into account in calculation of $(\sin(x)/x)^2$ by the following equation:

$$x = (\pi \sin(\theta)) / B \tag{2.2}$$

where θ is the azimuth angle from the peak of the beam and B is the 3-dB beamwidth.

The first plot in Figure 2.15 shows the basic $(\sin(x)/x)^2$ beamshape. The second graph shows the same data but with the units of amplitude given in decibels. The equation for converting amplitude in watts to decibels is as follows:

$$Amplitude(\text{dB}) = 10 \times \log 10(amplitude\ in\ Watts) \tag{2.3}$$

Decibels are the usual unit for expressing amplitude in ESM systems, and, as this is a log scale, it makes it easier to view amplitude profiles than if a linear scale was used: 1W = 0 dB and 0.5W (i.e., half the power) = –3 dB. An amplitude of 1 mW (–20 dB) can be plotted on the same graph as an amplitude of 1 MW (30 dB) so that both values are easily distinguished.

The default value of the first sidelobe for a (sin(x)/x)2 beam shape is –13.2 dB. However, to get more power into the main lobe and reduce the amount of power

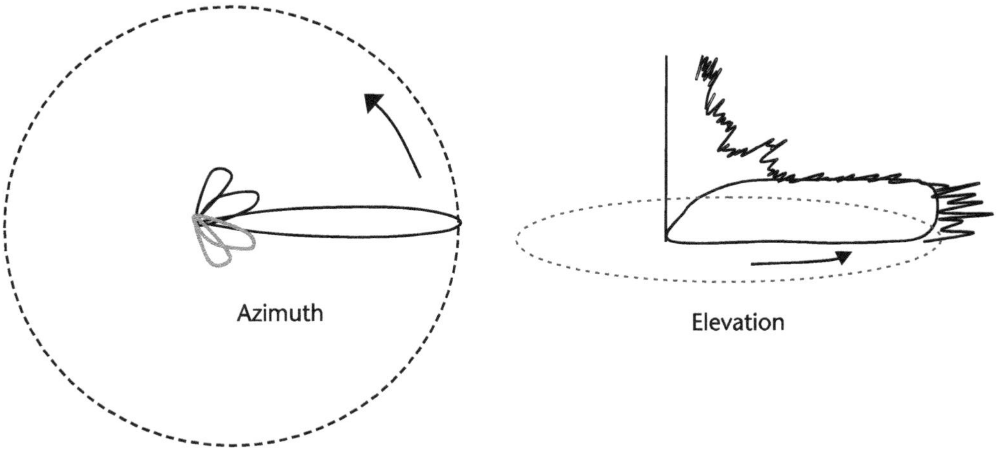

Figure 2.14 Radar azimuth and elevation beamshapes.

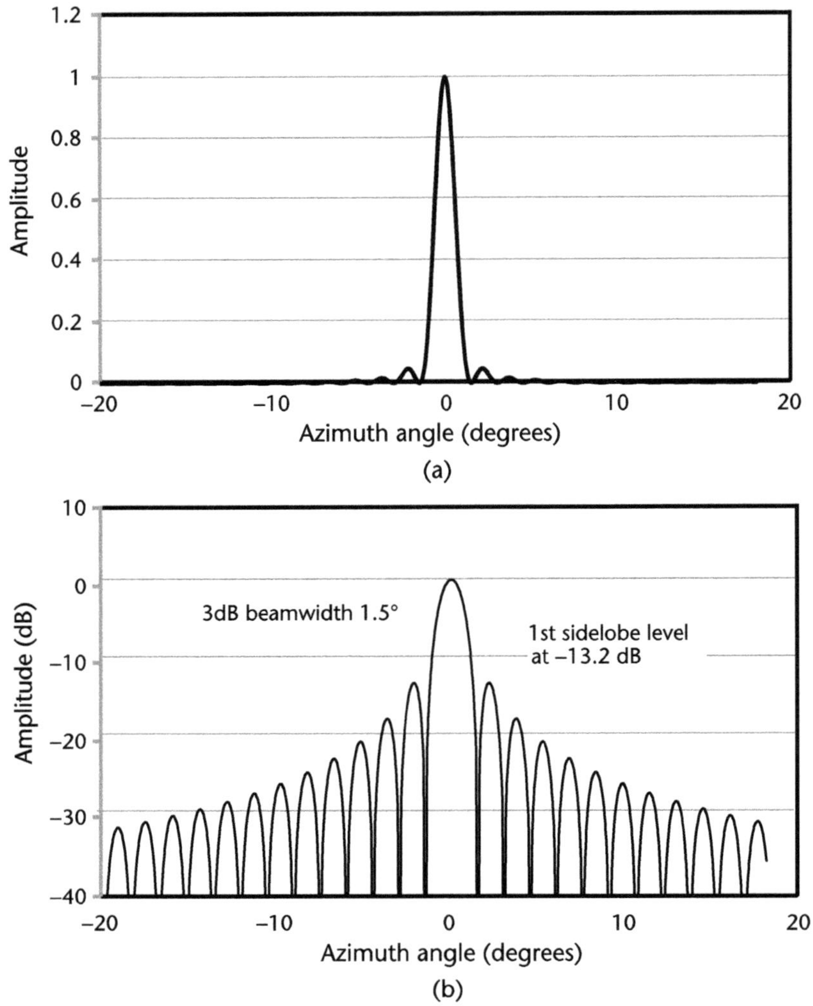

Figure 2.15 Basic azimuth beamshape (a) and in decibels (b).

going into the sidelobes, the designers of radar antenna beams lower the first sidelobe level. This is done by modifying the basic $(\sin(x)/x)^2$ equation as follows:

$$A_\theta = \left[\sin\left(u^2 - z^2\right)^{1/2} / \left(u^2 - z^2\right)^{1/2}\right]^2 \tag{2.4}$$

where A_θ is the amplitude at azimuth angle θ on the radar beam and

$$u = C\pi \sin(\theta) / B \tag{2.5}$$

C and z are found using a look-up table and correspond to the required sidelobe level [8] (see Appendix A for details).

When (2.4) is applied, a radar beamshape like the one in Figure 2.16 results. In this case, the 3-dB beamwidth is 1.5° and the first sidelobe level is at −23 dB.

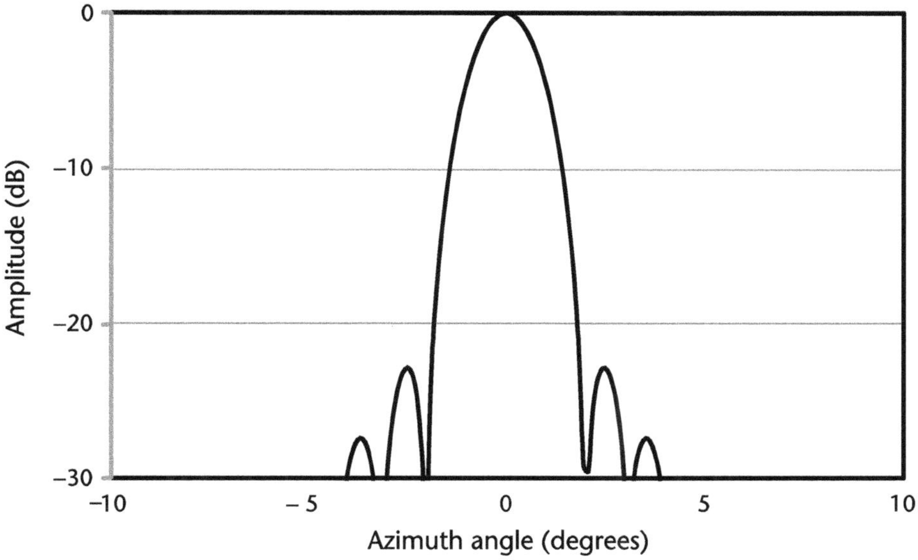

Figure 2.16 Typical azimuth radar beamshape.

Once the radar beamshape has been established, the amplitude of the pulses as the radar scans past the ESM platform can be plotted on the beamshape. The positions of the pulses will depend on two additional factors: the PRI and the scan period. Figure 2.17 shows the amplitude of all the pulses in the main beam and first sidelobes for a radar with PRI 1 ms and scan rate 3 seconds (see Appendix A for more details on the pulse positions).

Now that the pulses are plotted on the beamshape, the x-axis units on Figure 2.17 can be changed from degrees in azimuth to time, in this case in milliseconds. This conversion can be carried out as the scan period is included in the calculation

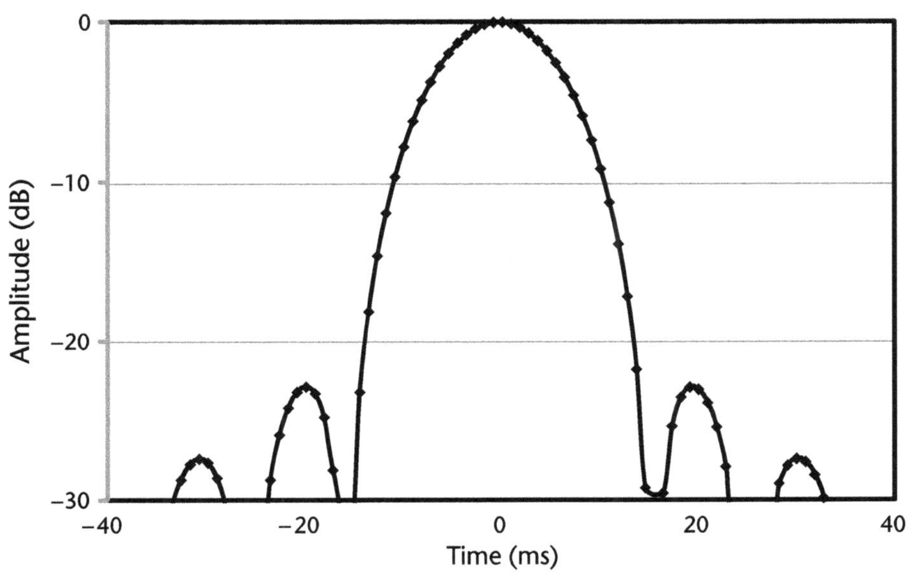

Figure 2.17 Typical azimuth radar beamshape with pulses.

of the position of the pulses on the beamshape. For example, a radar with a 3-second scan period and a 1-ms PRI will rotate by 0.12° from pulses to pulse. While neither the scan period nor the PRI affects the beamshape, they do affect where on the beam the pulses appear to be from the ESM system point of view.

2.8 Scan Patterns

Most radars operate with a circular scan and a rotating antenna. However, other scan types such as raster, spiral, conical, and sector are available to the radar developer and the way that an ESM would see each of the different types of scan is described below [9].

2.8.1 Circular Scan

This regular scan has a scan period of the order of a few seconds. The azimuth beamshape is usually $(\sin(x)/x)^2$ and the elevation beamshape is cosec^2. Coverage is 360° in azimuth and up to 70° in elevation. The amplitude of the pulses increases as the radar scans past the ESM receiver and then decreases once the peak of the beam has passed. Figure 2.18 shows the amplitude profile seen by an ESM system from a radar with a circular scan.

2.8.2 Sector Scan

The beamshape in azimuth is the same as for circular scan and in elevation the beamshape is cosec^2, but the radar covers a smaller azimuth range. Sector scans

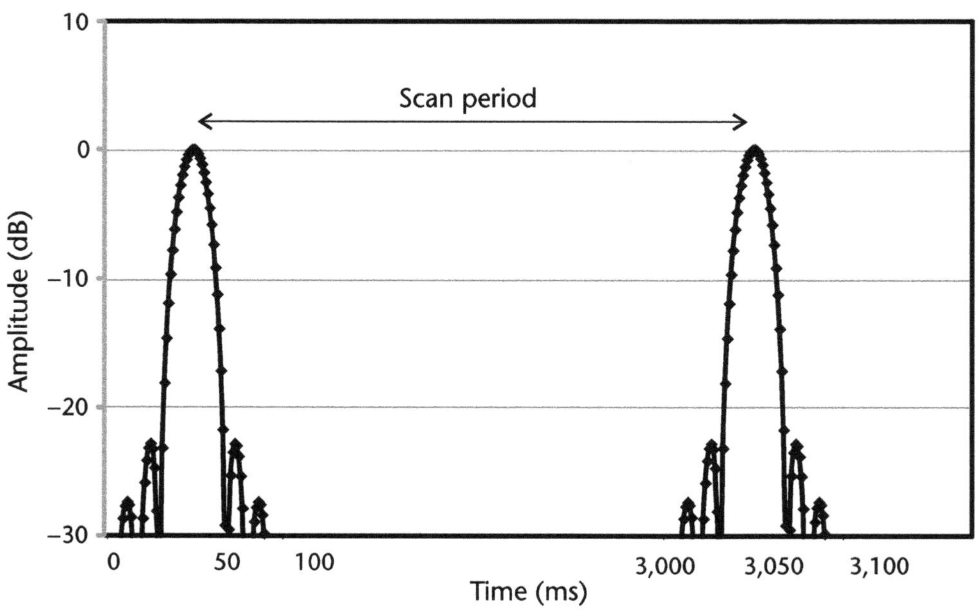

Figure 2.18 Amplitude profile from a circular scan radar.

are usually used to search in a particular direction or area, typically by target acquisition radars. There are three types of sector scan as follows.

- *Bidirectional sector scan:* Bidirectional sector scan radars transmit as the antenna moves in both directions through a fixed point. Typical of this application are weather or ground-mapping radars with the beam only being transmitted in the sector of interest. The ESM only sees pulses from bidirectional sector scan radars if it is within the azimuth extent of the sector. The amplitude profile of the pulses appears as two adjacent beamshapes, with the leading edge of the beam reversed on alternate scan peaks.
- *Unidirectional scan:* In this type of scan, the radar antenna is rotated through 360° in azimuth but is only transmitting when it passes through a particular segment in azimuth. The ESM only receives pulses from a radar that is transmitting a unidirectional scan if it lies within the segment that is being scanned, in which case the pulse amplitude profile appears to the ESM to be the same as that from a radar operating with a circular scan.
- *Vertical sector scan:* Vertical sector scans are used by height-finding radars. These have narrow elevation beamwidths and scan rates varying from 5 to 30 scans per minute. The ESM only sees pulses from a radar operating a vertical sector scan if the radar is transmitting in the direction of the ESM.

2.8.3 Raster Scan

Raster scans cover a limited search area rectangular in shape using a pencil beam. They may be used by air intercept (AI) radars to search a large volume of airspace in azimuth and elevation with good angle resolution. The beam scans in azimuth through the sector of interest and, at the end of each azimuth scan, the elevation of the beam is altered. Raster scanning can accurately fix a target position in range, azimuth, and elevation. Scan rates vary according to the angle of sector, but may generally be in the region of four to five scans per second.

Figure 2.19 shows the amplitude profile of a raster scan as seen by an ESM system. As the pencil beam becomes more centered in elevation on the ESM platform, the peak amplitude increases and more pulses are seen in a single scan.

2.8.4 Spiral Scan

The spiral scan antenna rotates outward by an angular step for each round using a pencil beam. The time to move from the outside to the center is the retrace time. Figure 2.20 shows the amplitude profile received by an ESM antenna from a radar using this type of scan.

2.8.5 Helical Scan

When using a helical scan, the antenna is continuously rotated in azimuth while it is simultaneously raised or lowered in elevation. It traces a helix in space and uses a

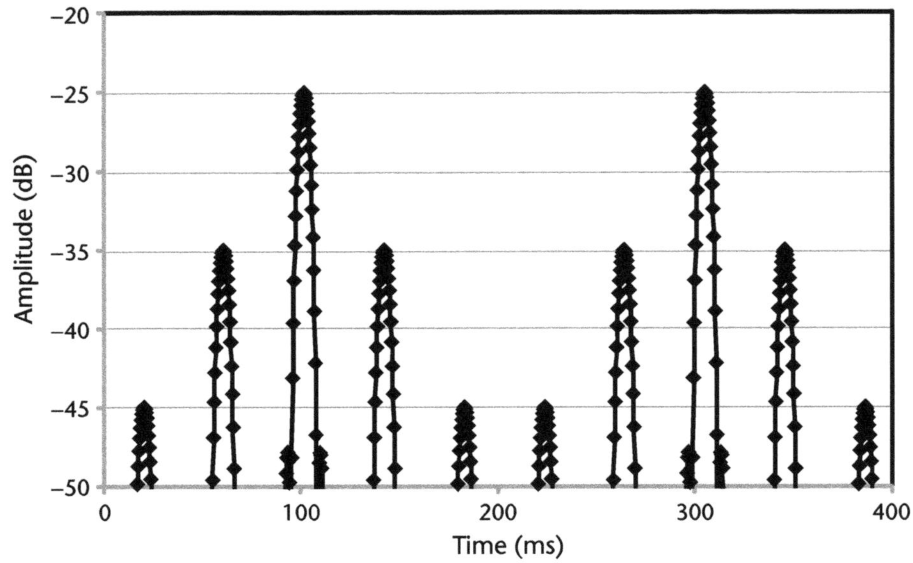

Figure 2.19 Amplitude profile from a raster scan radar.

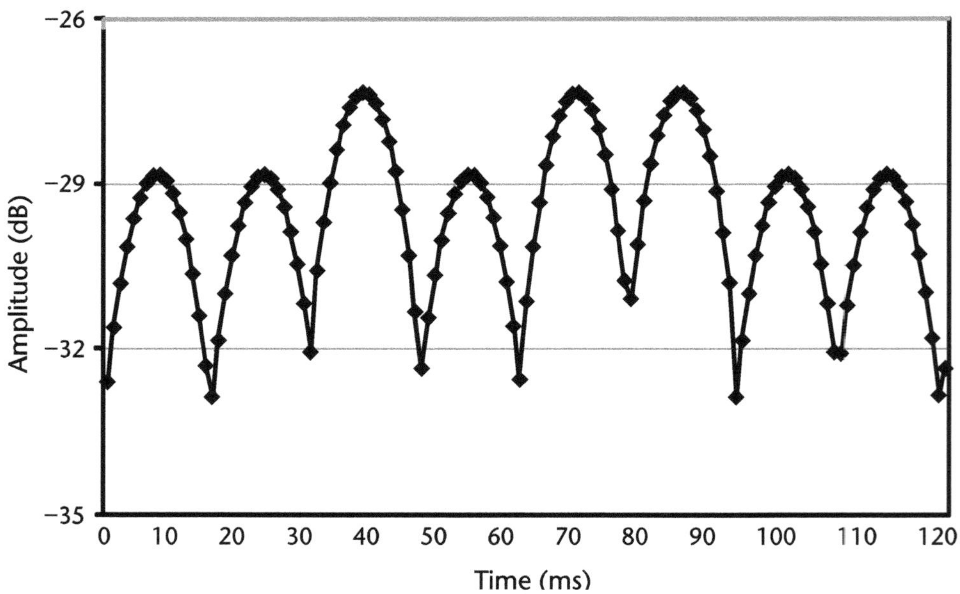

Figure 2.20 Amplitude profile from a spiral scan radar.

pencil beam. The amplitude profile of pulses seen from a radar using a helical scan is shown in Figure 2.21.

2.8.6 Conical Scan

In a conical scan, the radar pencil beam is rotated around a central boresight. A narrow beamwidth is used for good angle resolution. The angle between the axis of rotation and the axis of the antenna beam is called the squint angle. The squint

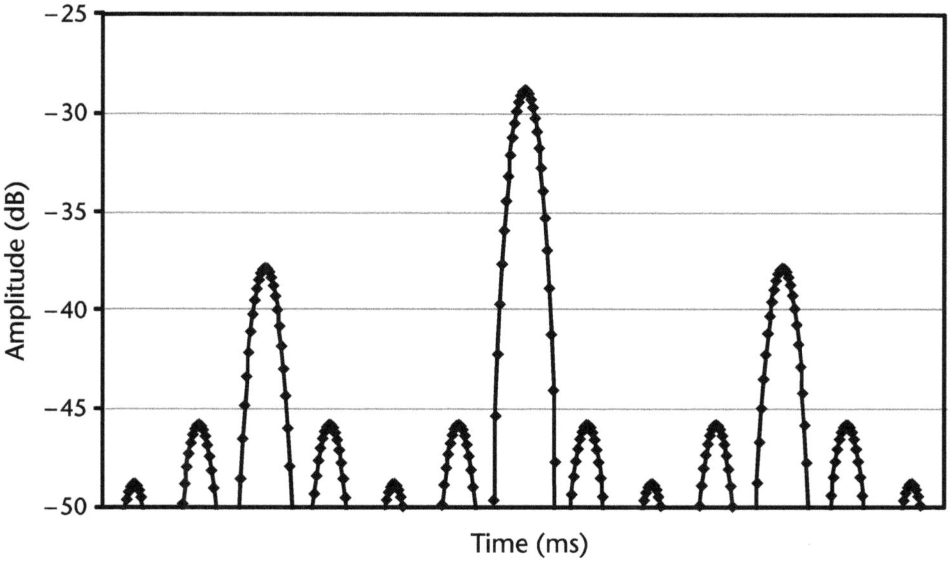

Figure 2.21 Amplitude profile from a helical scan radar.

angle is always less than half the beamwidth and the scan rates for conical scanning radars generally vary from 50 to 100 Hz.

Conical scans are used by tracking radars. A conical scan indicates that the radar is trying to lock onto a target by scanning conically around it to place it within the 3-dB beamwidth of the radar. The received amplitude profile, shown in Figure 2.22, is initially a sinusoid because the beam travels closer to and then farther from the ESM receiver.

As the lock improves, the amplitude of the sinusoid decreases. Eventually, when the radar is locked on completely and the target is within the 3-dB beamwidth, the

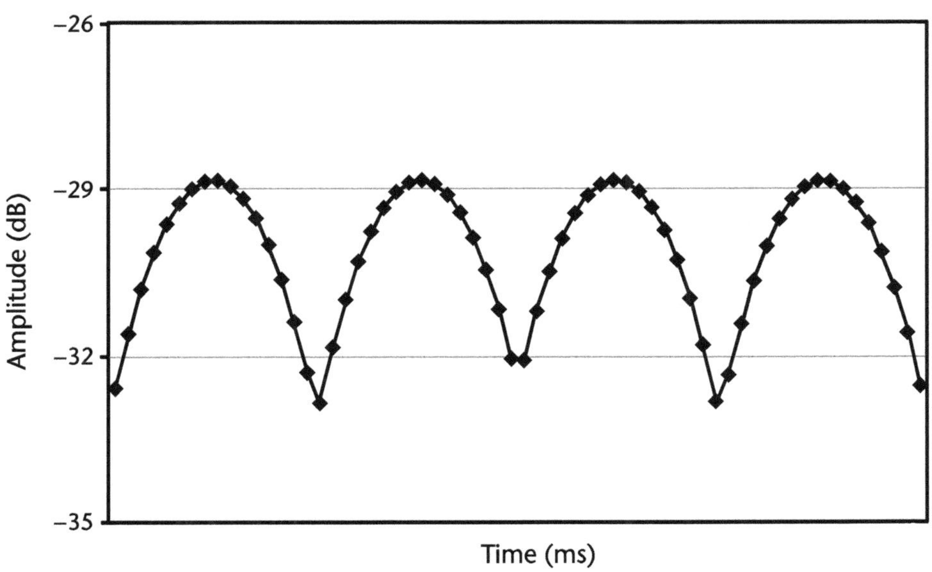

Figure 2.22 Amplitude profile from a conical scan radar.

pulse amplitude becomes constant and the amplitude profile seen by the ESM is that of a fixed scan [9].

2.8.7 Palmer Scan

This scan type consists of a rapid circular scan about the axis of the radar antenna, combined with a linear movement of the axis of rotation. When the axis of rotation is held stationary, the Palmer scan reduces to the conical scan. The Palmer scan is suited to a search area larger in one dimension than another. However, both the Palmer scan and the spiral scan suffer from the disadvantage that all parts of the scan volume do not receive the same energy unless the scan speed is varied during the scan cycle. The amplitude profile of pulses received from a Palmer scanning radar is shown in Figure 2.23.

2.8.8 Nodding Scan

Figure 2.24 shows the amplitude profile of a nodding scan. In this case, the basic scan is circular overlaid with a change in elevation of the pencil beam to obtain 360° azimuth coverage and 30° elevation coverage.

2.8.9 Lobe Switching

The lobe-switching scan positions the antenna beams in different squint angle locations to improve tracking accuracy of a tracking radar. It is similar to the conical scan, but instead of a constantly rotating beam, there are distinct positions. Typical configurations consist of two or four lobes. The amplitude profile seen for this type of scan is shown in Figure 2.25.

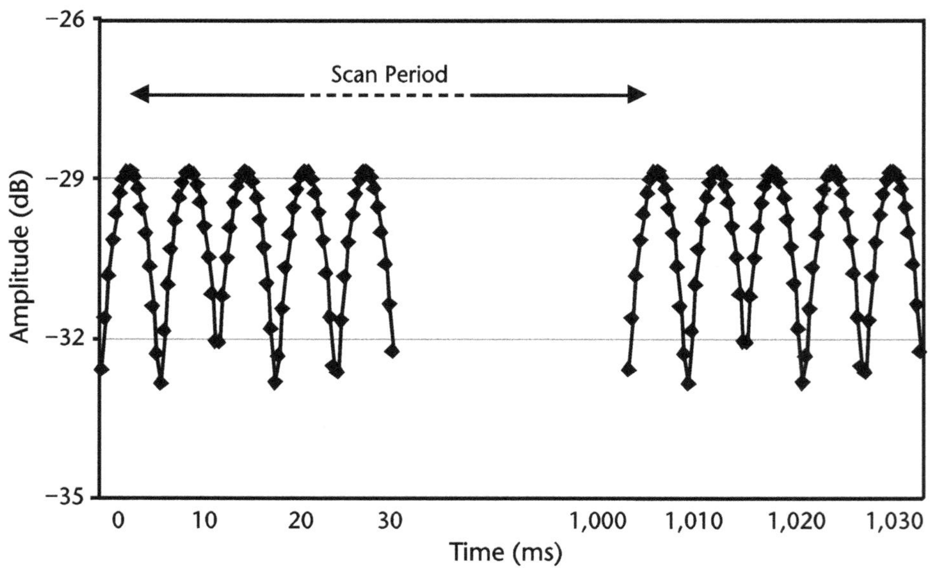

Figure 2.23 Amplitude profile from a Palmer scan radar.

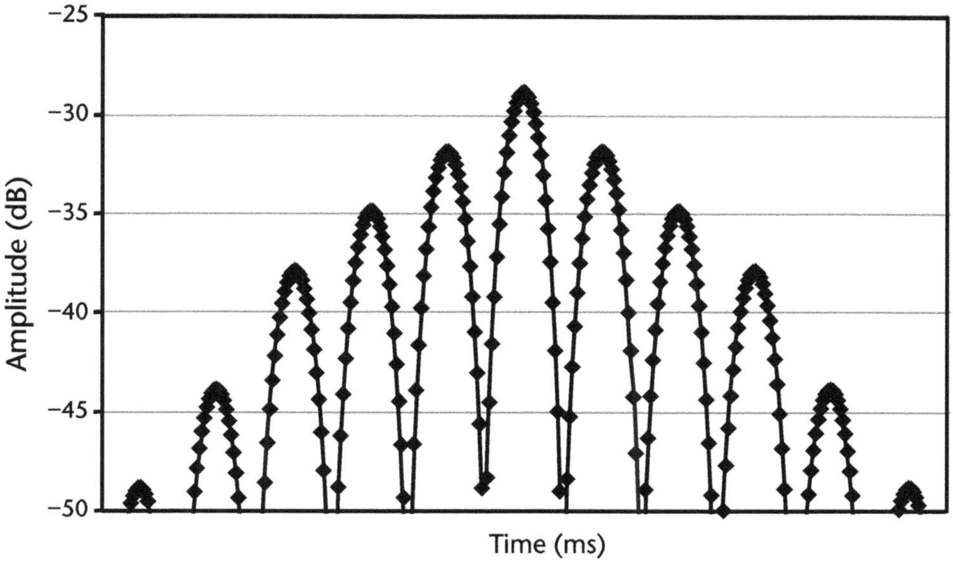

Figure 2.24 Amplitude profile from a nodding scan radar.

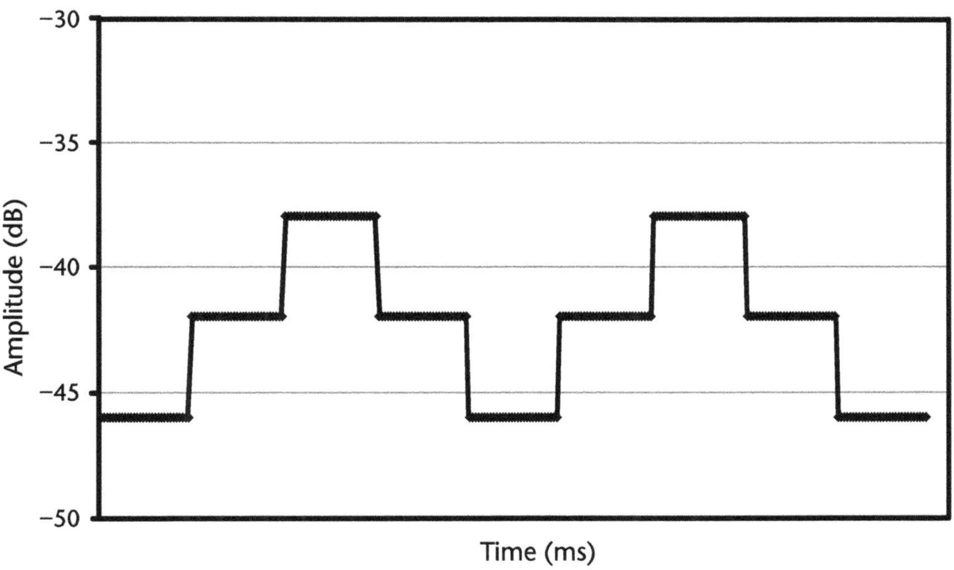

Figure 2.25 Amplitude profile from a lobe-switching scan radar.

2.8.10 Track-While-Scan

In this type of scan, the radar carries out a sector scan while using a separate beam to track a target. If the ESM receiver is within the sector being scanned then the amplitude profile seen by the ESM will be that of a sector scan. If the ESM platform is the target being tracked and then the pulses from the tracking beam all appear with equal amplitude to the ESM receiver in addition to the sector scan pulses.

2.8.11 Fixed (Locked On)

This type of scan is used by target tracking (TT) radars once they have locked onto a target. Each pulse is seen with equal amplitude by the ESM receiver.

2.8.12 Multiple Elevation Beam

Some radars operate on more than one elevation beam, as shown in the example in Figure 2.26. This radar emits each pulse on one of three different beams with elevations 10°, 30°, or 50°. The beam used for each pulse is predefined so that a pattern is seen in the amplitude profile for the radar.

The amplitude profile of the three beams can be superimposed on the overall scan profile, as shown in Figure 2.27, to show the beam on which each pulse is emitted.

2.8.13 AESA Radar Scan

AESA radars do not use a mechanically steered antenna but have an antenna array consisting of a matrix of small, solid-state transmit/receive modules (TRMs). Each TRM is capable of generating and radiating its own independent signal on individual frequencies. The amplitude profile seen from an AESA radar has a seemingly random pattern like the one simulated in Figure 2.28. This type of pulse sequence and amplitude profile are difficult for the ESM to process correctly. More details on AESA radars and the challenges they pose for ESM systems are given in Chapters 3 and 18.

2.9 Effective Radiated Power

Effective Radiated Power (ERP) is the power supplied to a radar antenna multiplied by the antenna gain in a given direction. If the direction is not specified, the

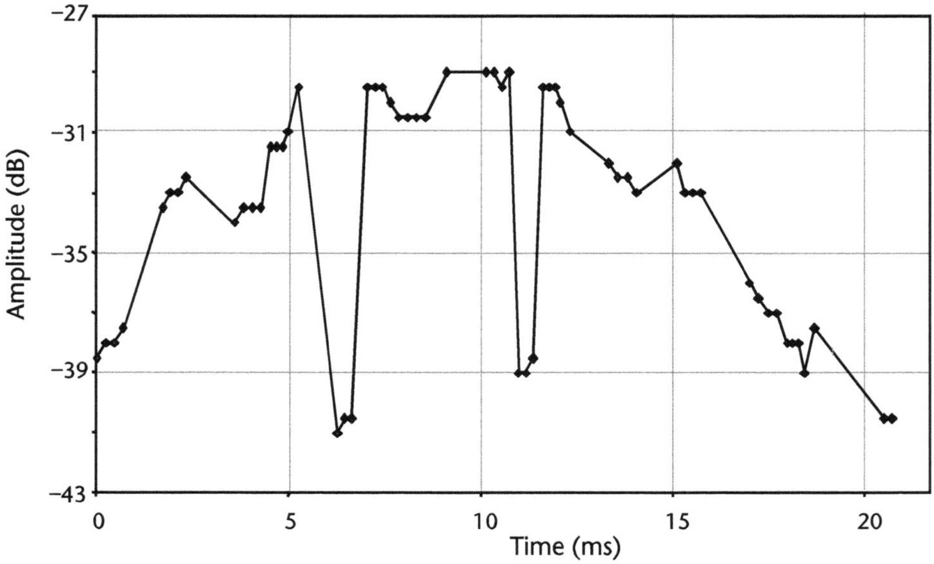

Figure 2.26 Amplitude profile of a single scan of a radar using three different elevation beams.

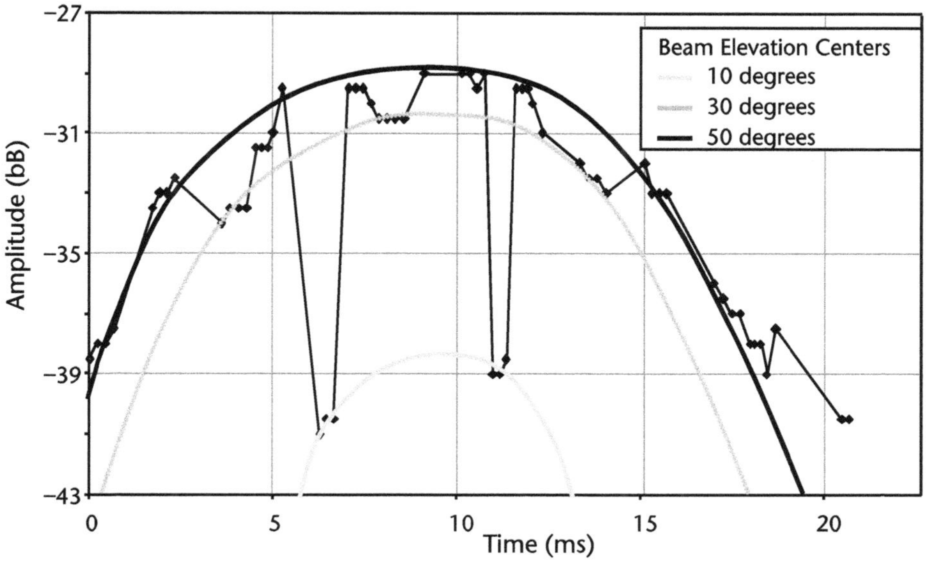

Figure 2.27 Amplitude profile of a single scan of a radar using three elevation beams with superimposed theoretical beam patterns.

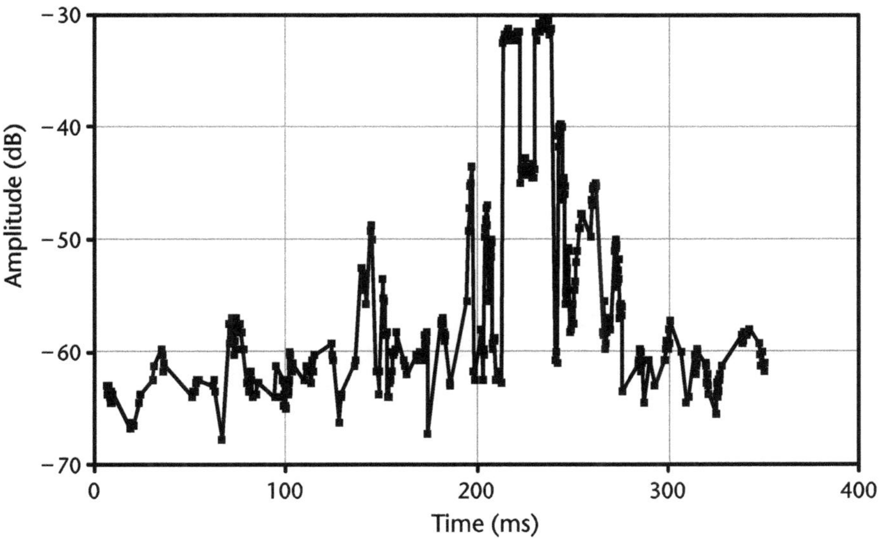

Figure 2.28 Amplitude profile from an AESA radar.

direction of maximum gain is assumed. As a matter of definition, dB is the accepted abbreviation for decibel(s). Further, dBW is the abbreviation for dB referenced to 1W. The decibel is used rather than arithmetic ratios or percentages because when circuits are connected in series, expressions of power level, in decibels, may be arithmetically added and subtracted. For example, in an RF transmission system, if a known amount of RF power, is fed into the system, and the losses, in dB, of each component (for example, lengths of feeder coax, connectors, duplexers) are known, the overall system loss may be quickly calculated with simple addition and subtraction.

The ERP of a radar is an important parameter for the ESM analyst and system designer. It is needed to predict the number and amplitude of pulses that will be received from a particular radar. Appendix A shows how ERP is used in these calculations. It is also used to determine when a radar will come into detection range of the ESM platform (i.e., it is used to find the probability of intercept for a particular radar). Typical ERP values range from 73 dBW for a harbor surveillance radar through 80 dBW for an ATC radar to more than 100 dBW for some threat radars.

2.10 Polarization

The radiation field of an antenna is composed of electric and magnetic lines of force. These lines of force are always at right angles to each other. The electric field determines the direction of polarization of the wave. When an antenna receives energy from a passing radio wave, maximum pickup will result when the antenna is oriented in the same direction as the electric field of the signal [10].

Two planes of linear polarization are usually used: vertical polarization, where the electric lines of force are in a vertical direction, and horizontal polarization, where the electric lines of force are in a horizontal direction.

Figure 2.29 illustrates the sense of travel of the radar wave for vertical and horizontal polarizations.

Linear polarization can actually take all planes but besides the horizontal plane and vertical plane only the positions: slant or sloping +45° and slant or sloping −45° have been assigned specific names.

Circular polarization arises when the electric lines of force rotate through 360° with every cycle of RF energy. Circular polarization can be right-handed or left-handed. Figure 2.30 shows an example of right-hand circular polarization.

Circularly polarized radars currently account for less than 10% of radars in total. However, they dominate the polarizations of threat radar functions such as missile guidance, beacon interrogation, and missile-mounted radars transmitting beacon signals.

In some cases, the orientation of the electric field does not remain constant. Instead, the field rotates as the wave travels through space. Under these conditions, both horizontal and vertical components of the field exist and the wave is said to have an elliptical polarization.

Polarization is important for the ESM system for two reasons: (1) the ESM system may not be able to receive pulses of all polarizations, and (2) the effects

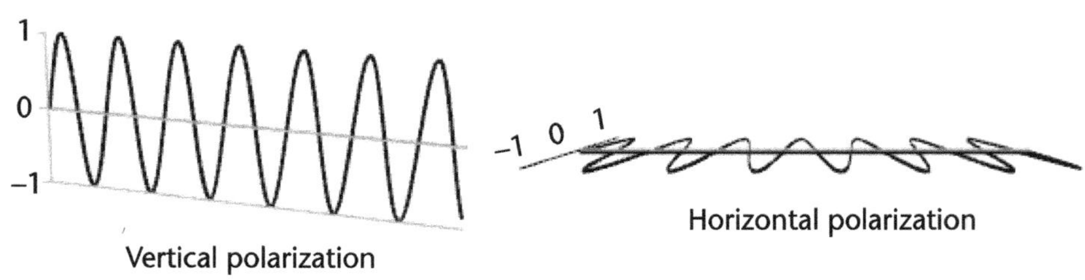

Figure 2.29 Vertical and horizontal polarizations.

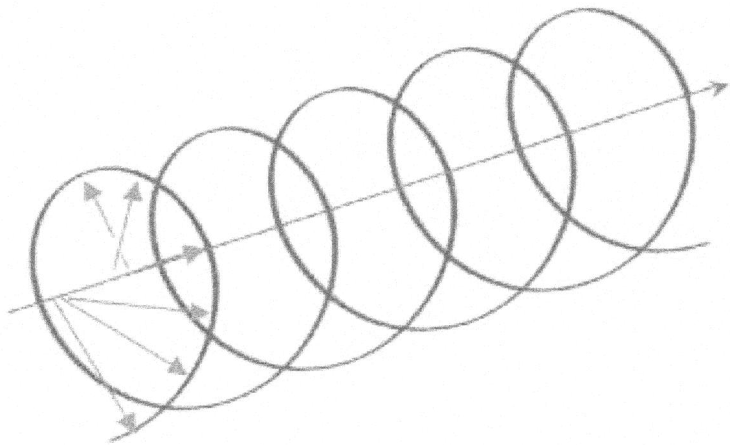

Figure 2.30 Right-hand circular polarization.

of multipath interference are different for different polarizations, as the reflection coefficient is different (see Appendix B).

ESM antennas are usually capable of receiving only one hand of circular polarization. The hand of the circular polarization changes at reflection causing problems in that the ESM receiver detects reflected pulses, but not direct path pulses when it is receiving the wrong hand of polarization. If a wrongly polarized antenna is used, then considerable losses arise, in practice between 20 and 30 dB. When the ESM antenna polarization is not matched to the polarization of the radar, pulses may be missed even when the aircraft is well within the range at which it should have been able to detect pulses from the radar.

References

[1} Barton, D. K., *Radars, Volume 7: CW and Doppler Radar*, Dedham, MA: Artech House, 1978.

[2} Belov, L. A., S. M. Smolskiy, and V. N. Kochemaso, *Handbook of RF, Microwave, and Millimeter-Wave Components*, Norwood, MA: Artech House, 2012.

[3} Laster, J., and W. L. Stuzmann, "Frequency Scaling of Rain Attenuation for Satellite Communication Links," *IEEE Transactions on Antennas and Propagation*, November 1995.

[4} Wiley, R. G., *ELINT: The Interception and Analysis of Radar Signals*, Norwood, MA: Artech House, 2006.

[5} Smith, S. W., The Scientist and Engineer's Guide to Digital Signal Processing, San Diego, CA: California Technical Publishing, 1997.

[6} Barker, R. H., *Group Synchronizing of Binary Digital Systems Communication Theory*, **Oxford, UK**: Butterworth, 1953.

[7} Skolnik, M. I., *Introduction to Radar Systems, 3rd ed.*, New York: McGraw-Hill, 2008.

[8} Skolnik, M. I., *Radar Handbook, 1st ed.*, New York: McGraw-Hill, 1970.

[9} Prathyusha, S., and R. D. Nageswar, "Radar Scan Pattern Synthesis and Implementation on FPGA," *International Journal of Scientific & Engineering Research*, June 2015.

[10} Seybold, J. S., *Introduction to RF Propagation*, New York: Wiley, 2005.

CHAPTER 3

The RF Environment

There is a popular misconception that ESM systems have to contend with pulse densities numbered in millions of pulses per second. In reality, an ESM system can only detect pulses from radars that are pointing towards it. For scanning radars, relatively few pulses are detected by the ESM system on each radar scan cycle. A tracking radar is seen constantly by the ESM and therefore has a higher received pulse density. However, even with a short PRI, of say, 50 μs, such a radar will only generate 20,000 pulses per second. Most of the time, the ESM will not be in the vicinity of a target tracking radar, so designing the ESM system with constraints on its operation to cope with very high pulse densities is likely to compromise its use as a provider of a good picture of the RF environment.

3.1 The Radar Range Equation and ESM systems

The radar range (R) is limited by the radar equation:

$$\text{Power Received} \propto 1/R^4 \tag{3.1}$$

Equation (3.1) applies to radars as the pulse must travel to the target and back to the radar to be of any use [1]. The ESM system only has to receive pulses that have been transmitted by the radar, so for the ESM:

$$\text{Power Received} \propto 1/R^2 \tag{3.2}$$

For example, a radar with a range of 60 nautical miles (nm) could be seen at a range of 3,600 nm by an ESM, if the curvature of the Earth did not limit the radar horizon.

There are two units used for power received by an ESM system. One of them, watts/m^2, does not depend on the frequency of the radar signal being received and the equation for the calculation of power received, P_r, is:

$$P_r = (P_t G_r G_t)/(4\pi R^2) \tag{3.3}$$

where P_t is the power transmitted by the radar in watts and R is the range to the radar in meters. G_t and G_r are the antenna gains of the transmitter and receiver, respectively, and can each be set to a value of 1. The result of (3.3) is expressed in dBW/m² as follows:

$$P_r = \left(dBW / m^2\right) = 10 \times \log 10\left(P_r\right)\left(Watts / m^2\right) \tag{3.4}$$

The second unit for power received is dBmi. This does depend on the frequency of the radar and the equation for the calculation of power received, P_r, is:

$$P_r = \left(P_t \lambda^2\right) / \left(16\pi^2 R^2\right) \tag{3.5}$$

In this case, P_r and P_t are expressed in milliwatts. λ is the wavelength of the radar in metres and is calculated using the radar frequency in the following equation:

$$\lambda = f / c \tag{3.6}$$

where f is the radar frequency and c is the speed of light (3×10^8 m/s).

P_r in (3.5) is expressed in dBmi using the 10 × log10 conversion. (Note: to convert from dBW to watts, use the equation Pr (dBW) = $10^{(\text{dBW})/10}$ watts.)

A comparison of the maximum ranges for receivers specified to have sensitivities of –50 dBmi and –50 dBW/m² gives 430 nm and 480 nm, respectively, for a 3-GHz signal transmitting 80 dBW of power. At other RFs, the maximum ranges for the two receiver sensitivities differ. More details on the derivation of required ESM sensitivity are given in Chapter 4.

3.2 Radar Horizon

The radar horizon increasing with altitude is the real governing factor for the range at which the ESM can see the radar. The curvature of the Earth means that there is a limit on the range that radar signals, traveling in straight lines, can be seen by an ESM system. The radar horizon depends on the height of the transmitting radar and also on the altitude of the ESM platform. Figure 3.1 is a simple diagram illustrating this point.

Simple equations for calculating radar horizon are [2]:

$$radar\,horizon = 1.23\sqrt{(h)} \tag{3.7}$$

where the h is the height of the ESM platform in feet and the radar horizon is in nautical miles and

$$radar\,horizon = 4.124\sqrt{(h)} \tag{3.8}$$

where the h is the height of the ESM platform in meters and the radar horizon is in kilometers.

Equations (3.7) and (3.8) are good enough approximations for use in ESM analysis and for the simulation of the radar environment. However, they do not

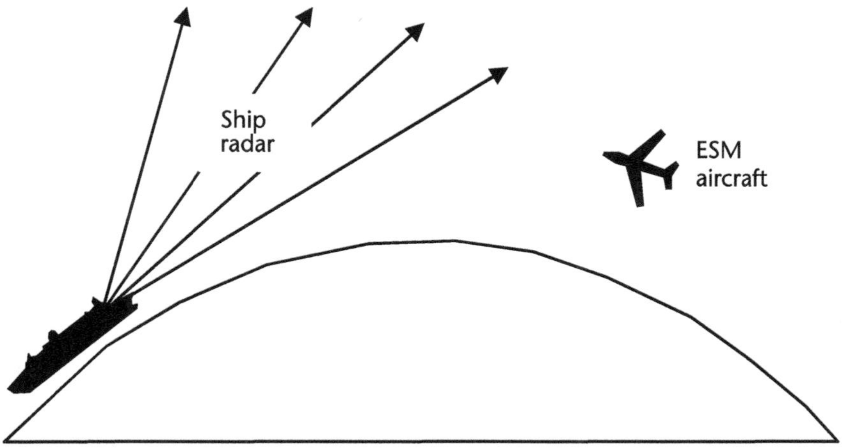

Figure 3.1 The radar horizon.

take into account the height of the radar. If this is known, then the following equations give more accurate measures of the radar horizon:

$$radar\,horizon\,(nm) = 1.23 \times \left(\sqrt{(h_{ESM})} + \sqrt{(h_{radar})}\right) \tag{3.9}$$

$$radar\,horizon\,(km) = 4.124 \times \left(\sqrt{(h_{ESM})} + \sqrt{(h_{radar})}\right) \tag{3.10}$$

In (3.9) the heights of the radar and ESM are in feet and in (3.10) both heights are in meters.

Table 3.1 shows the radar horizon in kilometers and in nautical miles for a variety of platform altitudes.

3.3 Types of Radar and Their Functions

The most common types of radar carry out functions such as air traffic control (ATC), weather detection and early warning, and coastal and harbor surveillance (HS). Ship navigation and aircraft transponders, altimeters, and airborne weather radars also make up substantial radar categories. The detection of threat radars is a

Table 3.1 Radar Horizon at Various Altitudes

Aircraft Altitude (ft)	*Radar Horizon (km)*	*Radar Horizon (nm)*
200	32	17
1,000	72	39
2,000	102	55
5,000	162	87
10,000	227	123
20,000	322	174

very important function of the ESM system, so an understanding of the parameters of radars carrying out functions such as target acquisition, tracking, and missile guidance is needed for the design of ESM systems.

3.3.1 ATC Radars

ATC radars monitor civil and military air traffic. This category of radar covers primary radar where aircraft are detected within a particular area and secondary radar relying on transponder responses from interrogated aircraft.

Primary ATC radars transmit pulses that are reflected by the target aircraft and then received by the same radar. They are usually located at fixed radar sites, either at an airport or at specific air traffic control centers monitoring large regions of airspace. Most ATC radars operate in E-band with staggered PRI. They often use dual RF and may have more than one PW and an ERP of at least 75 dBW. Operating with simple circular scans of a few seconds scan period, they have beamwidths of 1° or 2°, meaning that there are several tens of pulses available for detection in the main beam, assuming that the range to the ESM is adequate to allow for detection.

An example of the parameters of a primary ATC radar is given in Table 3.2. The radar emits pulses on its two RFs, in turn, with the same PRI sequence for each RF. The pulse on RF2 is emitted 20 μs after each pulse on RF1, as shown in the pulse sequence in Table 3.3.

In most RF environments, there are only a few examples of primary ATC radar so it should be possible for the ESM to distinguish between them. However, problems may arise if there are several of the same type of radar using the same PRI sequence. Chapter 14 discusses this in more detail.

Table 3.2 Parameters of a Typical Primary ATC Radar

RF (MHz)	*PRI Sequence (μs)*	*PW (μs)*	*Scan Period (seconds)*	*3-dB Beamwidth and First Sidelobe Level*	*ERP (dBW)*
2,780	1,020, 1,030	0.8	4	1.5°, –23 dB	82
2,930	1,060, 1,040, 1,050	12			

Table 3.3 Pulse Sequence for the ATC Radar

Pulse	*TOA*	*RF*	*TSLP**	*PW*	*Pulse*	*TOA*	*RF*	*TSLP**	*PW*
1	10:00:00.000	2,780		0.8	11	10:00:00.520	2,780	1,030	0.8
2	10:00:00.002	2,930	20	12	12	10:00:00.522	2,930	20	12
3	10:00:00.102	2,780	1,000	0.8	13	10:00:00.622	2,780	1,000	0.8
4	10:00:00.104	2,930	20	12	14	10:00:00.624	2,930	20	12
5	10:00:00.205	2,780	1,010	0.8	15	10:00:00.725	2,780	1,010	0.8
6	10:00:00.207	2,930	20	12	16	10:00:00.727	2,930	20	12
7	10:00:00.311	2,780	1,040	0.8	17	10:00:00.831	2,780	1,040	0.8
8	10:00:00.313	2,930	20	12	18	10:00:00.833	2,930	20	12
9	10:00:00.415	2,780	1,020	0.8	19	10:00:00.935	2,780	1,020	0.8
10	10:00:00.417	2,930	20	12	20	10:00:00.937	2930	20	12

*TSLP = Time since last pulse in microseconds.

Secondary ATC radar has two components, a ground transmitter and a transponder mounted on the airborne target. The ground unit, called the interrogator, transmits coded pulses towards the target aircraft using an RF of 1,030 MHz. The transponder on the airborne platform receives the pulse, decodes it, prepares a suitable coded response, and then transmits it back to the ground unit at an RF of 1,090 MHz. The data received from the aircraft is displayed on the display of the primary ATC radar.

There are several modes of interrogation, each indicated by the difference in spacing between two transmitter pulses, known as P1 and P3 [3] and shown in Figure 3.2. Each mode produces a different response from the aircraft. A third pulse, P2, is for sidelobe suppression.

A mode-A interrogation elicits a 12-pulse reply, including an identity number associated with that aircraft. The 12 data pulses are bracketed by two framing pulses, F1 and F2. A mode-C interrogation produces an 11-pulse response including aircraft altitude. The ground antenna is highly directional but cannot be designed without sidelobes. Aircraft could also detect interrogations coming from these sidelobes and reply appropriately. These replies cannot be differentiated from the intended replies from the main beam and can give rise to a false aircraft indication at an erroneous bearing.

To overcome this problem, the ground antenna has a second, mainly omnidirectional, beam with a gain that exceeds that of the sidelobes but not that of the main beam. A third pulse, P2, is transmitted from this second beam 2 μs after P1. An aircraft detecting P2 stronger than P1 (therefore in the sidelobe and at the incorrect main lobe bearing) does not reply

Mode S [4] has different interrogation characteristics. It comprises pulses P1 and P2 from the antenna main beam to ensure that Mode-A and Mode-C transponders do not reply, followed by a long phase-modulated pulse.

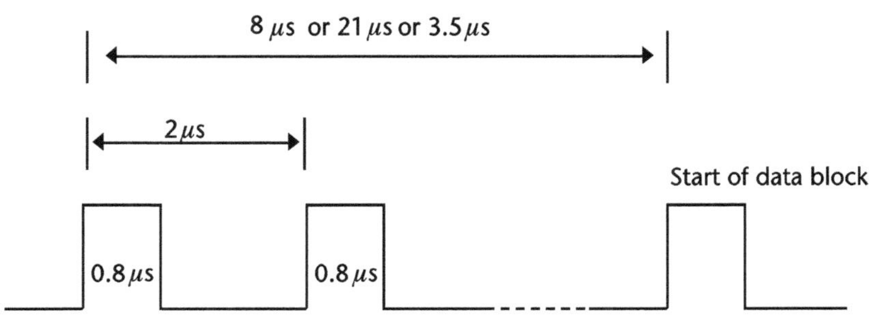

Mode	P1 – P3 Pulse spacing	Purpose
A	8 μs	identity
C	21 μs	altitude
S	3.5 μs	multipurpose

Figure 3.2 IFF modes.

It is difficult for an ESM system to make much sense of this type of signal as, in busy air traffic environments, there are likely to be many sets of pulses at the two RFs (1,030 and 1,090 MHz) with seemingly random PRIs. In general, it is useful for the ESM to have a filter to avoid the pulses at these RFs.

3.3.2 HS Radars

Traditional HS radars operate in I-band and have very narrow 3-dB beamwidths and fast scans, making them difficult to detect by an ESM system, despite usually having fixed RF and simple PRI and PW. The reason for this is that most of the pulses have powers well below that of the peak of the main beam. Section 3.7 discusses this further.

Modern HS radars operate in a variety of RF bands and often scan electronically in azimuth using passive electronically scanned arrays. The typical power output from such radars is only a few watts making them difficult to detect by ESM systems [5].

3.3.3 Airborne Radars

This category of radar is the most challenging for the ESM system. Since aircraft are fast-moving, it is difficult for the ESM to correlate intercepts from airborne radars into existing tracks. An example of the difficulty that an ESM system has in tracking an airborne radar is shown in Figure 3.3. In this plot, there are multiple tracks in existence for a single radar, a situation that is unfortunately common for current ESM systems.

Military aircraft radar cause many problems for ESM systems. This type of aircraft may not travel on a predicted path and has radars with more complexity than straightforward civilian airliner radars. The most modern of the airborne intercept radars are active electronically scanned arrays (AESA), discussed in more detail in Section 3.3.9 and Chapter 18.

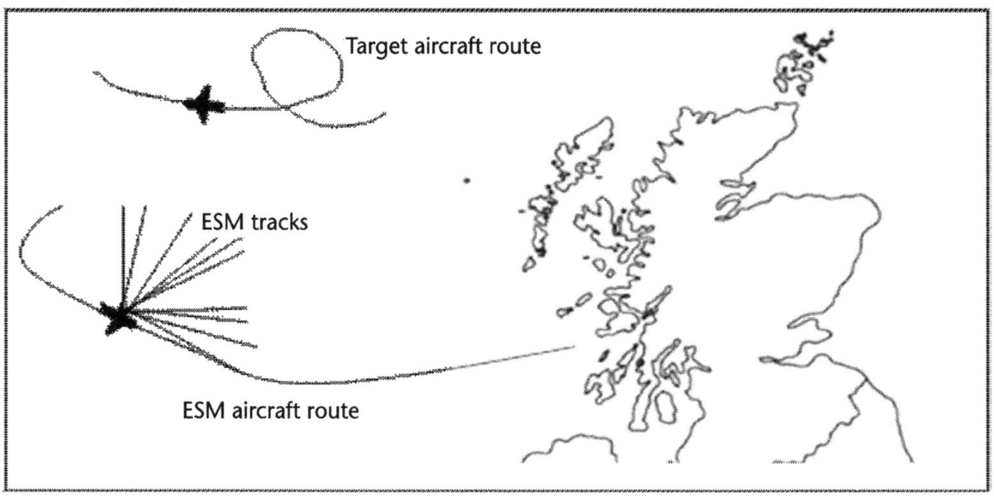

Figure 3.3 ESM tracks for a single airborne radar.

Another particular type of airborne radar is that seen on early warning aircraft such as Airborne Warning and Control System (AWACS). This radar operates in the E/F-band and has a nodding circular scan, giving rise to a distinctive amplitude pattern [6].

Radar altimeters form another category of airborne radar. These radars often operate in the RF range of 4.2 to 4.4 GHz, a radar band that is typically not used for any other purpose [7]. Radar altimeter antennas have a fairly large main lobe of about 80° so that at bank angles up to about 40° the radar can detect the range from the aircraft to the ground (specifically to the nearest large reflecting object).

3.3.4 Ship Navigation Radars

Ships over 300 tons and passenger ships are mandated to have an I-band (~9 GHz) radar and ships over 3,000 tons must also have an E/F band (~3 GHz) radar [8]. Traditional ship radars have simple RF with several PRI and PW modes matched to the range at which they are set to operate. Some examples of the parameters of ship radars are given in Table 3.4.

The pulses from these radars are easy for the ESM system to detect and, in a sparse RF environment, it should be possible to deinterleave the pulses from the individual radars. Later in this chapter, there are descriptions of different RF environments including the challenges for the ESM in dealing with this type of radar.

Newer versions of ship navigation radars are broadband radars, considered further in Chapter 18. These radars are of such low power that it is unlikely that they would be detected by an ESM system.

3.3.5 Weather Radars

Weather radars [9] are a type of radar used to locate precipitation, calculate its motion, and estimate its type (rain, snow, hail, and so forth). Modern weather radars are mostly pulse-Doppler radars, capable of detecting the motion of rain droplets in addition to the intensity of the precipitation. The most common RF range for weather radars is 5 to 6 GHz. They typically operate with a fixed PRI and PW and have circular or helical scans, with a relatively long scan periods of about 10 seconds. Table 3.5 shows the parameters of typical weather radars.

The RF band where these radars are found is relatively sparsely populated. The simplicity of the parameters of typical weather radars therefore makes them good candidates for ESM testing and trials.

Table 3.4 Examples of Ship Navigation Radar Parameters

Radar	*RF (MHz)*	*PRI Sequence (μs)*	*PW (μs)*
1	3,040	760, jittered ±10	0.8
2	3,060	980 fixed	0.3
3	9,380	1,020 fixed	0.7
4	9,410	330, 350, 370	0.3

Table 3.5 Examples of Weather Radar Parameters

Radar	*RF (MHz)*	*PRI Sequence (μs)*	*PW (μs)*
1	5,630	3,333	2.0
2	5,420	3,330	2.5
3	5,590	2,500	3.0

3.3.6 Threat Radars

Threat radars are those that are associated with weapons. Weapon systems typically have radars for target acquisition (TA), target tracking (TT), and missile guidance (MG). Other radar types associated with weapons are missile homing (MH), beacon interrogators (BI), target illuminators (TI), missile beacon (BN), and fire control (FC) radars.

A few notes are given here on these types of radar and their functions, but it is not possible to give much information about the parameters of threat radars as nation states are sensitive about the release of data concerning their weapon systems. However, there are some sources of information in the public domain [10–12].

TA radars usually have circular scans, although sector, conical, or raster scans are also used by some TA emitters. The scan periods are typically of a few seconds for circular scans but may be as long as a minute for raster scans. TA radar beams have small 3-dB azimuth beamwidths, often of only 1° or 2° and may have ERP up to 130 dBW. There is no one RF range that is particular to this radar function, and there are examples of TA radars in all radar bands.

TT radars lock onto a target and then emit pulses in the direction of the target. These are the only types of pulsed radar that may be received 100% of the time by the ESM system. If the ESM platform has been targeted by a TT radar with a PRI of 1 ms, the pulse density received will be 1,000 pulses per second (pps). Even if the radar has a very short PRI of 10 μs, then 100,000 pps will be received. In reality, the PRIs of such radars are unlikely to be as short as 10 μs.

MG radars often use pulse position modulation (ppm), meaning that there are pulse doublets or triplets used to convey guidance commands to a missile. An example of a ppm missile guidance signal is given in Figure 2.6. This type of signal is difficult for an ESM to deinterleave and correctly identify, so precise specifications of their parameters should be put into the ESM radar library to ensure that they can be identified.

A BI signal may appear in conjunction with a MG signal and be related in some way to the MG signal either by a known separation in RF or by synchronization of the BI signal with the PRI sequence of the related TT radar. BI signals usually have fixed RF, use either fixed PRI or simple stagger sequences of no more than 3 positions/3 elements and have short pulse widths of less than 1 μs.

TI radars usually transmit a CW signal to be used by a missile to show it the direction in which to travel in to reach the target. .Azimuth beamwidths are generally small, of only 1° or 2°; however, some TI radars have wider beamwidths than this and narrow their beamwidths as the missile gets near to its target.

BN signals are found in several RF bands with PRI values varying from CW to 2 ms. ERP values are not usually high, at most 30 to 40 dBW. Some BN signals are circularly polarized, which, coupled with low ERP, make these signals difficult to detect and identify.

MH signals are usually of high PRF with only a few microseconds between pulses and they may have RF agility. One of the other features that MH signals have in common with BN signals is that they are of relatively low ERP, typically about 50 dBW. Azimuth beamwidths range from 2° to 15° and they all operate with a tracking scan or vertical sector scan.

An FC radar provides target information (azimuth, elevation, range, and range rate) for use in the direction of a weapon. FC radars use pencil beams and are often operated in synchronization with a scanning (track-while-scan) radar. The FC function is increasingly being carried out by AESA radars with other radar functions carried out simultaneously by this type of radar.

3.3.7 Cellular/Mobile Phones

There are several systems for designating mobile phone signals including Global System for Mobile (GSM), Long Term Evolution (LTE), and Universal Mode Telecommunications System (UTMS). Apart from a few obscure network providers, all of the signals operate in the RF range 800 to 2,600 MHz. This frequency range encompasses secondary ATC radar and some threat radars, but does not interfere with the RF ranges of the majority of radars from which the ESM expects to receive pulses.

If mobile phone signals are seen by an ESM system, they are likely to appear as a stream of pulses with short PRI. This can have serious consequences for the ESM system such as filling up the pulse buffers with these unwanted pulses leaving the desired radars undetected. Once the frequencies used by mobile phone operators in the region in which the ESM is to operate are known, it should be possible to filter out pulses at the specific frequencies. For example, the mobile phone network frequencies used in Europe are given in Table 3.6.

With the advent of 5G mobile phone signals, there will probably be more RF ranges that could cause problems for the ESM system. The exact frequencies are not known at the time of this writing, but 3.4–3.8 GHz and 24.25–27.5 GHz have been discussed [13].

Table 3.6 Cellular (Mobile) Bands Used in Europe

Mobile Band	*Frequency (MHz)*	*Uplink (MHz) (Mobile to Base)*	*Downlink (MHz) (Base to Mobile)*
E-GSM-900 (2G and 3G*)	900	880.0–915.0	925.0–960.0
DCS-1800 (2G and 4G)	1,800	1,710.2–1,784.8	1,805.2–1,879.8
IMT (3G)	2,100	1,920–1,980	2,110–2,170
IMT-E (4G)	2,600	2,500–2,570	2,620–2,690
EU Digital Dividend† (4G)	800	832–862	791–821

*G refers to generation

†This is the frequency spectrum freed up when digital TV was introduced.

3.3.8 Low Probability of Intercept Radars

Low probability of intercept (LPI) radars [14] are those whose agile parameters make them difficult to detect by ESM systems. For example, RF agility, a complex PRI sequence, long chirped pulses, and transmitting pulses on several elevation beams would make it difficult for an ESM system to correlate pulses into a single intercept for a radar and also increase the difficulty of correlating intercepts into existing ESM tracks.

Constructing a radar beam so as to minimize the sidelobes and back lobes may also reduce the probability of intercept. However, when a radar is sweeping a large volume of space for targets, it is likely that the main beam will point at the ESM platform frequently enough for ESM detection to occur. Reducing sidelobes and back lobes is desirable as it makes the radar more difficult to identify.

LPI radars using parameter and power manipulation have been in service for many years and careful design of deinterleaving and identification processes can allow an ESM to cope with standard LPI radars. However, the latest type of LPI radar, the AESA radar, is likely to cause problems with existing ESM systems.

3.3.9 AESA Radars

Conventional radars, with rotating antennas, operating on fixed frequencies or with limited frequency agility usually perform only one function, with separate units required for surveillance, tracking, and targeting. However, AESA radars are capable of carrying out several functions simultaneously. The AESA radar antenna does not move but consists of a matrix of small, solid-state transmit/receive modules (TRMs). Each TRM is capable of generating and radiating its own independent signal allowing the AESA to produce radar pulses on different frequencies with interleaved pulse streams carrying out several functions simultaneously [15].

The use of multiple frequencies creates difficulties for ESM systems. The AESA can change its frequency with every pulse and generally does so using a random sequence so it is very difficult for an ESM system to deinterleave pulses from a specific radar. The AESA does not need to have a fixed PRI, making an AESA a low-probability of intercept radar. AESA radars provide surveillance and dedicated tracking of hundreds of targets with a range of 400 km.

More information about this type of radar is given in Chapter 18.

3.4 Radar Pulse Density

Understanding the radar pulse density and the range at which radar detection is possible are fundamentally important to ESM analysis and design. A typical radar has a PRI of about 1 ms, so it emits 1,000 pulses per second. However, the ESM does not detect 1,000 pulses per second from the radar for two reasons; the shape of the radar beam and the scanning characteristics of the radar. As described in Chapter 2, the radar beam pattern has a narrow 3-dB beamwidth with low sidelobes and scans past the ESM platform with a scan period of a few seconds.

Figure 3.4 shows radars scanning past an ESM system. The purpose of showing this picture is to emphasize that most of the time the ESM cannot detect the radars

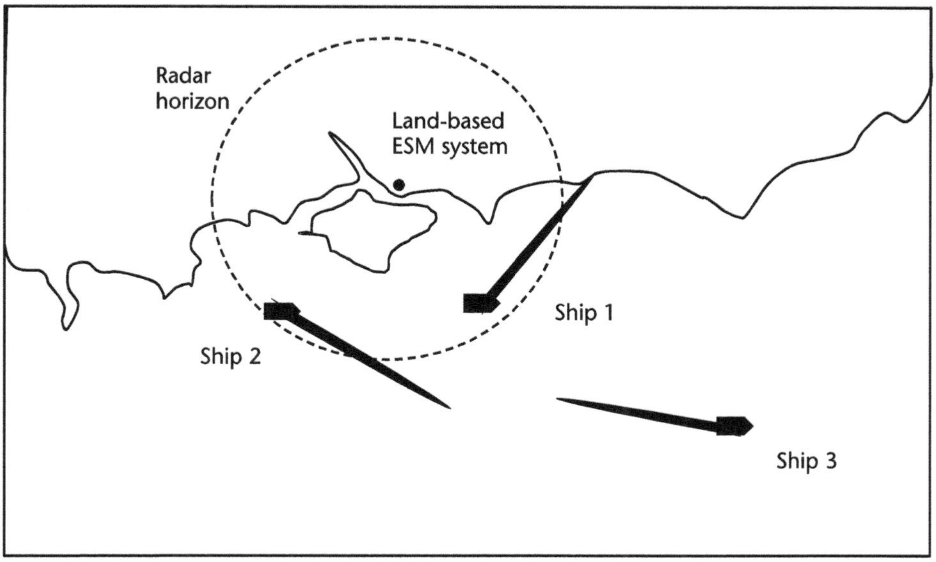

Figure 3.4 Land-based ESM and scanning radars.

as they are pointing away from the ESM location. At the instant of time shown in Figure 3.4, the radars mounted on ships 1 and 2 are not detected by the ESM system as the main beams of the radars are pointing away from the ESM location. They will only be detected by the ESM when the radar beam scans past the ESM. The radar on ship 3 shown on the picture will not be seen at all by the ESM as it is outside the radar horizon.

A radar is only seen for a percentage of time that the radar is emitting: so, for a relatively small duty cycle. In Figure 3.5, the amplitude profiles of an ATC radar operating at various ranges to the ESM system are shown. This simulated radar has a scan period of 3.5 seconds, an average PRI of 900 μs, 3-dB beamwidth of 1.5°, and a first sidelobe level of –24 dB.

At long range, the ESM system sees the radar for less than 1% of the time that it is transmitting. At medium ranges (25 to 50 nm), the radar is still not detected for at least 95% of the time and it is only at these ranges that the sidelobes are detected. It is only when the ESM platform is at short range to the radar (5 nm) that the radar is seen for 20% of its scan cycle. In this case, there are many sidelobe pulses seen.

3.5 Example of Low-Pulse Density Environment: South Australia

South Australia has a relatively low radar pulse density, so it has been chosen as the first example to examine radar pulse density. In the vicinity of Adelaide, the capital of the state of South Australia there are 6 land-based radars. Three of them are ATC radars and the other three are weather radars. Their locations are shown on the map in Figure 3.6. At any one time, there are likely to be 8 ships contributing to the radar pulse density in this environment. Figure 3.6 shows the location of the ships chosen to be representative of the marine traffic seen off the coast of South Australia.

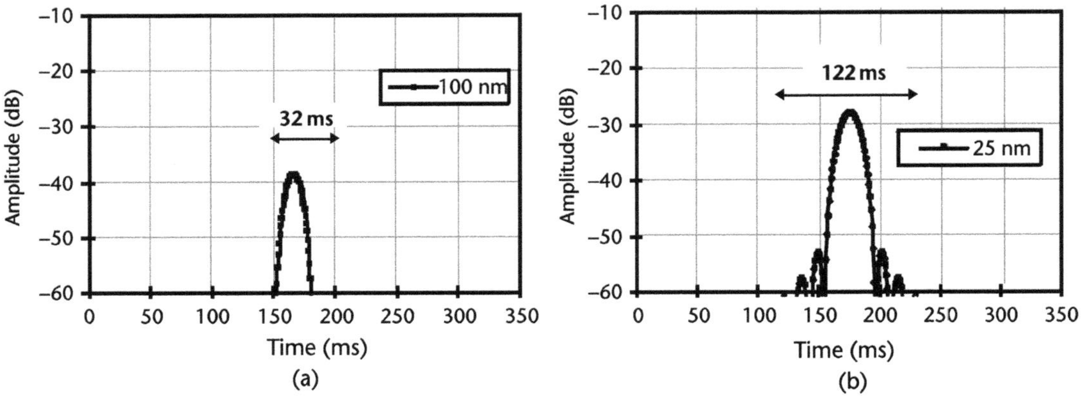

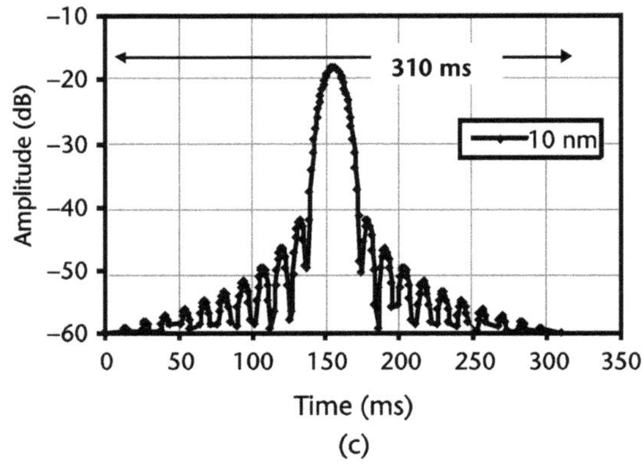

Range (nm)	Duration of Pulses (ms)	Number of Pulses	% of Radar Scan Time
200	28	32	0.8
150	30	34	0.9
100	32	36	0.9
50	52	52	1.4
25	122	90	3.5
15	202	150	5.8
10	310	221	8.8
5	633	450	18.1

(d)

Figure 3.5 Amplitude profiles of a scanning radar at various ranges.

Studies have shown that shipping generally follows a repeatable pattern [16] so that the same ships can reliably be predicted to be in a particular area. The ships shown in Figure 3.6 are therefore very typical of those seen in this area of ocean.

When considering the radar pulse density, the limiting factor for the ESM is the radar horizon. Figure 3.7 shows the map of ground truth for the Adelaide area but this time with radar horizon rings superimposed on it. All the radar horizon rings are centered on Adelaide itself and all the radar locations lie within a radar horizon of 350 km produced for an altitude of 20,000 ft.

Each radar is considered in turn and the number of pulses likely to be detected on each scan is calculated using the radar parameters such as 3-dB beamwidth, PRI, and scan period. The range to the radar is also needed, as is the ERP of the radar. The graph on Figure 3.8 is generated to show the time for which pulses are detected from each radar on each scan cycle.

In conjunction with the creation of the timeline plot of pulses for each radar, a plot of the amplitude profile for the radars is generated in Figure 3.9. There are 5,000 pulses from all the radars in the 20-second time period shown on Figure 3.9, making a pulse density of only 250 pps. The ATC radars contribute most of the pulses with 155 pps and the weather radars contribute on average only 9 pps.

At this level of pulse density, the ESM system should be able to deinterleave the pulses from the various radars and should produce a near perfect picture of the RF environment.

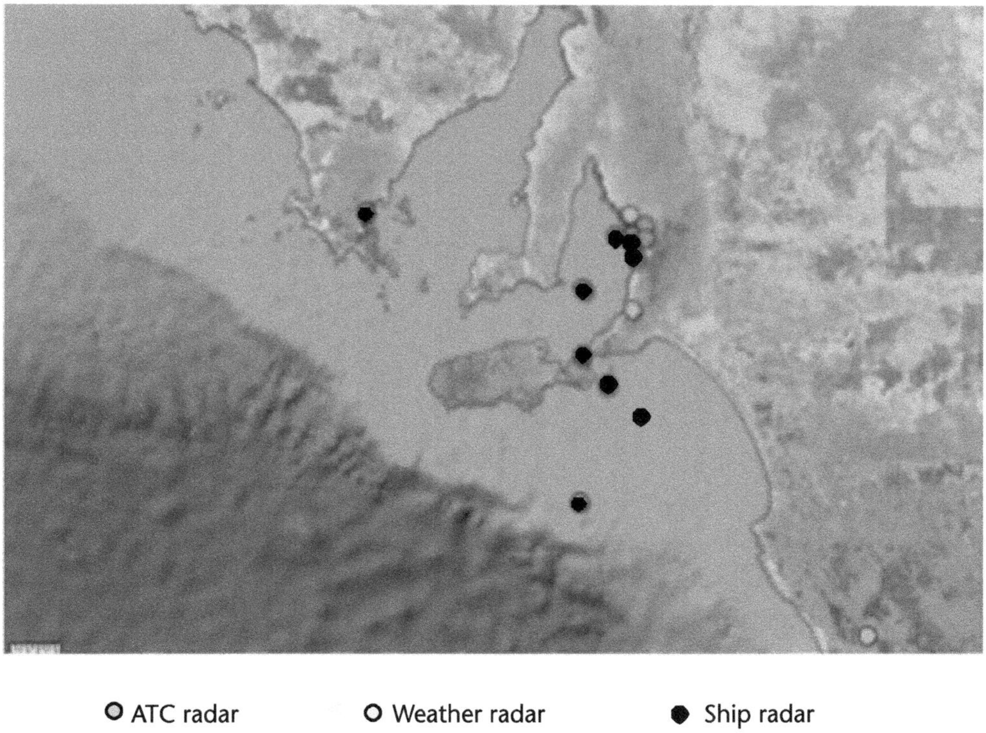

Figure 3.6 Map of radar ground truth for the Adelaide area.

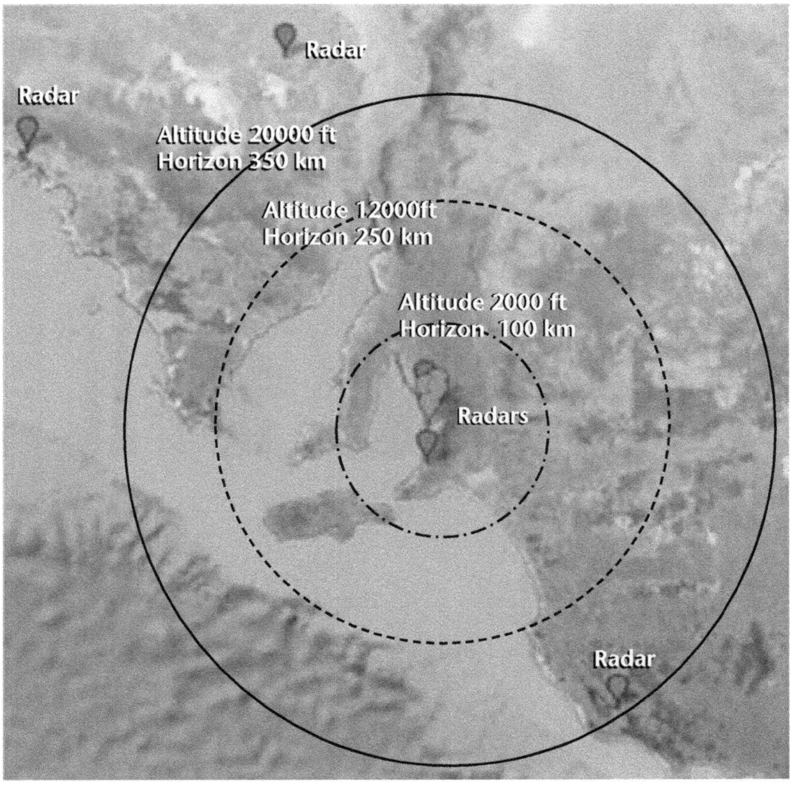

Figure 3.7 Map of radar ground truth with the radar horizon for various altitudes.

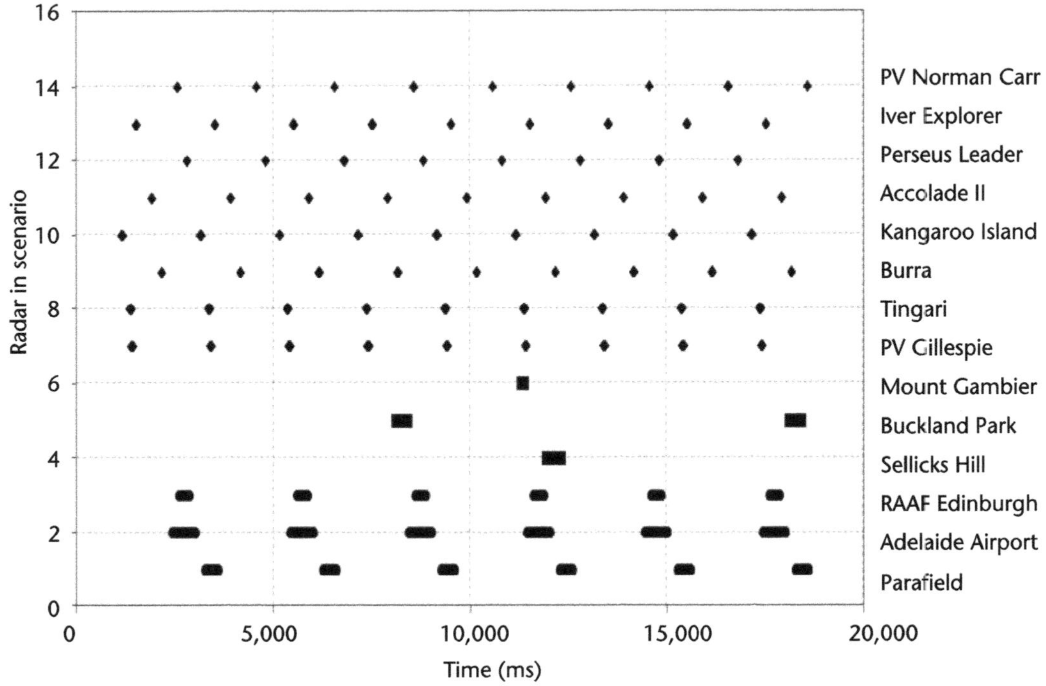

Figure 3.8 Timeline of pulses seen by an ESM platform 20,000 ft above Adelaide.

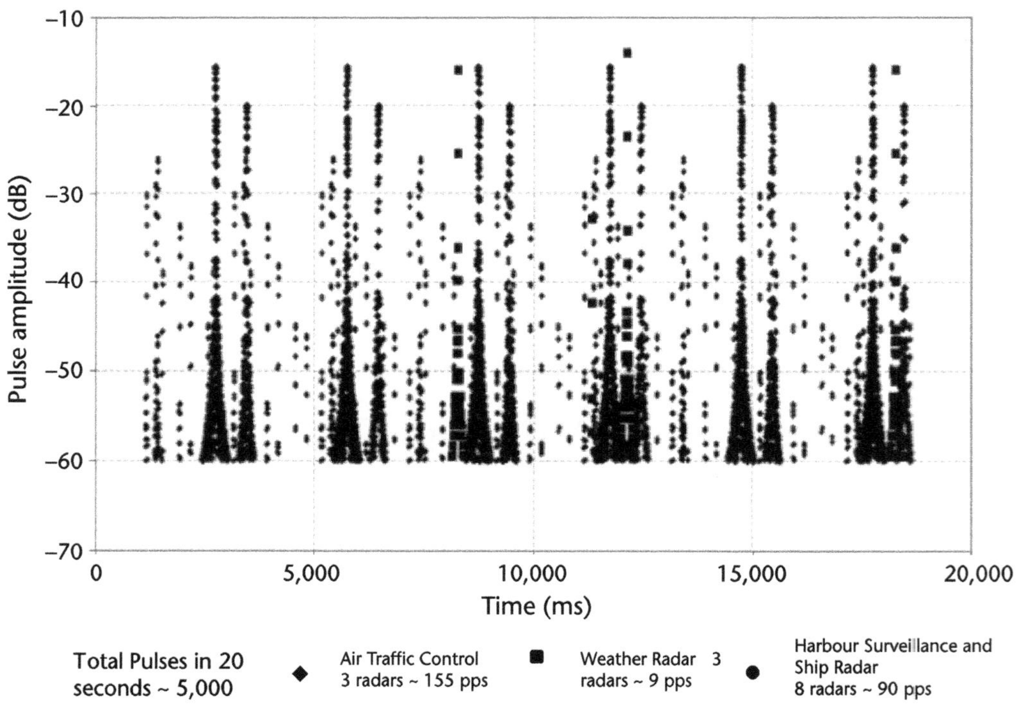

Figure 3.9 Amplitudes of pulses seen by an ESM platform 20,000 ft above Adelaide.

3.6 Example of High-Pulse Density Environment: Straits of Malacca

At the other end of the scale of pulse density environments lies the Straits of Malacca. This is one of the busiest shipping areas in the world with 10% of the world's large ships visiting the area every year. The map in Figure 3.10 shows more than 200 ships, each with the mandatory two radars, one of which is in the RF band at around 3 GHz and the other in the 9-GHz RF band. Figure 3.10 is actually a map of targets transmitting Automatic Identification System (AIS) data [17]. The picture shown is that detected by satellite AIS and made available on a number of commercial web sites [18, 19].

With 200 radars emitting pulses in this environment, there are pulses being received continuously by an ESM system. Figure 3.11 shows the amplitude profiles of a simulation of the pulses from all the radars. In the 2-second time period shown on this plot the scan peaks from all 200 radars can be seen.

With all these ship radars transmitting it might be thought that the ESM would experience a very high pulse density. However, as described above, the ESM can only see these radars when they are transmitting in the direction of the ESM, so the situation is not as bad as might be thought. There are only actually 4,000 pulses shown on Figure 3.11, so the pulse density is 2,000 pulses per second (pps).

Figure 3.12 shows the timeline of pulses emitted by ship radars in the Malacca Strait. Pulses from all 200 radars are shown here.

However, as these radars are only seen for a small percentage of the time that they are transmitting, there are only pulses from a small number of radars interleaved at any one time.

To illustrate this point, Figure 3.13 is a plot of the amplitude of pulses seen in a time period of 180 ms. There are pulses about 400 pulses from 16 radars.

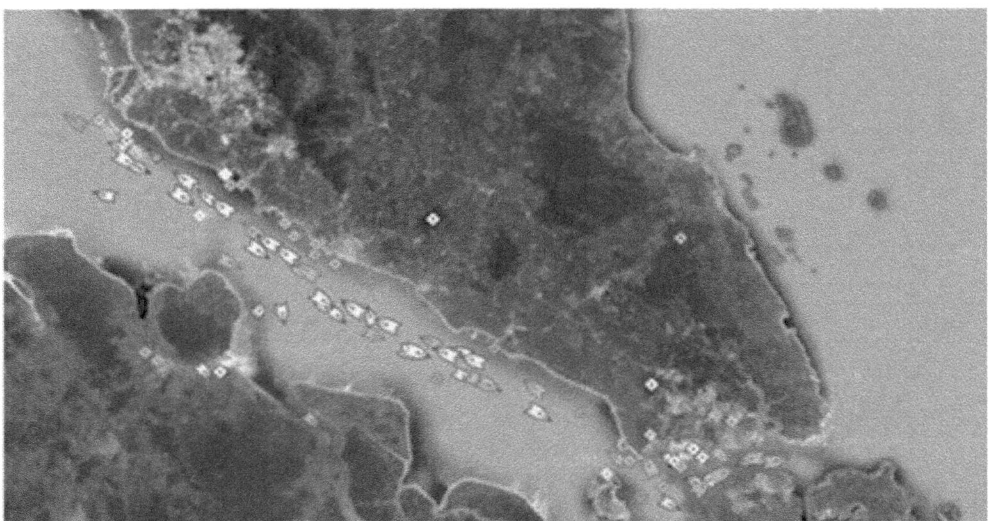

Figure 3.10 Shipping in the Straits of Malacca.

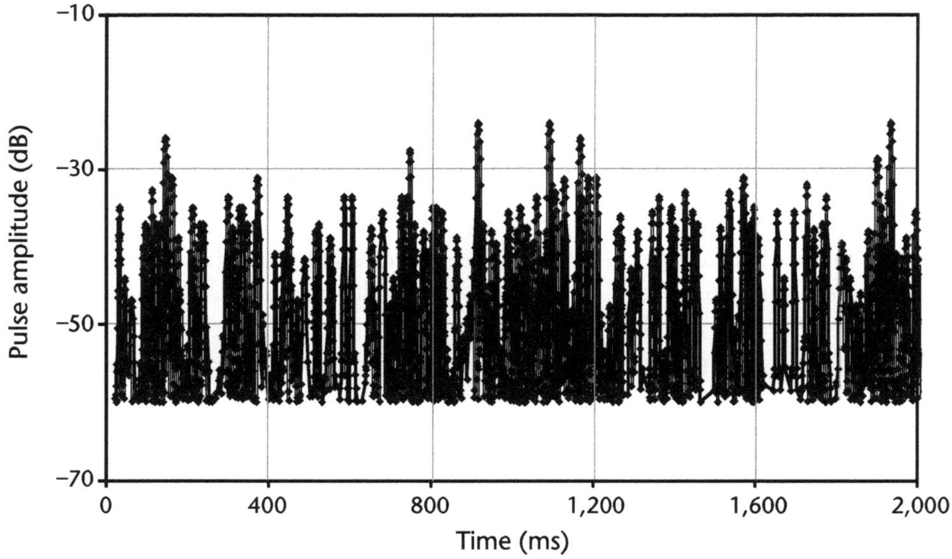

Figure 3.11 Amplitude of pulses from 200 radars in the Straits of Malacca.

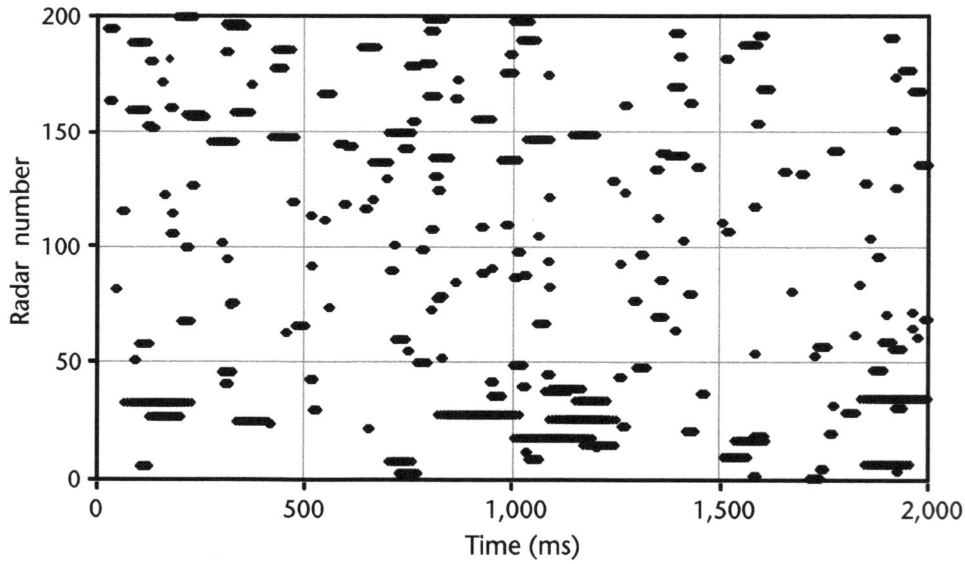

Figure 3.12 Timeline of radar pulses in the Malacca Straits.

At any one time it looks as if there are a maximum of five radar pulse sequences interleaved.

Figure 3.14 shows the pulse number in a sequence of 400 pulses together with the radar number. On this plot, there are actually never more than three radars with interleaved pulses and often only two radars have interleaved pulse sequences.

To conclude, even in a very busy RF environment, the radar pulse density is of the order of a few thousand pulses per second, and there are never going to be more than a handful of radars whose pulses are interleaved. This reality of the radar pulse environment should be taken into account when designing an ESM system.

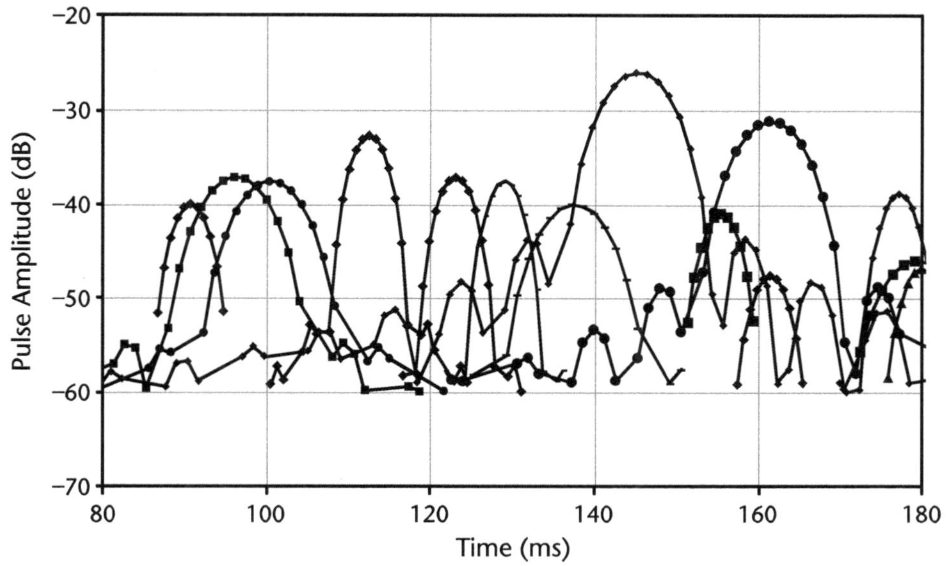

Figure 3.13 Detail of predicted amplitude profiles.

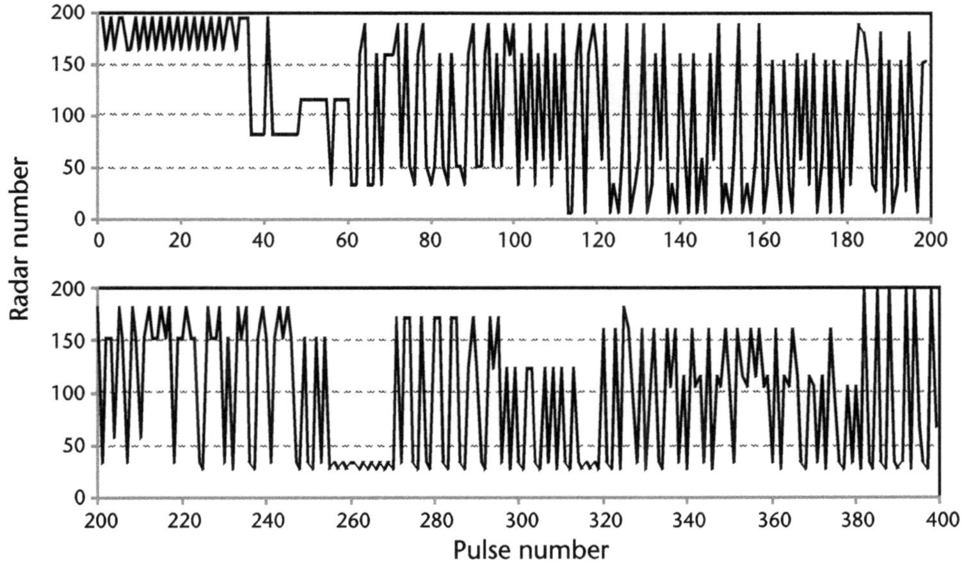

Figure 3.14 Graph showing interleaved pulses from radars in the Straits of Malacca.

3.7 Number of Pulses Needed for ESM Detection

The designer of the ESM system must decide on the minimum number of pulses that the ESM receiver should detect before reporting an intercept from a particular radar. Typically, six pulses are needed to allow the for parameter extraction from the intercept.

Some ESM manufacturers have expressed the desire to use a larger number of pulses for intercept creation, but this is not likely to prove reasonable. The characteristics of current scanning radars carrying out functions such as target acquisition for weapons systems mean that detecting more than six pulses at a safe range for the ESM platform is difficult. In reality, ESM systems are likely to have to be able to make use of fewer pulses per scan in the future.

The amplitude profile of a typical ATC radar is shown in Figure 3.15 with each pulse shown separately. The main beam and sidelobes of this radar comprise 75 pulses, with 15 pulses lying within the 3-dB beamwidth. This radar has a beamwidth of 1.5°, a scan period of 4 seconds, and a PRI of 1.2 ms.

The radar beam profile shown in Figure 3.15 has a relatively wide 3-dB beamwidth. However, there are many radars with narrow beamwidths and short scan periods, for example, the amplitude profile of a HS radar shown in Figure 3.16. The main beam and both the sidelobes of this radar comprise only 20 pulses, with only three pulses between the 3-dB beamwidth.

The radars that absolutely must be detected by an ESM system at adequate range are those associated with weapons systems. The number of pulses to make an intercept depends on the detection requirement for those radars that have the narrowest amplitude profiles and are considered to be threats to the ESM platform.

The number of pulses seen in a single scan of a scanning radar depends on the beamshape as well as the scan period and PRI of the radar. Figure 3.17 shows examples of the detection of nine pulses from a radar with azimuth 3-dB beamwidth of only 0.8° by an ESM system of sensitivity of –55 dBmi. The same ESM system would see 40 pulses from a radar with identical parameters except for a beamwidth of 3.5°, also shown in Figure 3.17.

The radar with the 0.8° beamwidth required a dynamic range of 17 dB above the ESM sensitivity threshold for the detection of nine pulses. However, the radar with 3.5° beamwidth needs only 0.5 dB above the threshold for nine pulses to be seen, as shown in Figure 3.18. The PRI for both the radars shown in the plot is 1 ms and the scan period is 2 seconds.

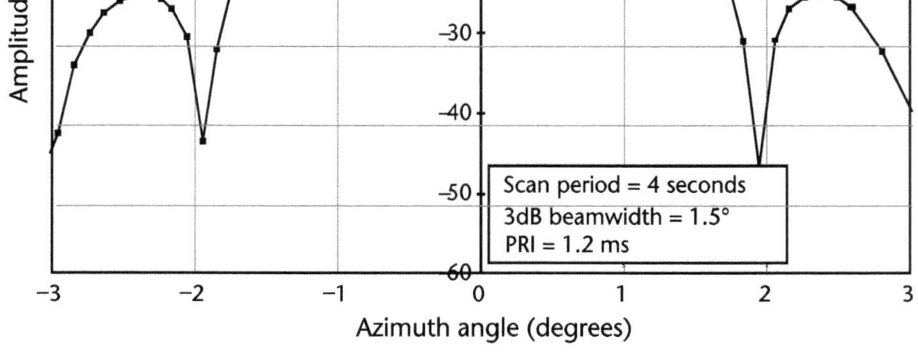

Figure 3.15 Example of an amplitude profile of an ATC radar.

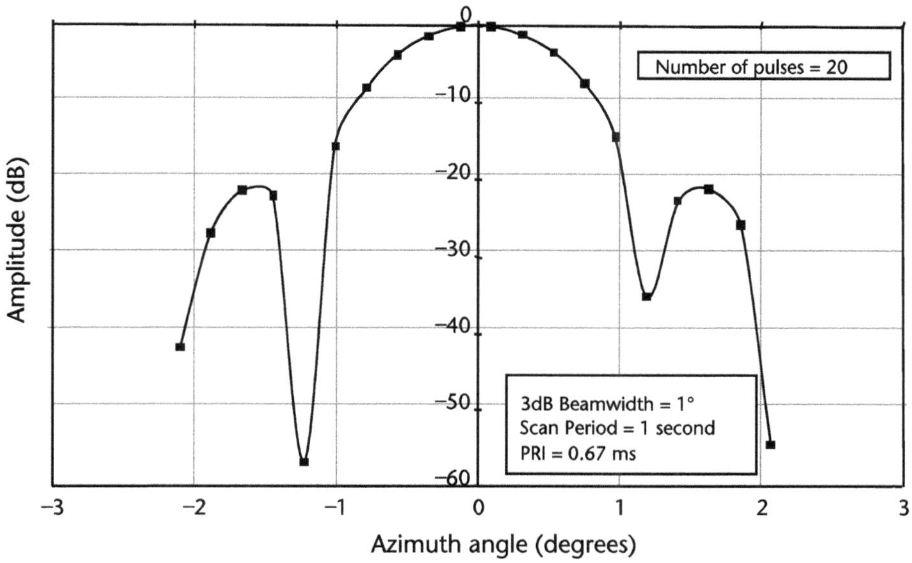

Figure 3.16 Example of an amplitude profile of a HS radar.

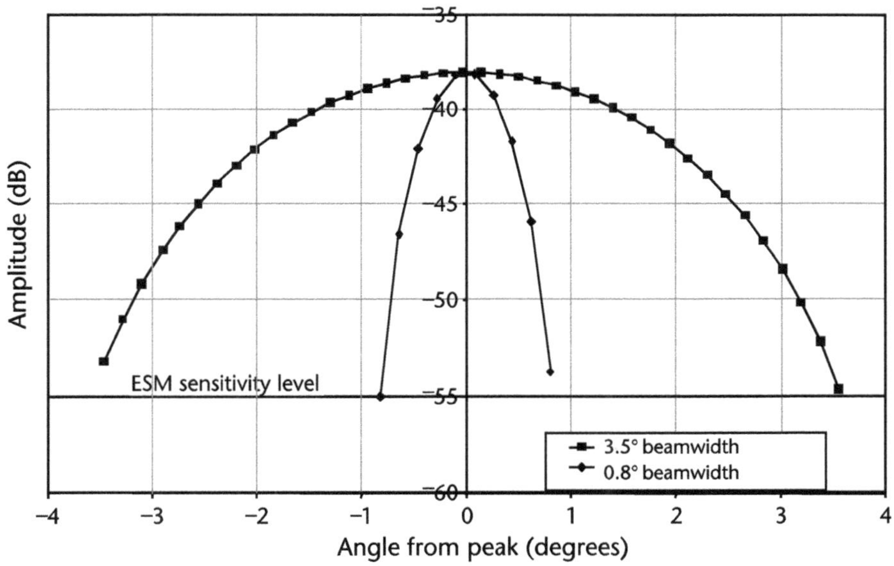

Figure 3.17 Detection of radars with azimuth beamwidths of 0.8° and 3.5° by an ESM system of sensitivity –55 dBmi.

The PRI is also an important factor in governing the amplitude profile of a scanning radar. In Figure 3.19 the effect of PRI on amplitude profile is shown for radars with PRI of 100 μs to 3 ms; almost all target acquisition radars have PRI values within these limits.

The effects of the final factor that significantly affects the amplitude profile of a scanning radar, the scan period, are shown in Figure 3.20. Four amplitude profiles are shown here, generated for radars with parameters identical in every respect except for scan period.

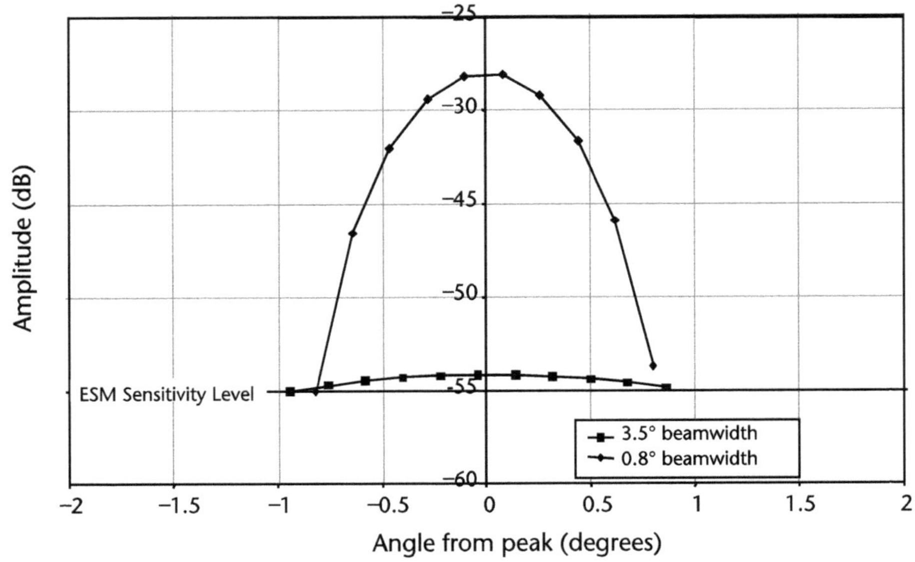

Figure 3.18 Detection of radars with azimuth beamwidths of 0.8° and 3.5° at the limits of detection of an ESM system of sensitivity –55 dBmi.

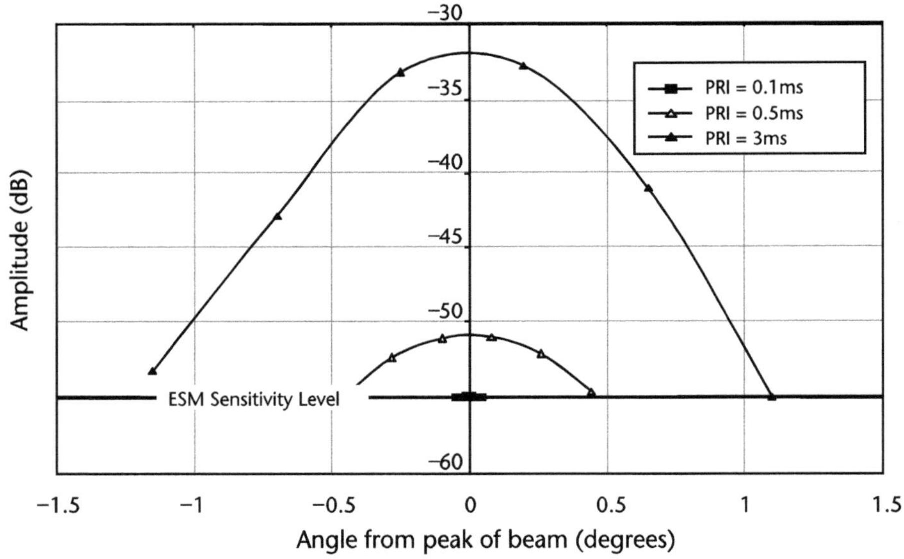

Figure 3.19 Effect of PRI on the amplitude profile of a scanning radar.

The radar in Figure 3.20 having a scan period of only 1 second requires a dynamic range of 9 dB (i.e., the number of decibels above the sensitivity level) for the detection of 10 pulses whereas all 10 pulses from the radar with the 10-second scan period are detected at amplitudes within 0.1 dB of the peak pulse.

It can be seen from the above analysis that radars with small azimuth beamwidths, long PRI, and short scan intervals will be the most difficult to detect due to their narrow amplitude profiles. Ideally, the number of pulses needed for production of an intercept should be chosen to allow radars with the most extreme combination of these parameters to be detected.

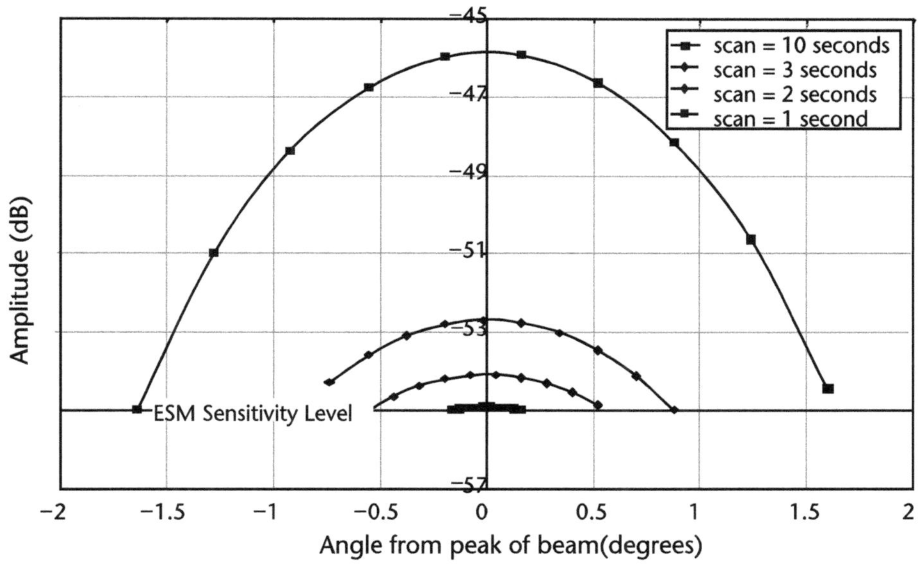

Figure 3.20 Effect of scan period on the amplitude profile of a scanning radar.

References

[1] O'Neil, S., *Electronic Warfare and Radar Systems Engineering Handbook*, U.S. Naval Air Warfare Center Weapons Division, 2013.

[2] https://msi.nga.mil/MSISiteContent/StaticFiles/NAV_PUBS/RNM/310ch1.pdf.

[3] Illman, P., *The Pilot's Radio Communications Handbook*, 5th ed., New York: McGraw-Hill, 1998.

[4] Binns, C., *Aircraft Systems: Instrumentation, Communications, Navigation and Control*, Hoboken, NJ: Wiley-IEEE, 2018

[5] Radford, M., "Coastal and Harbor Security White Paper," www.blighter.com/images/pdfs/white-papers/coastal-and-harbour-security-white-paper.pdf.

[6] "AWACS Surveillance Radar," www.northropgrumman.com/Capabilities/AWACSAPY2/Documents/AWACS.pdf.

[7] Sandwell, D., "Radar Altimetry," http://topex.ucsd.edu/rs/altimetry.pdf, 2011.

[8] Nav radar/www.marineinsight.com/marine-navigation/marine-radars-and-their-use-in-the-shipping-industry/.

[9] Doviak, R., and D. Zrnic, *Doppler Radar and Weather Observations*, 2nd ed., San Diego CA: Academic Press, 1993.

[10] Kopp, C., "Surface to Air Missile Systems and Integrated Air Defence Systems," *Air Power Australia*, www.ausairpower.net/sams-iads.html.

[11] Kopp, C., and Wise, J., "HQ-9 and HQ-12 SAM System Battery Radars," *Air Power Australia Technical Report APA-TR-2009-1201*, 2012.

[12] https://missilethreat.csis.org.

[13] Rogerson, J., https://5g.co.uk/guides/5g-frequencies-in-the-uk-what-you-need-to-know/.

[14] Deng, A., "Detection and Jamming of LPI Radars," Thesis, U.S. Naval Postgraduate School, 2006.

[15] Robertson, S. M., "Advantages of AESA Radar Technology" *Naval Forces*, No. V, 2016.

[16} Robertson, S. M., "Ship Fingerprinting for Maritime Security," *Proc. MAST Conference*, 2013.

[17} www.nmea.org/Assets/nmea%20collision%20avoidance%20through%20ais.pdf.

[18} www.marinetraffic.com.

[19} www.aishub.net.

CHAPTER 4

ESM Equipment

This chapter describes the aspects of ESM equipment that affect the measurement and resolution of radar parameters with the basic components of the ESM system shown in Figure 4.1.

Antennas are the eyes of the ESM system. They feed signals into front-end receivers (FER), which begin the task of processing the electromagnetic (EM) energy and turning the signals received into pulse descriptor words (PDW).

The parameter measurement unit (PMU) produces data on individual pulses and stores them in a buffer ready for the next stage of processing. A buffer may contain several thousand pulses that are deinterleaved to produce intercepts for the different radars in the EM environment. Intercepts are correlated to existing ESM tracks or are used to create new tracks if there is no relevant track already in existence (see Chapters 2, 8, and 9).

Many ESM systems attempt to locate the radars, generating a fix with an associated error ellipse (see Chapter 11). An identification process then makes use of data from a radar library to assign an identity to the radar track (see Chapter 10). Most ESM systems also have recording facilities so that PDW, intercept, and track data can be analyzed after the mission (see Chapter 10).

4.1 ESM Antennas

The most commonly used antenna type in ESM systems is the spiral antenna. These can be used in all types of ESM systems including in arrays to allow phase measurements for phase comparison systems. Other types of ESM antennas are sinuous antennas and horn antennas, with spinning antennas being used for narrowband RF reception. Reference [1] gives a good overview of antennas for EW applications.

4.1.1 Spiral Antennas

Spiral antennas are particularly useful for direction-finding applications, such as ESM systems [2]. This type of antenna is shaped as a two-arm spiral, as shown in Figure 4.2, although more arms may be used [3].

Spiral antennas belong to the class of frequency-independent antennas as they can operate over a wide range of frequencies. The fractional bandwidth of a spiral antenna may be as much as 30:1. This means that if the lower frequency is 1 GHz,

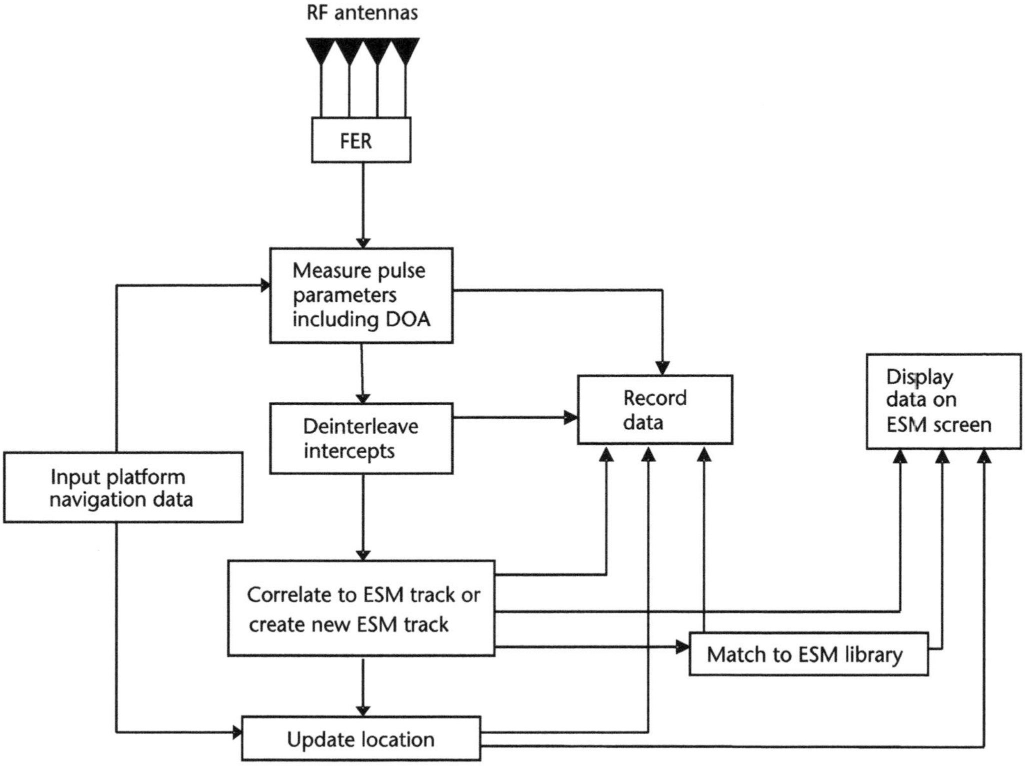

Figure 4.1 ESM system components.

Figure 4.2 A spiral antenna.

the antenna would still be efficient at 30 GHz and at every frequency in between. Polarization and the antenna radiation pattern are unchanged over the frequency range [4].

The spiral antenna radiation pattern typically has a peak radiation direction perpendicular to the plane of the spiral with a 3-dB beamwidth of 70° to 90°. Lossy cavities are usually placed at the back of the spiral to eliminate back lobes, so the antenna radiation pattern appears as in Figure 4.3 for a single antenna. Figure 4.3 also shows the radiation pattern for a set of four antennas covering the whole 360° azimuth range.

4.1.2 Sinuous Antennas

A limitation of spiral antennas is their inability to receive circularly polarized signals with the opposite sense of circular polarization. Sinuous antennas are capable of simultaneously receiving signals of either sense of circular polarization and all senses of linear polarization [5].

The sinuous antenna is a dual orthogonal linear polarized radiator with an output connector for each linear polarization. By combining the signals from these two outputs with varying amplitude and phase ratios, any polarization can be created.

4.1.3 Horn Antennas

These antennas have large azimuth and elevation beamwidths for instantaneous signal interception. Biconical horn antennas are linearly polarized but can be integrated with a suitable polarizer to enable them to receive all polarizations [1]. Antennas covering a frequency range from very high frequency/ultrahigh frequency (VHF/UHF) to microwave frequencies have been developed for shipborne and vehicle-based platforms. These antennas can also be used for direction-finding (DF) applications based on time difference of arrival (TDOA), resulting in very high DF accuracy.

4.1.4 Antennas for Phase Comparison Systems

An interferometer array consists of several individual antenna elements, either single polarization spiral antennas or dual-polarized sinuous antennas. Phase tracking of the antenna elements is needed to achieve a precision DF array. For maximum precision, all the antennas can be etched on a single sheet with spacing between antennas positioned to a tolerance of ±0.001 inch.

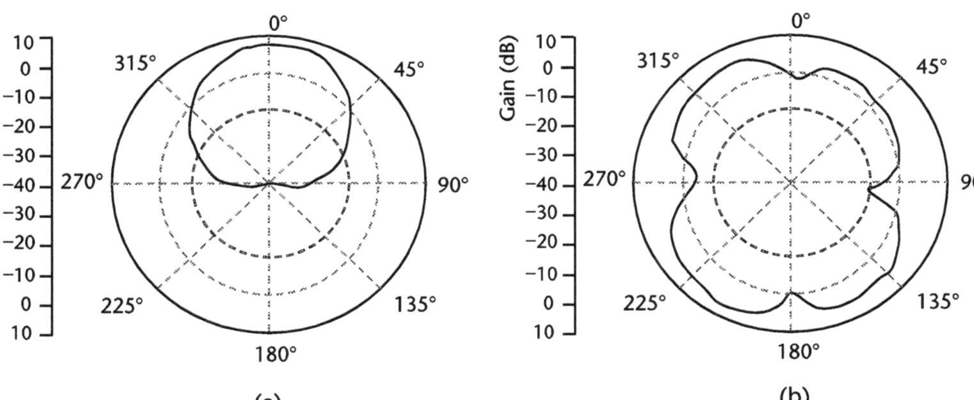

Figure 4.3 Spiral antenna azimuth radiation patterns for a single antenna and for 360° coverage.

The length of the interferometer array is determined by the accuracy of the DOA requirement and the phase resolution of the receiver (see Chapter 7). To realize high-precision DOA accuracy, the interferometer must be populated with enough antenna elements to resolve ambiguities in the DOA measurement.

4.1.5 Spinning Antennas

A spinning antenna provides focused, high gain detection. It is especially useful for detecting signals of high frequency (millimeter wavelengths) that suffer significant path loss through the atmosphere. They search a narrow RF band, typically 10 to 20 MHz at a time with spinning antennas having approximately 20-dB extra gain over traditional spiral antennas. Spinning antennas have a viewing angle of between 12° and 15° in azimuth and speeds of up to 240 revolutions per minute (RPM). The beam dwell time, t_d, is calculated as follows:

$$t_d = 1/\left(RPM/60 \times 360/\theta\right)(ms) \tag{4.1}$$

where θ is the viewing angle in azimuth.

The beam dwell in milliseconds for a viewing angle of 15° is plotted against the RPM in Figure 4.4.

Figure 4.4 shows that, for a dwell in any 15° azimuth segment to last for more than 40 ms, the RPM must be less than 60. The time for which a scanning radar is visible to the ESM depends on a number of factors including the radar scan period. At a range of 100 km, a typical Air Traffic Control (ATC) radar is detected for 200 ms on each scan cycle by an ESM system with a sensitivity of –70 dBmi (typical sensitivity of a spinning antenna) (i.e., for about 5% of its scan cycle). The probability of intercept (POI) for this type of radar is therefore low unless the spinner rotation speed is slow.

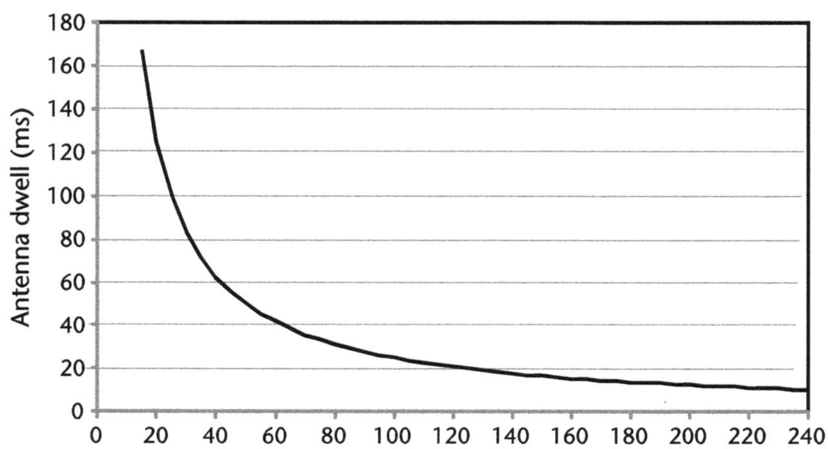

Figure 4.4 Spinning antenna beam dwell.

4.2 ESM Receivers

Once the EM energy has been detected by the ESM antenna array, the next stage of processing takes place. There are several types of ESM receiver; the choice depends on the RF range and sensitivity required for the ESM system.

4.2.1 Superheterodyne Receivers

A superheterodyne receiver linearly transforms an input signal to a lower frequency, known as an intermediate frequency (IF). It is relatively easy to phase match between two identical superheterodyne receivers because of their narrow bandwidth so they are commonly used in phase comparison systems.

The two basic microwave components of the superheterodyne receiver are the mixer and the local oscillator (LO). A basic superheterodyne receiver is shown in Figure 4.5.

Since a mixer is a nonlinear device, it generates many other frequencies besides the desired frequency. All these undesired frequencies are called spurious frequencies or, simply, spurs. The frequencies of these spurs are predictable and can be written as:

$$f_{out} = Mf_{in} + Nf_{IF} \tag{4.2}$$

where f_{in} is the signal and f_{IF} is the LO frequency. M and N are integers, either positive or negative, and the output frequency, f_{out}, is a positive value.

A limitation of a superheterodyne is its relatively narrow bandwidth. To cover a wider input bandwidth, the receiver can be made to scan a certain bandwidth at a fast rate. This kind of receiver is known as a scanning superheterodyne. The fastest scan rate achievable without sacrificing the sensitivity of the receiver is B^2, where B is the final bandwidth of the IF filter (e.g., if the IF bandwidth of the receiver is 10 MHz, the fastest scan rate is 100 MHz/μs). If the input bandwidth is 10 GHz, it will take the receiver 100 μs to scan across the whole band. A scanning superheterodyne can therefore be used to intercept long pulsed or CW signals, but has a low POI for short pulses from low duty cycle signals.

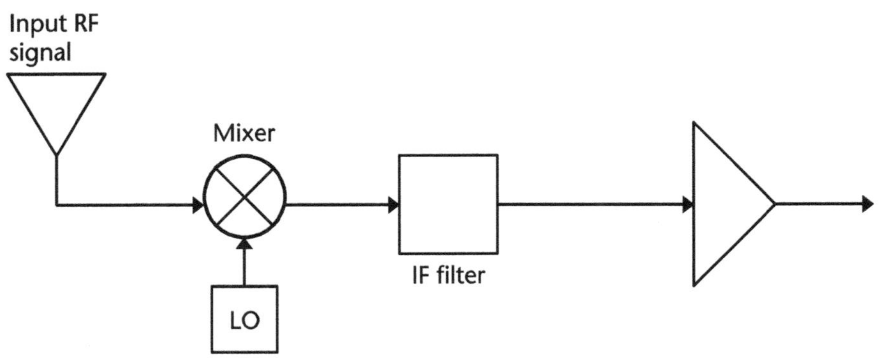

Figure 4.5 A basic superheterodyne receiver.

4.2.2 Crystal Video Receivers

This type of receiver can cover a very wide frequency bandwidth, but it cannot determine the frequency of the input signals. Since the RF bandwidth is wide, the sensitivity of the receiver is relatively low. The receiver cannot separate input signals by frequency, so it cannot detect the existence of simultaneous signals.

The key component of the crystal video receiver is the crystal detector. A simple crystal video receiver is shown in Figure 4.6. The output of an antenna feeds directly to a detector. Following the detector is a video amplifier that amplifies the detector output and passes the video signal to the input of a comparator.

The crystal video receiver measures pulse amplitude, PW, TOA, and DOA.

4.2.3 Instantaneous Frequency Measurement Receivers

An instantaneous frequency measurement (IFM) receiver divides the incoming RF signal into two parts, one having a delay of a known length and the other with no delay. The signal passing down the delay line then has a phase shift relative to the signal with no delay. This phase shift is a function of the input frequency.

The two signals are applied to a phase correlator and envelope detector that forms two video signals; one (V_c) is proportional to the cosine of the phase shift, and the other (V_s) is proportional to the sine of the phase shift. The phase angle ϕ is given by:

$$\phi = \omega\tau_{delay} \tag{4.3}$$

where ω is the signal angular frequency and τ_{delay} is the fixed delay time.

The phase angle is found by taking $\tan^{-1}(V_s/V_c)$. The intercepted frequency, f_{in}, is then found from the measured phase angle and the known time delay:

$$f_{in} = \phi / \left(2\pi\tau_{delay}\right) \tag{4.4}$$

The output of the frequency discriminator repeats itself each time the input frequency changes by an amount equal to the electrical length of the delay line.

$$\Delta f = 1/\tau_{delay} \tag{4.5}$$

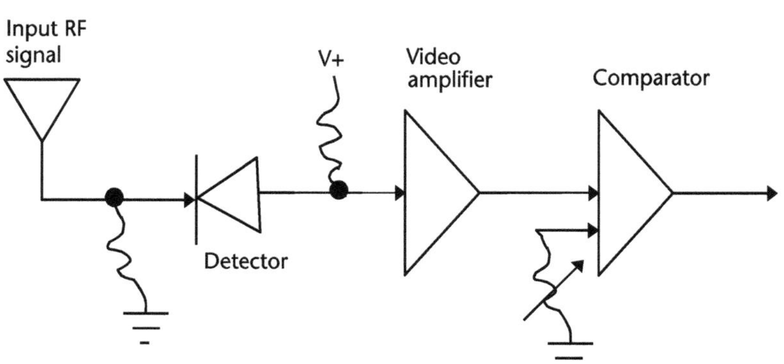

Figure 4.6 A basic crystal video receiver.

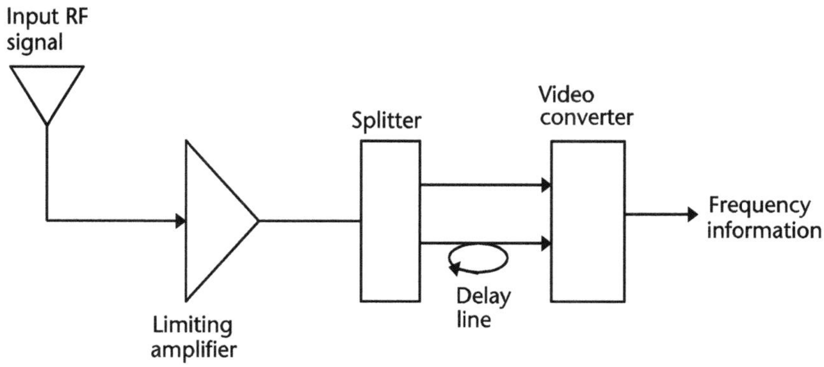

Figure 4.7 An IFM receiver.

Thus, the wider the input bandwidth to be covered, the shorter the delay line needs to be to prevent ambiguities. However, as the delay line length is shortened, the amount of phase shift associated with a given frequency resolution is also reduced.

Most practical IFM receivers are implementations of a bank of frequency discriminators or correlators with the longest delay line corresponding to the desired frequency resolution, and the shortest delay determined by the highest frequency to be measured (Figure 4.7). The length of the longest delay line must be shorter than the minimum pulse width the receiver is designed to measure. Otherwise the delayed and undelayed portions of the input signal never overlap and the receiver is never able to measure the frequency of the signal.

Advantages of IFM receivers are their wide bandwidth and ability to accurately measure radar frequency. However, a serious limitation is their inability to measure more than one signal at a time. If two signals occur simultaneously, either only the strongest one is measured by the IFM, or if the signals are similar in power level, then the frequency measured is an average of the frequencies of the two signals. If a continuous signal is seen by the IFM receiver, then any subsequent weaker signals may be lost altogether.

4.2.4 Channelized Receivers

A channelized receiver consists of a large number of parallel narrowband receivers with adjacent frequencies. The RF band over which the receiver is required to operate (for example, 2 to 18 GHz) is typically divided into contiguous bands each 1 or 2 GHz wide. These bands are further divided into subbands, perhaps 200 MHz wide. The subbands are further divided into channels of width 20 MHz or so to give the frequency resolution of the receiver.

A problem for channelized receivers is the activation of multiple channels by high power, narrow pulses. Another problem for this type of receiver is that the outputs from all the filters do not resemble the input, which is flat in amplitude. The output signals from the filters have larger amplitude at the leading and trailing edges of the pulse, referred to as transients. This transient effect is critical in the design of channelized receivers. Generally, the narrower the filter bandwidth and the more sections that the filter has, the more pronounced the transient effect. To have less transient effect the filter should have relatively wide bandwidth and fewer sections. These requirements contradict each other, and, in practice, the choice of filters depends on the frequency resolution and the minimum pulse width expected. Figure 4.8 shows a typical channelized receiver configuration.

4.3 Parameter Measurement, Accuracy, and Resolution

RF, PW, TOA, and amplitude are measured by the ESM receiver. Pulse repetition interval (PRI) and DOA are calculated by the ESM processing system.

4.3.1 RF

For receivers that can measure RF (superheterodynes and IFMs), the frequency measurement takes place by processing the IF signal. A fine frequency (FF) clock typically runs at 1 GHz and acts on the IF. The frequency is created by finding the zero crossings of the IF near the start of the pulse. This is done by first amplifying the IF of the pulse and then passing the amplified signal through a limiter as depicted in Figure 4.9 to approximate a square wave for more accurate sampling. The FF clock samples the energy and registers a 1 for positive and 0 for negative. A transition from 0 to 1 indicates the beginning of a complete sine wave or one cycle of the IF within the pulse. By counting the number of cycles within the pulse duration or up to a limit of a few microseconds, whichever occurs first, an FF value is reached. The IF is reached by a series of upconversions and downconversions from the original RF to get to the system IF, typically 250 to 500 MHz. Once the FF is known and the reverse calculations are made, the original RF can be found.

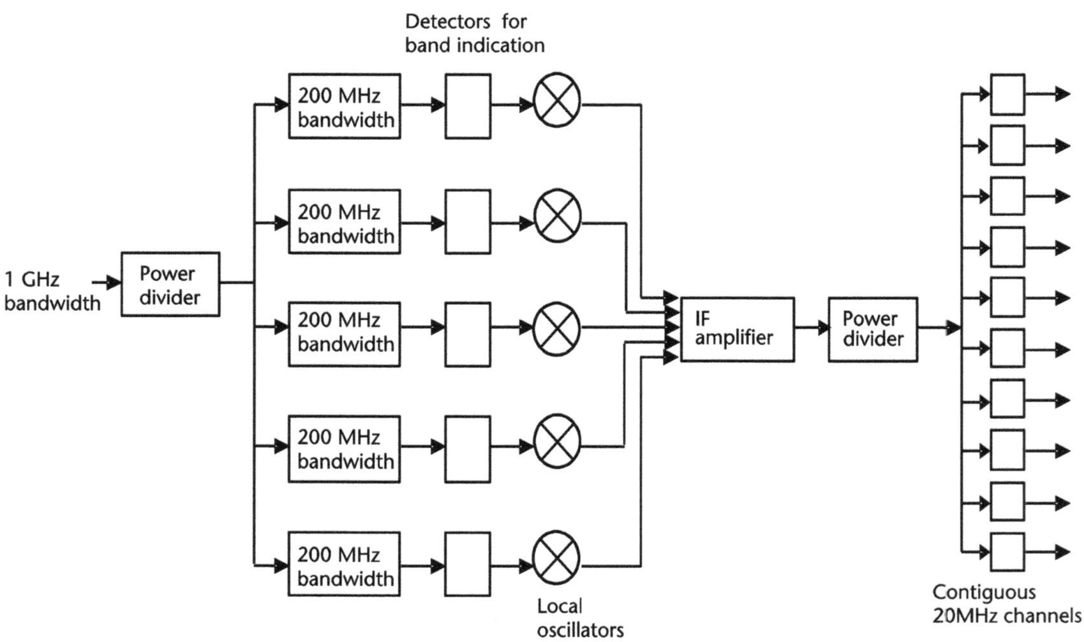

Figure 4.8 A channelized receiver

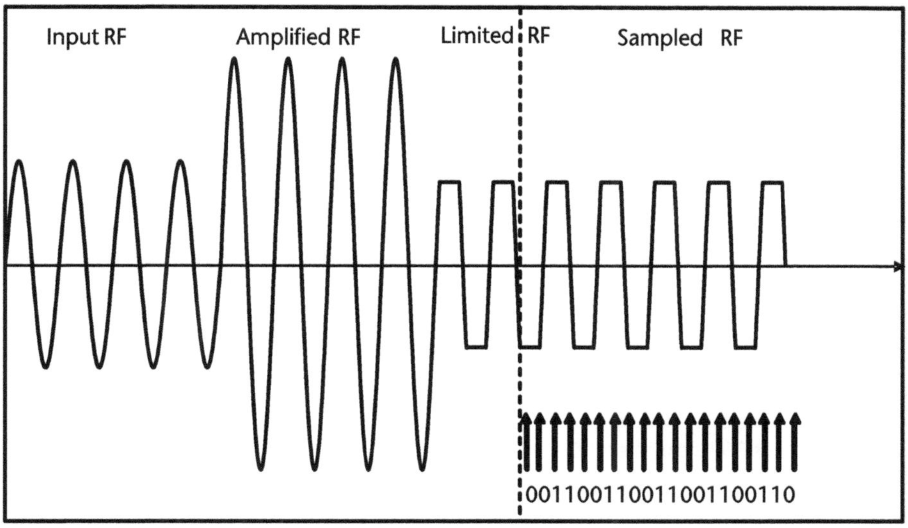

Figure 4.9 Frequency estimation process.

In general, the resolution to which RF is measured is within ±1.5 MHz of the average RF of a radar, but the spread in RF may be up to 5 MHz on either side of the average. When using a filter to select pulses from a particular radar for post-mission analysis, a filter of ±5 MHz should capture all the pulses from the radar. However, in a busy RF environment, this should be reduced to ±2 or ±3 MHz to avoid selecting pulses from multiple radars.

4.3.2 Timing

The accuracy to which TOA must be measured for the purposes of PRI calculation is typically to the nearest microsecond, a standard that is easily achievable by ESM systems. A resolution of 100 ns is sometimes reported, but this level of accuracy is not needed for the calculation of PRI by the usual methods.

However, in TDOA systems, a much more accurate TOA is needed. In these systems, the TOA should be measured to at least 1 ns and, if possible, to an accuracy of 0.1 ns.

There are several clocks running in TDOA ESM equipment. A 1-GHz clock is used to detect the leading edge of the pulse, to measure the TOA to an accuracy of 1 ns. A 100-MHz clock is likely to be used to measure PW to an accuracy of 10 ns. Figure 4.10 shows the different clocks that measure pulse timing.

A noise rider detects energy and triggers the TOA clocks. A 1-GHz or 1-ns rate samples the leading edge of the pulse for TOA. A separate clock may sample frequency to produce RF. The PW clock operating at 100 MHz samples the pulse leading to the PW measurement. The lower the noise rider is set, the more sensitive the system. However, if it is set too low, it will trigger on noise spikes and create false pulses.

Figure 4.11 shows a pulse arriving at each of four antennas in a TDOA system. In this example, the pulse is seen first on the starboard antenna (TOA_s) and then on the aft ($TOA_s - TOA_A$) and forward ($TOA_s - TOA_F$) antennas. The pulse is not detected by the port antenna in this example.

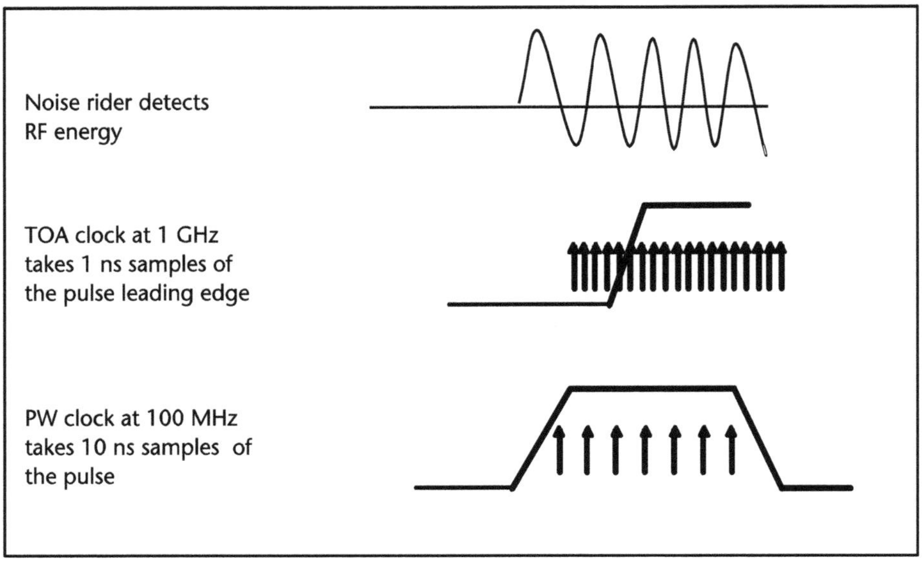

Figure 4.10 Timing measurements.

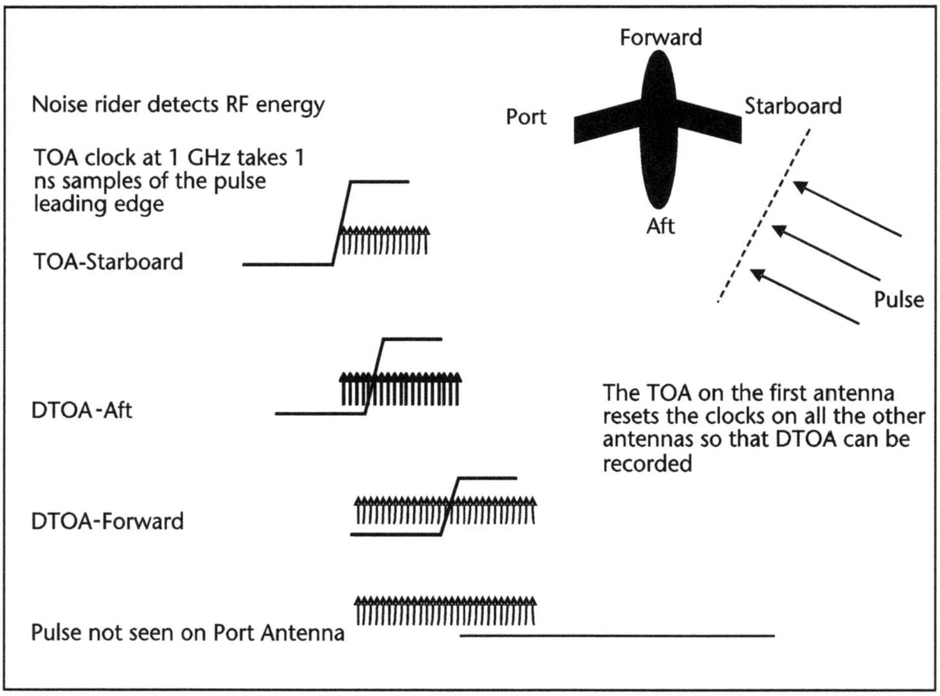

Figure 4.11 Radar pulse arriving at separated antennas in a TDOA system.

The TOA clock, triggered by the noise rider, registers the arrival of the pulse in the first antenna/receiver, and this is the value used in PRI calculations. At this point, all TDOA receiver clocks start running and DTOA is measured for each subsequent antenna and made available for recording into the PDW.

4.3.3 PW

PW measurement begins along with other measurements when the noise rider is activated. The PW clock typically runs at 100 MHz or a 10-ns sample rate and runs until the end of pulse is flagged or until the buffer allocated to store the samples from the pulse clock is full. In practice, the maximum PW that can be accommodated is usually about 500 μs. The end of the pulse is measured when the pulse amplitude drops from the maximum, usually by 3 dB, as shown in Figure 4.12.

A minimum PW is set to prevent noise spikes taking up space in the pulse buffer. This minimum PW is usually 50 ns. As there are currently no radars with PW less than this value, it is a good compromise to prevent the detection of noise while allowing genuine pulses to be detected.

4.3.4 Amplitude

In some ESM systems, a measurement of amplitude is used to estimate the range to the radar. However, amplitude cannot be reliably measured due to effects such as multipath that cause constructive interference to the pulse (see Section 11.8). Cross-polarization, where the ESM antennas are not matched in polarization to the radar, also significantly affects the received pulse amplitude. Table 4.1 shows the reduction in amplitude seen due to cross-polarization [6].

In almost all cases, even when the ESM and radar are matched in polarization, there is a difference of at least 1 dB between the predicted amplitude for a particular radar range and that measured by the ESM system. The resolution to which amplitude is reported is typically 0.5 dB, a sufficient resolution given the limitations in the measurement of amplitude described above.

4.3.5 PRI Calculation

PRI is one of the most important parameters used in the identification of radars by ESM systems. There are several techniques for determining PRI:

- *Time of arrival difference histogram (TOADH):* This is the most usual method of deinterleaving. Pulse TOA differences are used to form a histogram giving peaks at PRI values.

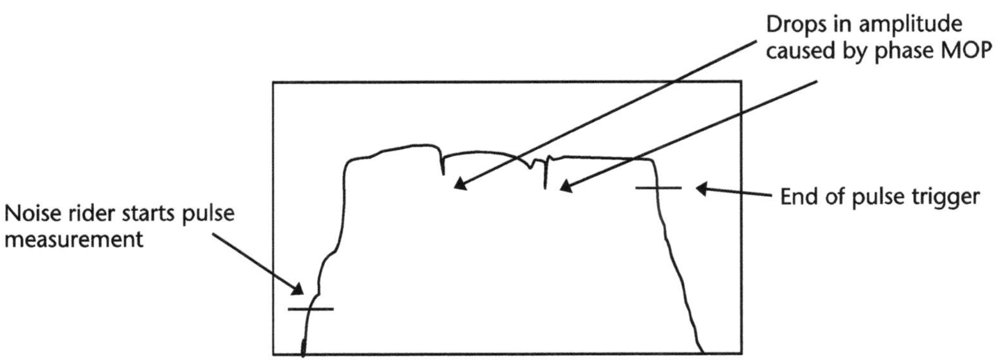

Figure 4.12 PW measurement.

Table 4.1 Polarization Loss for Combinations of Transmit/Receive Polarizations

Transmitter Polarization	*Receiver Polarization*	*Ratio Received Power to Maximum Power*
Vertical	Vertical	1
Vertical	Slant	0.5
Vertical	Horizontal	0
Vertical	Circular right-hand or left-hand	0.5
Horizontal	Horizontal	1
Horizontal	Slant	0.5
Horizontal	Circular right-hand or left-hand	0.5
Circular right-hand	Circular right-hand	1
Circular right-hand	Circular left-hand	0
Circular right-hand or left-hand	Slant	0.5

- *Predictive gates:* This techniques looks for the next pulse in a chain, once PRI limits for a particular radar have been established.
- *Graph theory deinterleaver:* This pulse association process is based on the use of Graph theory and nonlinear optimization. It attempts to determine if one or more pulse chains exist within a cluster.

These methods of PRI calculation are discussed in detail in Chapter 8 on deinterleaving.

4.3.6 DOA Calculation

There are three methods of DOA calculation used by ESM systems:

- *Amplitude comparison:* DOA is determined by measuring the relative amplitude of the signal received by two or more antennas whose boresights are squinted relative to each other. Theoretical calculations suggest that amplitude comparison systems give DOA accuracies of between 3° and 20° (see Chapter 5).
- *TDOA:* DOA is determined by the measurement of relative TOA of the pulse received by two separated antennas. TDOA can be used for high-accuracy direction finding (HADF) if the antenna separations (i.e., baselines) are long (typically >10m) and the time-difference measurements of the receiver are precise (e.g., of the order of 1 ns). Multiple baselines are required to provide the necessary spatial coverage (see Chapter 6).
- *Phase comparison (interferometry):* DOA is determined by the measurement of relative phase difference of the pulse received by two or more antennas arranged in an array. This approach can provide HADF (e.g., 1° rms) [7]. To avoid ambiguities in DOA measurement, multiple antennas (typically 3 to 5) are required to provide the necessary spatial and frequency coverage (see Chapter 7).

Another method of determining DOA is the use of a spinning dish antenna. However, this is not a commonly used technique as the POI against scanning radars is poor.

4.4 ESM Sensitivity

The main factor in determining the sensitivity requirements of an ESM system are the ranges at which each type of radar must be detected. A start is made by considering the range at which a single pulse from the main beams of radars with ERPs from 45 dBW to 100 dBW can be detected by ESM systems of various sensitivities as shown in Figure 4.13.

From Figure 4.13, it can be seen that a radar of ERP of 60 dBW or less is not detected at a range of more than 200 km by an ESM system with sensitivity less than –70 dBmi. However, a radar of ERP of 80 dBW can be seen at a range of 200 km by an ESM system of sensitivity of –50 dBmi, and the detection range for this radar increases to greater than 1,600 km for an ESM with a sensitivity of –70 dBmi.

There are several factors that reduce the maximum range at which a radar can be detected. The first is the beamshape of the ESM antenna beam. Even with a 3-dB beamwidth as wide as 70°, there is a reduction in sensitivity away from the peak of the ESM antenna beam (the boresight of the antenna), as shown in Figure 4.14.

This reduction in sensitivity is particularly important for systems using only four antennas separated by 90° in azimuth. Figure 4.10 shows how the ESM sensitivity changes in the 360° around an aircraft using four antennas separated by 90°. The right graph in Figure 4.15 shows how the ESM antenna beamshapes overlap when there are six antennas separated by 60°.

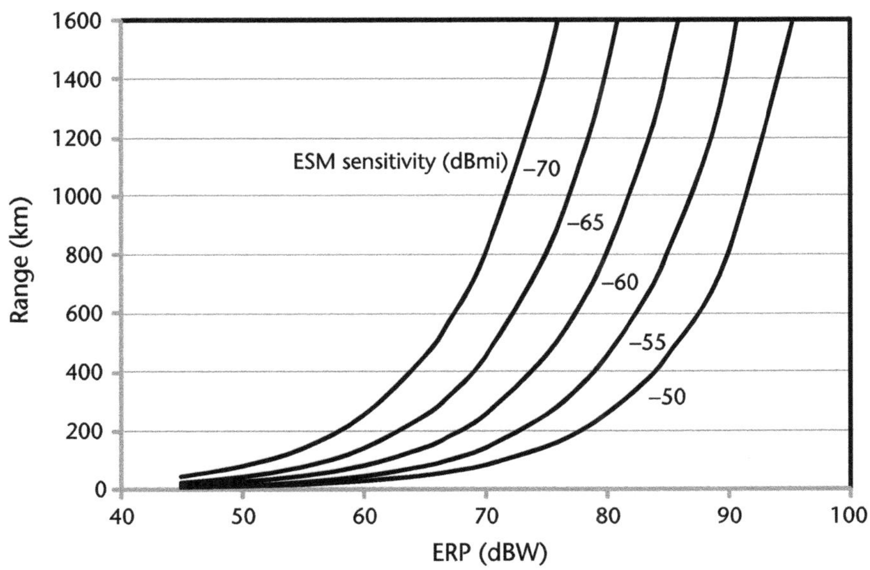

Figure 4.13 ESM detection range versus various ESM sensitivity levels for a 9-GHz radar.

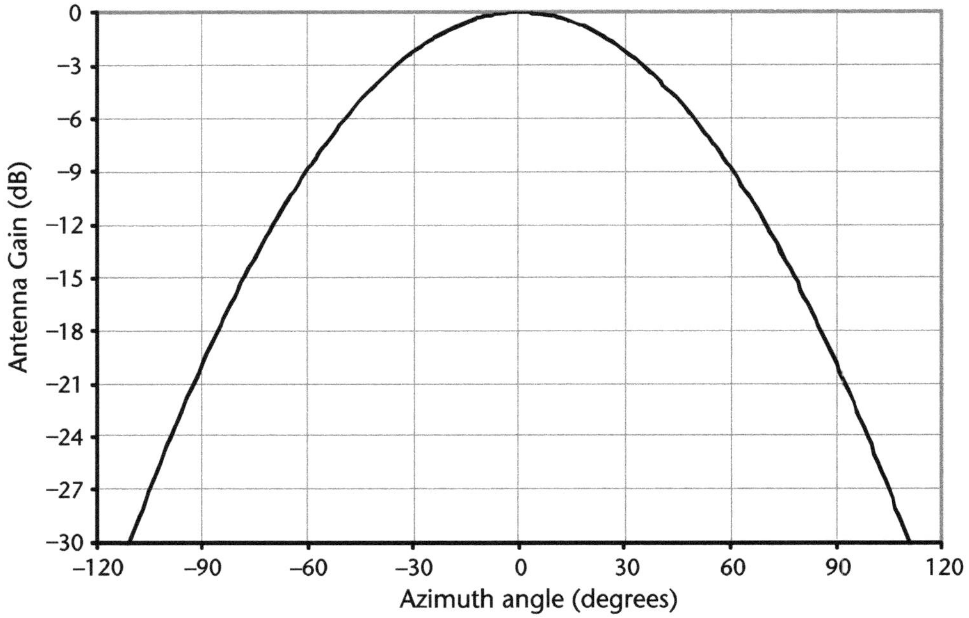

Figure 4.14 The ESM antenna beamshape.

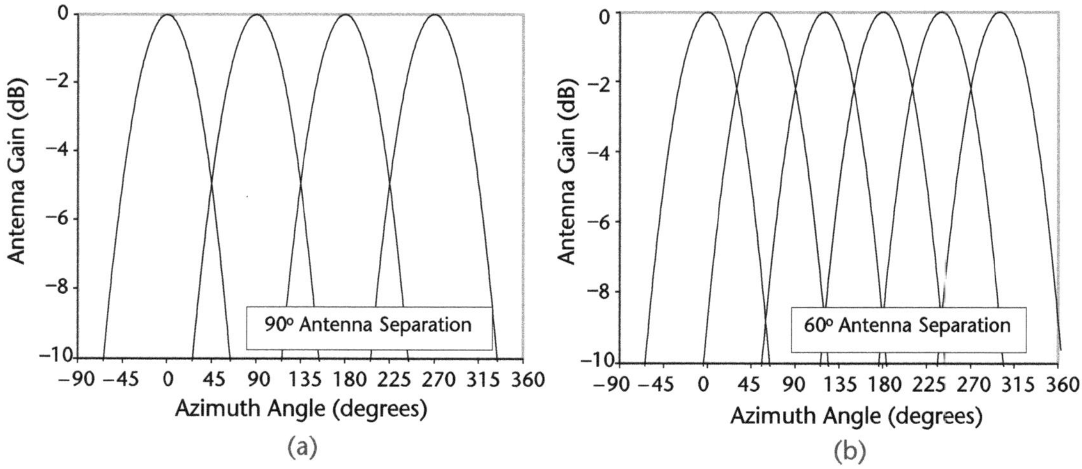

Figure 4.15 ESM sensitivity in azimuth for antenna separations of 90° and 60°.

It can be seen from Figure 4.15 that the ESM antenna sensitivity is at its peak at the boresight of the antennas and is reduced at points exactly between the antennas. The reduction in sensitivity is more than 4 dB in the case of the installation with four antennas and is 2 dB when six antennas are configured to provide 360° azimuth coverage. However, there is a further complication when considering amplitude comparison systems where the pulse must be detected by at least three antennas and also for TDOA systems where the pulse must be seen on more than two antennas for unambiguous DOA to be calculated.

In the case of an amplitude comparison system with six antennas, a pulse arriving at a DOA of 20° from the peak of antenna 1 is received at significantly different levels on each of three relevant antennas, as shown in Figure 4.16.

The pulse is seen at −1.5 dB from the peak of the beam on antenna 1, at −4 dB from the peak of antenna 2, and at −25 dB from the peak of antenna 3. If the amplitude of the pulse itself is low, then there is a possibility that it will not be detected by antenna 3 and the pulse will be rejected by the three antenna amplitude comparison process.

Another factor that reduces the maximum detection range is the number of pulses needed for detection, as discussed in detail in Chapter 3. Usually six pulses are needed before an intercept is declared for a radar. Typical ATC radars have a negligible reduction in amplitude between the pulse at the peak of the beam and two or three pulses either side of the peak. However, harbor surveillance (HS) radars with much narrower beamwidths and faster scans can have a reduction in amplitude by as much as 15 dB for the detection of 6 pulses (see Section 3.7).

Although it is desirable to be able to detect all radars, it is essential that the ESM system has sufficient sensitivity to detect threat radars. Most of the threat radar range requirements depend on the range of the weapon with which they are associated. Target acquisition radars are the only threat function where the actual radar range is used in the specification of required ESM detection range and therefore ESM sensitivity.

A graphical method for determining the sensitivity requirement for threat radars is shown in Figure 4.17.

The curved line on Figure 4.17 shows the range (R_{ESM}) at which the radar can be detected by ESM systems of various sensitivities given by the equation:

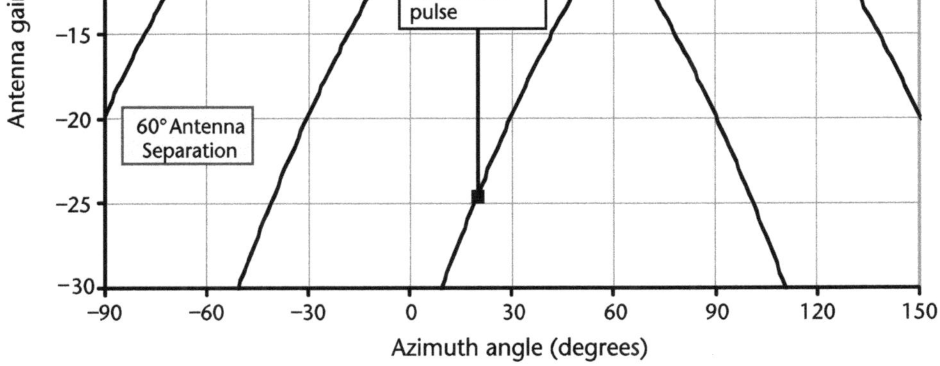

Figure 4.16 A pulse arriving at DOA of 20° in a three-antenna amplitude comparison system.

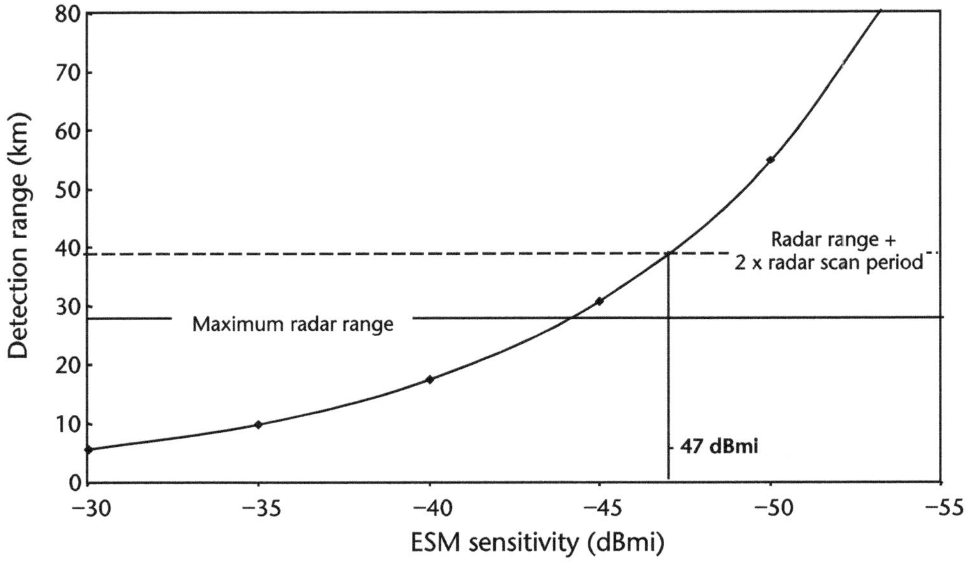

Figure 4.17 Graphical method for determining ESM sensitivity requirement.

$$R_{ESM}{}^2 = \left(P_t\lambda^2\right)/\left(16\pi^2 P_r\right) \tag{4.6}$$

where P_t is the radar transmitted power, P_r is the power received by the ESM (sensitivity) and λ is the radar wavelength.

The solid, horizontal line shows the maximum detection range of the radar (R_{radar}). This is the range at which the radar can detect the ESM platform. A simple form of the radar range equation is:

$$R_{radar} = \left[\left(P_t G^2\lambda^2\sigma\right)/\left(\{4\pi\}^3 P_r\right)\right]^{1/4} \tag{4.7}$$

where P_t = transmitted power in watts, G = radar antenna gain, λ = radar wavelength in σ = target cross-section in m^2, and P_r = received power in watts.

The radar cross-section of the ESM platform (σ) is an important parameter in the calculation of the radar detection range as aircraft and ships have very different radar cross-section values. The cross-section of an aircraft such an F-16 is 5 m^2, whereas stealth aircraft such as F-22 have quoted cross-sections as small as 0.0001 m^2 [8]. Ships can have cross-sections as large as 100,000m^2 [9] meaning that the radar can detect ships at much greater range than it can detect aircraft.

The dashed straight, horizontal line on Figure 4.17 shows the minimum detection range that the ESM must achieve in order to be safe from the threat. In this case this is specified as the maximum detection range of the radar plus a margin for the ESM platform to detect the radar before it is in detection range of the radar. For threat radars with an associated weapon the required ESM detection range may be the associated weapon range + a margin of 20% or 50%, depending on the radar type.

A brief description of the required ESM detection range for the various threat radar functions is given below.

4.4.1 Target Acquisition

The range requirement for target acquisition (TA) radars is typically that they should be detected at the radar range + 2 times the distance that would result from the target aircraft traveling directly towards the TA radar for two of the radar scan periods. If a radar has a long scan period then the distance from the aircraft to the radar can decrease by several kilometers between scan peaks.

For example, if a radar operating at 3 GHz has an ERP of 80 dBW its detection range is about 47 km for a target of cross-section 1 m^2 with a radar antenna gain of 30 dB and a minimum detectable power of −90 dBmi. If the radar has a scan period of 5 seconds and the ESM aircraft is traveling at 400 knots (200 m/s) then the distance traveled by the aircraft for 2 of the radar scan periods is 2 km, assuming that the aircraft is traveling directly towards the radar. The required detection range for this radar is therefore (47 + 2) = 49 km. For the ESM system to detect a radar of ERP 80 dBW at a range of 49 km a sensitivity of at least –46 dBmi is needed, found by reordering (4.6) as follows:

$$P_r = \left(P_t\lambda^2\right) / \left(16\pi^2 R_{ESM}{}^2\right) \tag{4.8}$$

4.4.2 Target Tracking

When target tracking (TT) is taking place, the scan of the threat radar is usually locked on to the target. Ideally, the target aircraft should remain at least 1.5 times the range of the weapon during this phase of the engagement. Data on specific radars cannot be quoted here due to security classification, but analysis has shown that an ESM system with a sensitivity of –55 dBmi can detect current TT radars at the required range.

4.4.3 Missile Guidance and Target Illumination Radars

A suggested range requirement for missile guidance (MG) radar detection is 1.2 times the weapon range. MG radars do not have particularly high ERP, so a sensitivity of –55 dBmi is probably needed for the detection of current MG radars at adequate range. However, the parameters of future threats may change this requirement.

The detection range requirement for target illumination (TI) radars is also ideally 1.2 times the weapon range, although they may not be switched on until just before the weapon is launched, in which case they are not available for detection at a range greater than that of the weapon. TI radars are not particularly high power, so a sensitivity of −55 dBmi is needed for detection of this type of radar.

4.4.4 Beacon Interrogators and Missile Beacon Signals

There are few radars that carry out the beacon interrogator (BI) function. The detection requirement for this threat function is usually 1.2 times the weapon range on the basis that the threat radar will not carry out this function until the missile has been launched or possibly just prior to its launch.

Missile beacon (BN) radars are usually low power, below 60 dBW. It is not therefore reasonable to impose a detection range criterion for this radar function.

4.4.5 Missile Homing Systems/Seekers

It is desirable that missile homing (MH) radars are detected at the weapon range as they are not likely to be in operation until at or after the weapon launch. These radars are of relatively low ERP, so highly sensitive ESM receivers are needed for their detection, typically from −65 dBmi to −75 dBmi. Reducing the detection range requirement for all the missile homing emitters to 50 km would mean that most of the MH radars can be detected with an ESM system of sensitivity less than –70 dBmi.

4.5 RF Scan Strategies

For ESM systems that are narrowband (i.e., not capable of receiving radar signals over their entire RF range simultaneously), it is necessary to implement an RF scan strategy to have an acceptable POI for each RF band. Most traditional methods for generating scan strategies use short dwells for each band. The use of such strategies does not guarantee that all the radars in the environment are detected. The only way to ensure that all the bands are properly explored is to remain in a particular band for the duration of the scan cycles of all the radars of interest in that band and then move on to the next band.

Before describing the ideal RF scan strategy, it is worth considering what happens when a traditional RF scan strategy is used. An example of such a strategy is shown in Figure 4.18. This scan strategy has dwells of equal duration (300 ms) separated by 2 seconds between dwells at a particular band.

A simulation of the arrival time of pulses from consecutive scans of 10 radars in the RF band of the dwell is shown on Figure 4.18. The filled-in symbols show those radar scans that would be detected and the hollow symbols show those scans that would be missed. Only five of the radars are seen at all by the ESM using this scan strategy, and two of those are only seen on one radar scan, so half the radars are missed altogether using this scan strategy.

The reason that the radars are missing is due to the small percentage of its scan time that a radar is available for detection by the ESM system. This is discussed in more detail in Chapter 3, but a reminder is given in Figure 4.19 where a theoretical scan pattern for a typical ATC radar at a range of 150 km is plotted.

Only the main beam is seen for the radar in Figure 4.19 and even when the main beam is detected, pulses are only seen for 30 ms. This means that an RF scan strategy such as the one shown in Figure 4.18 is unlikely to be searching in the correct band for this radar during the short time period when the radar could have been detected.

An RF scan strategy that will ensure that all radars are detected, even those at long range, consists of dwells that are long enough to encompass the whole scan periods of all the radars of interest in a mission. Figure 4.20 shows such a scan strategy with dwells of 3 seconds in the RF band of interest separated by 6 seconds (during which time there are dwells in other RF bands).

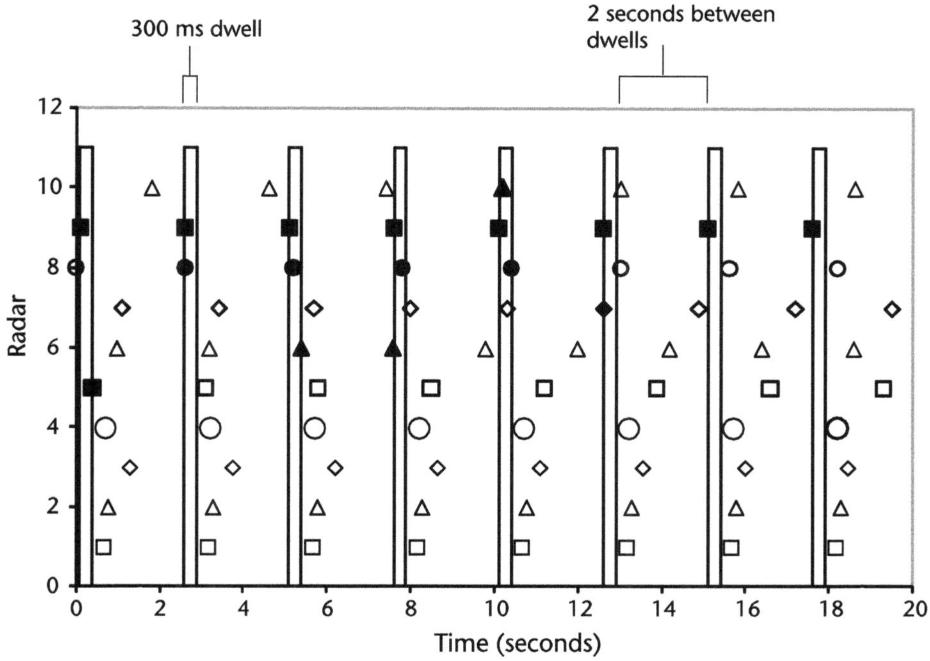

Figure 4.18 Traditional RF scan strategy.

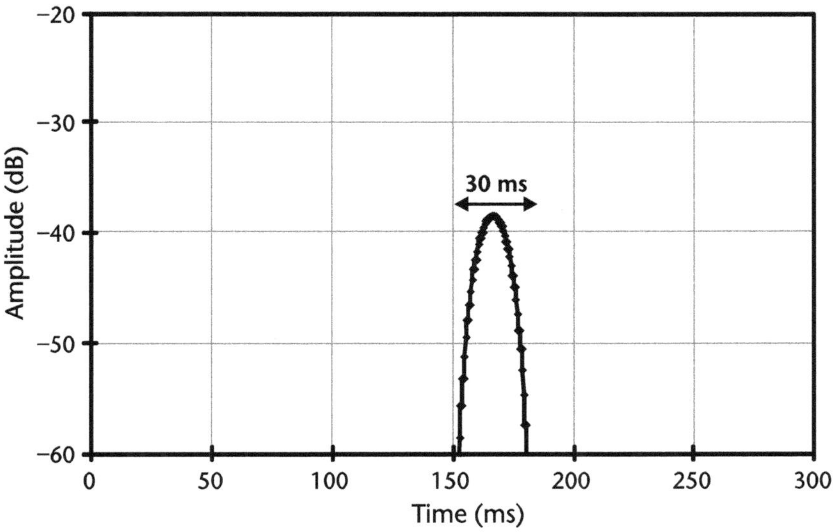

Figure 4.19 Theoretical scan pattern for an ATC radar at a 150-km range.

The same simulated radars that appeared in the traditional RF scan strategy (Figure 4.18) are plotted on the graph of the more effective scan strategy to show that all the radars are detected using this scan strategy (Figure 4.20). Each radar is detected during the dwell and then is missed for a few radar scan cycles while the ESM is looking at different RF bands. This RF scan strategy allows a track to be created for each radar and then updated at intervals of not more than three or four

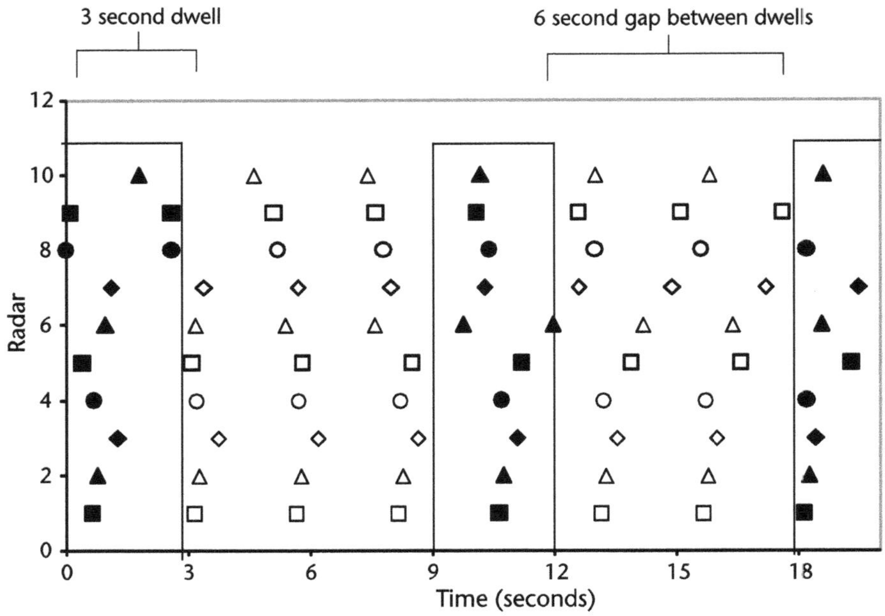

Figure 4.20 Effective RF scan strategy.

scans. This update rate should be acceptable for radars in the environment that are not threat radars.

Using an RF scan strategy like this will probably mean that several contiguous dwells will be needed in each band before moving on to the next band, as dwell durations are only likely to be of the order of a few hundred milliseconds. The reason for this limit in the dwell time is the size of the pulse buffer that is available. Typically, there is a limit of only a few thousand pulses that can be recorded in a single buffer/dwell in current ESM systems.

The use of this RF scan strategy means that it is certain that the main beam from each radar within the radar horizon will be seen by the ESM. The exact duration of the blocks of dwells is determined by considering the scan periods of all the types of radar in all RF bands of interest in a particular mission.

The scan strategy described above should be used for medium-range to long-range ESM operation. When the area of interest is of short range (< 25 nm), it is preferable to use dwells interleaved in each RF band. This is because pulses from sidelobes can be received at short range, so the time that elapses between the generation of intercepts and the updates to ESM tracks can be reduced and more intercepts are available for radar location fixing.

When there are threat radars likely to be in the environment, it is necessary to modify the scan strategy to optimize the search for the threat radars. In this case, a dwell at the threat radar band should be interleaved with the dwells at all the other bands. If there is one band that has a particularly long dwell, then this long dwell should be interrupted with a threat band dwell. This leads to a loss of detection capability in the band with the split dwell, but it is a compromise that cannot be avoided in the operation of narrowband ESM systems.

References

[1] Singh Bakshi, H., et al., "Microwave Antennas in Defence for Electronic Warfare Application," *International Journal of Electrical, Electronics and Data Communication*, Vol. 2, 2014.

[2] Lipsky, S. E., *Microwave Passive Direction Finding*, New York: SciTech Publishing, 2004.

[3] Johnson, R. C., and H. Jasik, *Antenna Engineering Handbook*, 2nd ed., New York: McGraw-Hill, 1961.

[4] Mayes, P. E., "Frequency-Independent Antennas and Broad-Band Derivatives Thereof," *Proc. of the IEEE*, Vol. 80, No. 1, 1992.

[5] Johnson, R. C., *Antenna Engineering Handbook*, 3rd ed., Ch. 14, New York: McGraw-Hill, 1993.

[6] Griffiths, H. D., and W. J. Bradford, "Digital Generation of High Time Bandwidth Product Linear FM Waveforms for Radar Altimeters," *Proc. of the IEE*, Vol. 139, No. 2, April 1992.

[7] Wiley, R. G., *ELINT: The Reception and Analysis of Radar Signals*, Norwood, MA: Artech House, 2006.

[8] https://www.globalsecurity.org/military/world/stealth-aircraft-rcs.htm.

[9] https://www.emcos.com/images/Applications/App_Ships4_RCS_Problems_in_Ship_Table.gif.

CHAPTER 5

Amplitude Comparison ESM

A technique commonly used by ESM systems is to compare the ESM antenna amplitude activation levels at separate ESM antennas to determine the DOA of a radar pulse. This is based on the assumption that the radar pulse arrives with equal amplitude at all receiving antennas. The DOA is found by looking at the amplitude differences of the receiver antenna beam patterns and assuming that they are due to the direction of arrival of the pulse at each antenna.

5.1 DOA determination in Amplitude Comparison Systems

A typical ESM system has m receiver antennas with an azimuth angle of $\theta° = 360°/m$ between the boresight of adjacent antennas. The beamshape for the antennas can be represented by a Gaussian model [1] with a nominal 3-dB beamwidth of θ_b, typically about 70°. The simplest form of amplitude comparison algorithm uses the activation levels of two ESM antennas to calculate the DOA (θ) of a signal.

Figure 5.1 shows the beamshapes of a 2-port ESM system with the ESM antennas separated in azimuth by 90°.

In a 2-port ESM system, the estimate of DOA [θ] is the one that has been created using the smallest amplitude difference of $|(A_{m-1} - A_m)|$ or $|(A_{m+1} - A_m)|$. To scale the measured amplitude to the signal direction, it is necessary to know the tracking slope for the system. This is equal to the change in the estimate of θ seen for a 1-dB difference in amplitude between two antennas. Gaussian-shaped beamwidths have the property that the logarithmic output ratio slope in decibels is linear as a function of an AOA [2]. A look-up table can therefore be used to determine the angle directly from the difference in activation levels between two antennas. Both the antenna beamwidth and the boresight squint angle vary with frequency over the RF range typical of ESM systems. Measurements in an anechoic chamber determine the true characteristics of the antennas [3]. However, a typical 2-port ESM system may have a 3°/dB tracking slope, as shown in Figure 5.2.

The measured signal strength A_m at the mth antenna due to a signal of amplitude A_0 incident on the antenna at angle θ is approximated by:

$$A_m = A_0 - k(\theta - m\theta_0)^2 \, (in\, units\, of\, dB) \tag{5.1}$$

where

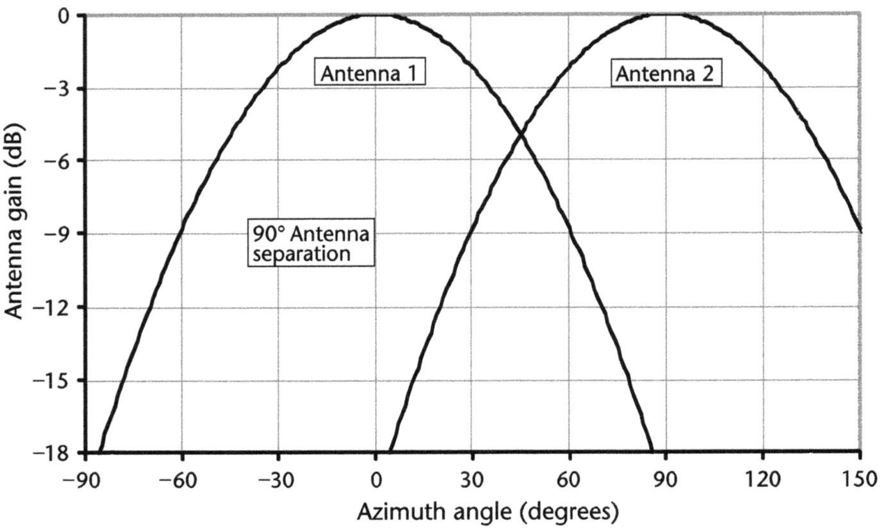

Figure 5.1 Two-port ESM antenna beamshapes.

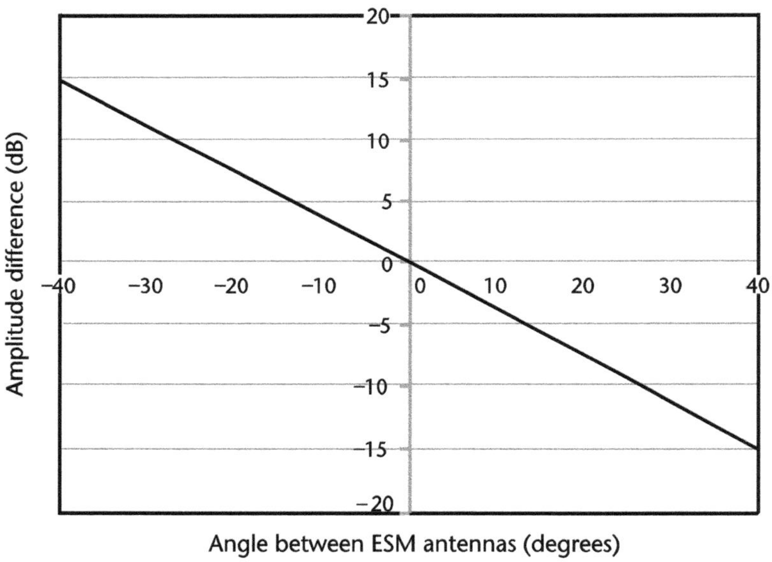

Figure 5.2 Tracking slope for a 2-port ESM system.

$$k \, is \, the \, tracking \, constant = 12(\theta_b)^2 \left(dB / degree^2\right) \tag{5.2}$$

θ_0 and θ_b are the separation in degrees of the antenna boresights and the 3-dB beamwidth of the ESM antennas, respectively.

When only two antennas are used, there are discontinuous jumps in the DOA at the boresights of the antennas, so it is usual to employ three antennas in the amplitude comparison calculation to avoid this situation. Figure 5.3 shows the profiles of 3 Gaussian beams of the 3-dB beamwidth of 70°, separated by 60°.

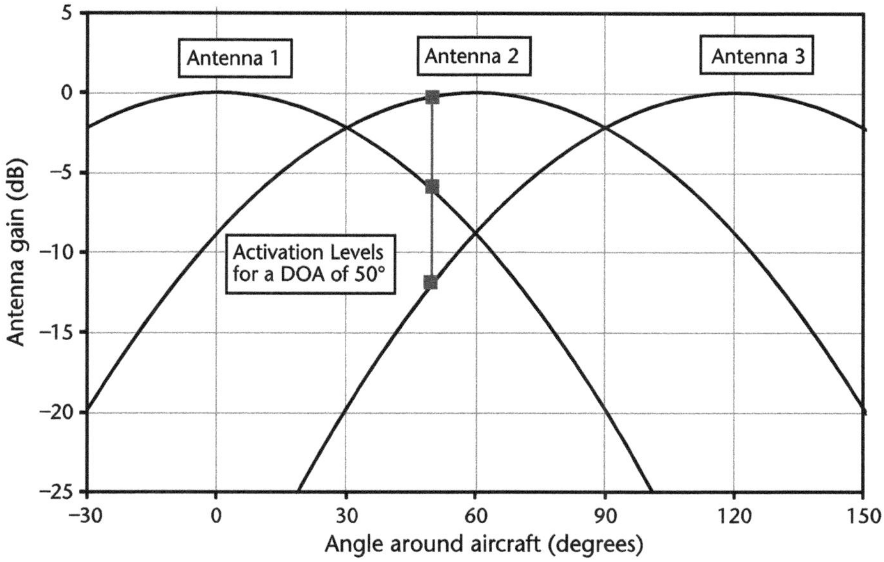

Figure 5.3 Typical antenna beam configuration for a 3-port ESM system.

The activation levels of each antenna for a pulse of DOA 50° are shown in Figure 5.3. In this representation of the amplitude comparison system, the activation levels lie in a straight line for a correct DOA calculation. The input amplitudes from the pulse to each of the antennas are equal in this case.

The following derivation of the DOA uses the activation levels of three ESM antennas.

A pulse arrives at the ESM antennas at an azimuth angle of θ such that the antenna having the largest activation level is antenna A_2 with the next largest activation levels for antennas A_1 and A_3. The antennas are separated in azimuth by θ_0 with antenna A_1 having a boresight at 0°. The following equations can be set up:

$$A_1 = A_0 - k(\theta - (0)\theta_0)^2 + e_1 \, (boresight\, at\, 0°) \tag{5.3}$$

$$A_2 = A_0 - k(\theta - (1)\theta_0)^2 + e_2 \, (boresight\, at\, 60°) \tag{5.4}$$

$$A_3 = A_0 - k(\theta - (2)\theta_0)^2 + e_3 \, (boresight\, at\, 120°) \tag{5.5}$$

where A_0 is the input amplitude, k is the tracking constant, and e_m is the measurement error for the mth antenna. Two activation level differences (D_L and D_R) can be calculated leading to two values of θ, θ_L and θ_R.

$$\begin{aligned} D_L &= A_2 - A_1 = A_0 - k(\theta - \theta_0)^2 - \left(A_0 - k(\theta)^2\right)(Assume\, e_m = e_{m-1} = e_1) \\ &= -k(\theta - \theta_0)^2 + k(\theta)^2 \end{aligned} \tag{5.6}$$

$$(D_L / k) = -\theta^2 - \theta_0{}^2 + 2\theta\theta_0 + \theta^2 \tag{5.7}$$

$$(D_L / 2\theta_0 k) = 0.5\theta_0 + \theta_L \tag{5.8}$$

$$\theta_L = -0.5\theta_0 + (D_L / 2k\theta_0) \tag{5.9}$$

A similar process using A_2 and A_3 yields and equation for θ_R:

$$\theta_R = 0.5\theta_0 - (D_R / 2k\theta_0) \tag{5.10}$$

Removing k from the equation and since $\theta_L = \theta_R$, the estimate of θ_a can be made.

$$\theta_a = \theta_0 (D_L - D_R) / 2(D_R + D_L) \tag{5.11}$$

A correction must be made to θ_a to take account of which antennas have been used in the calculation. For example, if the antenna that has the strongest activation level is the one with a boresight at 60°, then 60° must be added to θ_a to arrive at the correct AOA. The final part of the DOA calculation is the addition of the aircraft heading to the AOA to find the DOA of the radar pulse.

Problems arise for amplitude comparison systems when the amplitude incident at each antenna is not equal, a situation that can occur due to factors such as the physical separation of the ESM antennas, the radar beamwidth, and multipath interference of the emitted signal. Multipath effects are considered in detail in Chapters 15, 16, and 17.

5.2 Typical Amplitude Comparison Antenna Configurations

The basis of operation of an amplitude comparison system is that the received pulses must be seen on more than one antenna. To achieve 360° coverage in azimuth around an aircraft, the ESM antennas must be distributed around the platform.

A typical helicopter ESM antenna installation might consist of six ESM antennas located at the front (forward), rear (aft), and sides of the aircraft separated by a maximum of 20m. The antennas are orientated 60° apart giving 360° azimuth coverage, with two forward antennas, two aft antennas, and one on either side. Figure 5.4 shows a diagram of the helicopter ESM antenna installations. The other ESM installation shown in Figure 5.4 is for a fixed-wing aircraft with a separation of 30m between two sets of three antennas, one on the nose and the other at the tail of the aircraft.

A major assumption made in the implementation of amplitude comparison systems is that the aircraft can be thought of as a point in space. If the aircraft is far enough away from the radar, it can be thought of as a point, even though there is a separation of several meters between the ESM antennas. It is usually assumed

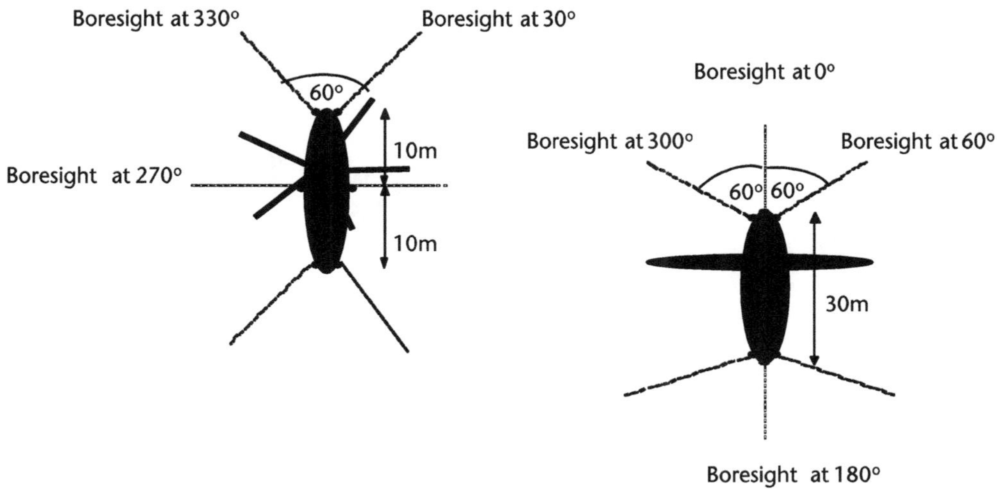

Figure 5.4 Typical ESM antenna configurations for a helicopter and a large aircraft.

in the design of ESM systems that even at short ranges this approximation to the point receiver is good enough.

However, many radar types have narrow azimuth 3-dB beamwidths, of 1° or 2° or less and sidelobes that are at least 20 dB down from the peak of the main beam. For radars with narrow main beams, different parts of the radar beam, at different amplitude levels, will be seen by the antennas at each end of the aircraft at the same time and DOA errors will result from this effect.

The main effect of the antenna separation is expected to be a slope in the pulse DOA profile as the radar beam scans across the aircraft (i.e., a change in the DOA error from pulse to pulse). However, larger DOA errors are expected at the edges of the main beam and when sidelobe detection occurs, for reasons explained next.

5.3 Effect of the ESM Antenna Separation on DOA Determination

The effect of antenna separation on the picture seen by the ESM system is shown in Figure 5.5. This shows an aircraft flying a racetrack route past a radar. The number of ESM tracks that are in existence at each of five points on the aircraft route is shown.

When the aircraft is at position 1, there is only one track formed. As the aircraft flies past the radar with the radar on the starboard side of the aircraft at an azimuth angle of 90° to the aircraft, extra tracks appear at position 2. When the aircraft has flown past the radar, the extra tracks disappear at position 3. The ESM is reset and a run past the radar with the radar on the port side of the aircraft starts. As the aircraft passes the radar on its port side with the radar at an azimuth angle of 270° from the aircraft nose, there is initially only one track at position 4. Extra tracks appear again at position 5 and then disappear as the aircraft ends the flight leg.

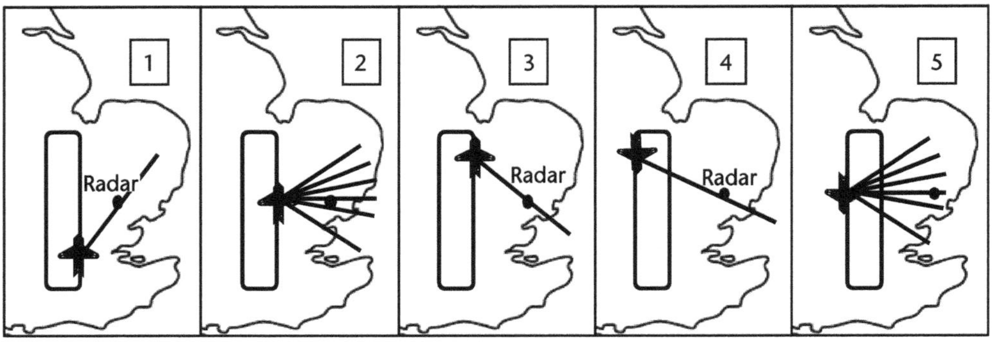

Figure 5.5 The effect of ESM antenna separation on ESM track formation.

A plot of DOA of the pulse data, Figure 5.6, shows that the reason for the multitracking is the increased spread in measured pulse DOA as the aircraft flies past the radar. As the aircraft draws level with the radar, the spread in DOA increases from ±5° to ±15°. The DOA spread decreases to ±5° as the aircraft flies beyond the radar.

Figures 5.7 shows simulations of amplitude and DOA profiles of an individual radar scan when the aircraft is alongside the radar. A slope in the DOA profile can be seen, so pulses from the leading edge of the main beam have a positive DOA error, those at the peak of the beam have no DOA error and those on the trailing edge of the main beam have a negative DOA error.

The slope that is seen across the main beam is dependent on the ESM antenna separation and the radar azimuth 3-dB beamwidth but is also dependent on the radar range and the orientation of the aircraft with respect to the radar. The slope is greatest at the edges of the main beam.

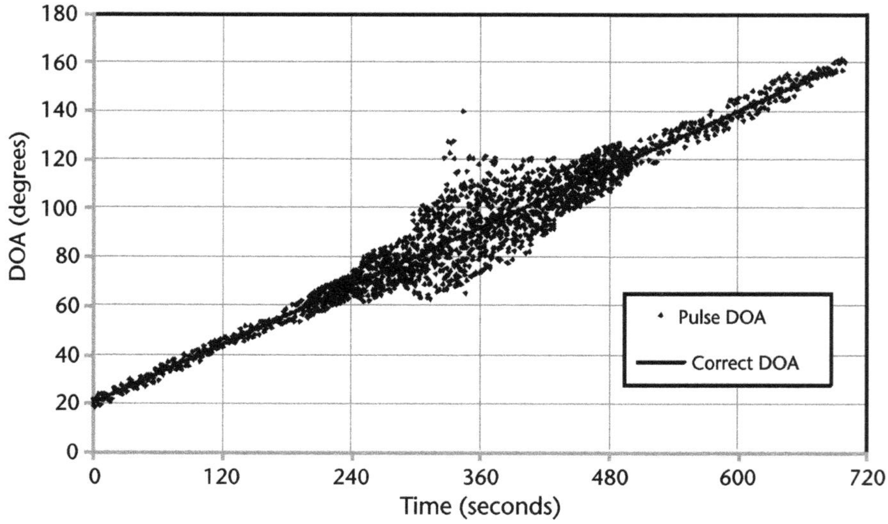

Figure 5.6 Radar pulse data received by an ESM system with separated ESM antennas.

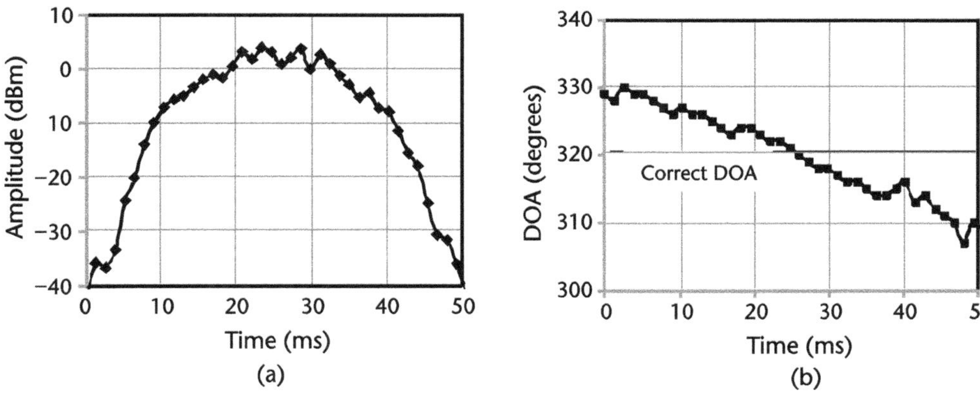

Figure 5.7 Single radar scan showing the slope in DOA across the main beam due to the separation of the ESM antennas.

5.4 DOA Errors Due to ESM Antenna Separation

To determine the DOA profile expected from a radar, a graphical method has been devised by the author. Figure 5.8 shows a simulation of a radar beam pattern with the azimuth angular extent of the aircraft drawn to scale on the same diagram. Vertical lines are drawn from the ends of the aircraft to the amplitude profile of the radar. The ESM antennas are situated at the ends of the aircraft.

The points at which vertical lines drawn from each end of the aircraft intercept the radar amplitude profile show the amplitude incident at each end of the aircraft.

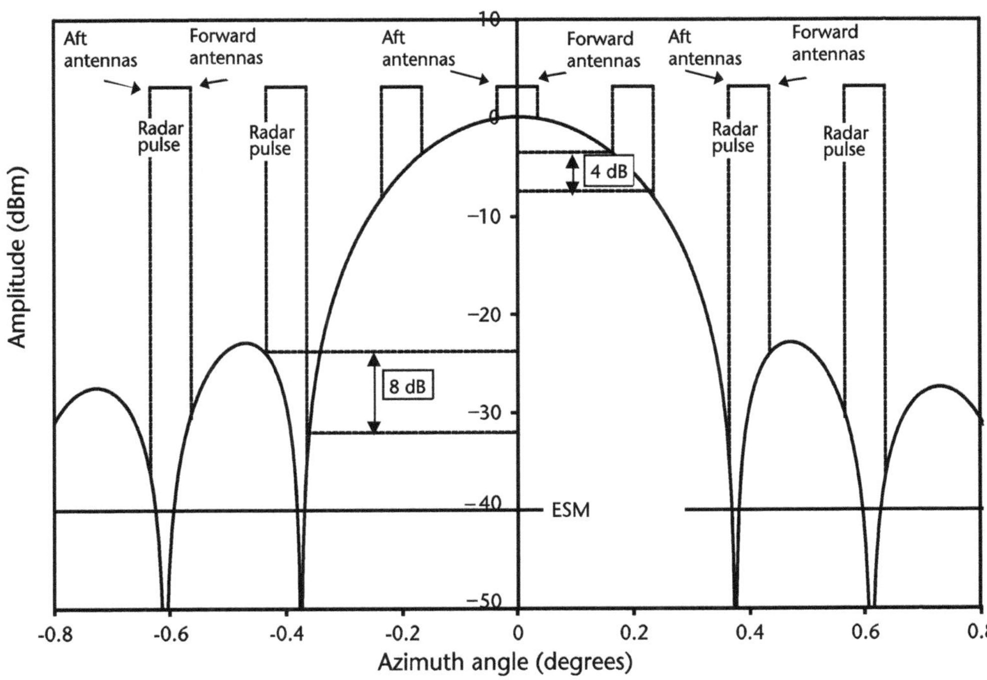

Figure 5.8 Method for determining DOA error due to antenna separation.

Horizontal lines drawn from the beam intercept points to the amplitude axis show the difference in radar amplitude seen by the forward and aft ESM antennas.

In the real world, the radar beam sweeps across the aircraft with little aircraft movement from pulse to pulse. However, in Figure 5.8 the aircraft is moved across the radar beam, creating the same effect of amplitude difference and allowing the angular distribution to be depicted linearly.

Figure 5.8 shows an extreme example where the range from the radar to the aircraft is short and the aircraft subtends a large azimuth angle to the radar beam to illustrate the method of defining the DOA error. A more realistic situation is shown in Figure 5.9 where the aircraft is at a range of 10 nautical miles (18.52 km) from the radar. As the antennas at the forward and aft of the aircraft are separated by 30m, the aircraft subtends an angle of 0.1° to a circle centered on the radar. The radar beamshape is typical for an air traffic control (ATC) radar with a 3-dB beamwidth of 1.5°.

As the radar beam begins to scan across the aircraft, the forward antennas will receive a larger input amplitude than the aft antennas and after the peak of the beam the situation is reversed. Thus, at positions A and B, the forward antennas will see a larger amplitude than the aft antennas, and at positions D and E the situation is reversed.

At the peak of the radar beam, position C, equal input amplitudes are seen by forward and aft antennas, so no error should result in the DOA calculation.

On the leading edge of the radar main beam, DOA errors of up to +36° are calculated as forward and aft antennas see different parts of the radar beam with an input amplitude difference of 12 dB. On the trailing edge of the main beam, there are DOA errors of up to −24° in this example due to input amplitude differences

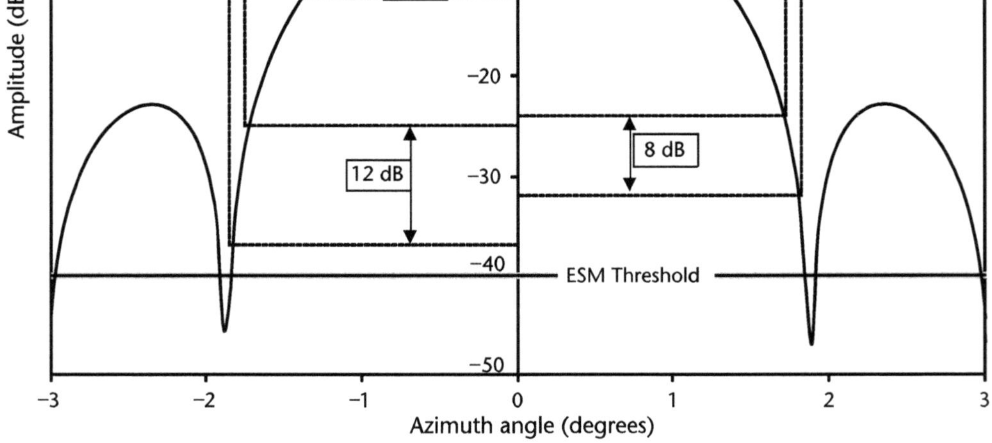

Figure 5.9 ATC antenna radiation pattern relative to an aircraft at a 18-km range.

of 8 dB between the forward and aft antennas. The effect of these amplitude differences leads to the slope in the DOA across the main beam as shown in Figure 5.7.

Although most ESM track construction software should be able to cope with the slope in the DOA across the main radar beam, some challenging effects may be seen at the edges of the main beam. For example, one cluster of antennas may see the main beam and the other cluster may see the first sidelobe, as shown in Figure 5.10.

A DOA error of –9° results from the difference in input amplitude at the forward and aft antennas for the pulse on the leading edge of the main beam in Figure 5.10. However, as this pulse is on the leading edge of the main beam, a positive DOA error would have been expected.

Other problems emerge when the pulse expected to be received by one antenna is lost altogether, as shown for the pulse on the trailing edge of the main beam in Figure 5.10 where the pulse is not seen at all on the aft antenna.

Figure 5.11 shows the effect of ESM antenna separation on sidelobe pulses and pulses at the edges of the main beam of a single scan of a radar. The DOAs of pulses at the edge of the main beam do not follow the same slope as that across the main beam as the errors are larger due to the shape of the radar beam. In the sidelobes, there may be positive or negative DOA errors because one antenna may receive the radar main beam and the other may receive the radar sidelobe. If one of the antennas does not receive the pulse at an input amplitude above the ESM threshold, then the pulse will be missed altogether.

The antenna separation effect is independent of the peak amplitude of the radar and depends on the antenna separation and the range and orientation of the aircraft. The slope across the main beam depends on these three factors and also on the radar beamshape, as described in Section 5.5.

The slope in the DOA error increases as the range to the radar decreases and increases as the radar 3-dB beamwidth decreases. However, the slope decreases

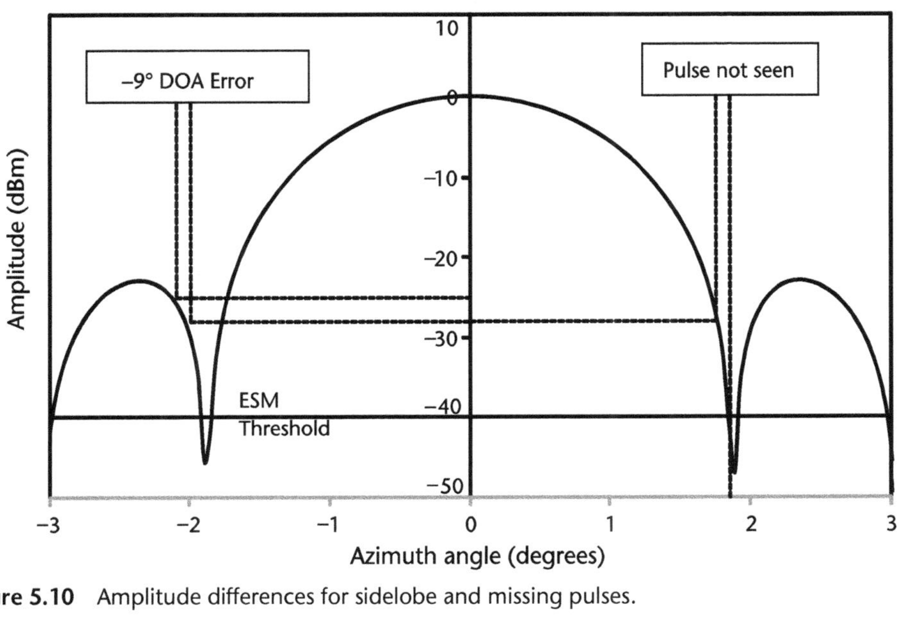

Figure 5.10 Amplitude differences for sidelobe and missing pulses.

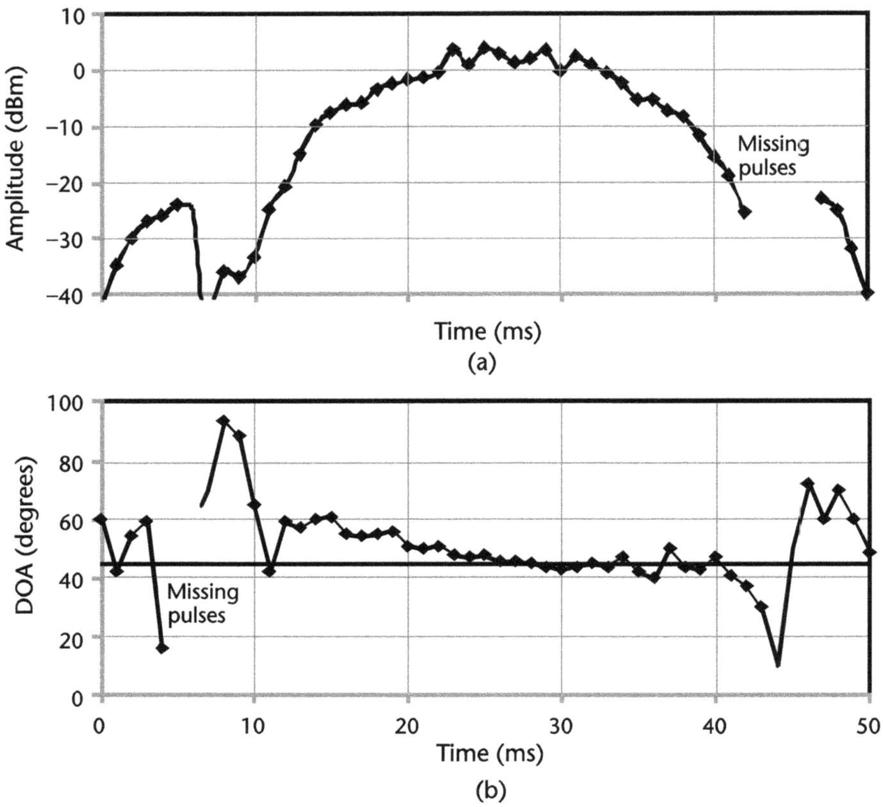

Figure 5.11 Single radar scan showing large DOA errors for sidelobe pulses.

with a decrease in effective antenna separation due to orientation. Figure 5.12 illustrates this point as the aircraft is drawn at three different orientations with respect to the radar.

The first (solid aircraft line) on Figure 5.12 shows the case where the azimuth angle is zero (i.e., the aircraft is perpendicular to a line drawn to the radar.) The difference in amplitude seen by the forward and aft antenna clusters is 5.5 dB, which equates to a DOA error of 16.5°. The dashed aircraft line is at an angle of 30° to the radar and the difference in amplitude seen by the forward and aft antennas is reduced to 3.7 dB or a DOA error of about 11°. The third aircraft (dotted line) is at an angle of 60° and in this case the amplitude difference is 1.5 dB with the corresponding DOA error being only 4.5°.

Figure 5.13 shows graphs of the maximum DOA error seen at the edge of the radar main beam for the large aircraft antenna configuration and also for the helicopter ESM antenna installation (from Figure 5.4). Two radar beamwidths are shown on Figure 5.13. The first is typical of an ATC radar with a 1.5° 3-dB beamwidth, and the second graph shows a radar of beam width 0.4°, typical of a harbor surveillance (HS) radar.

When a radar of 1.5° beamwidth is seen by a large aircraft ESM antenna configuration, there is a maximum error of 25° at the edge of the main beam at a range of 10 km. This error increases to 80° for the same aircraft ESM configuration when the radar beamwidth is 0.4° at a range of 10 km. For both radar types, the helicop-

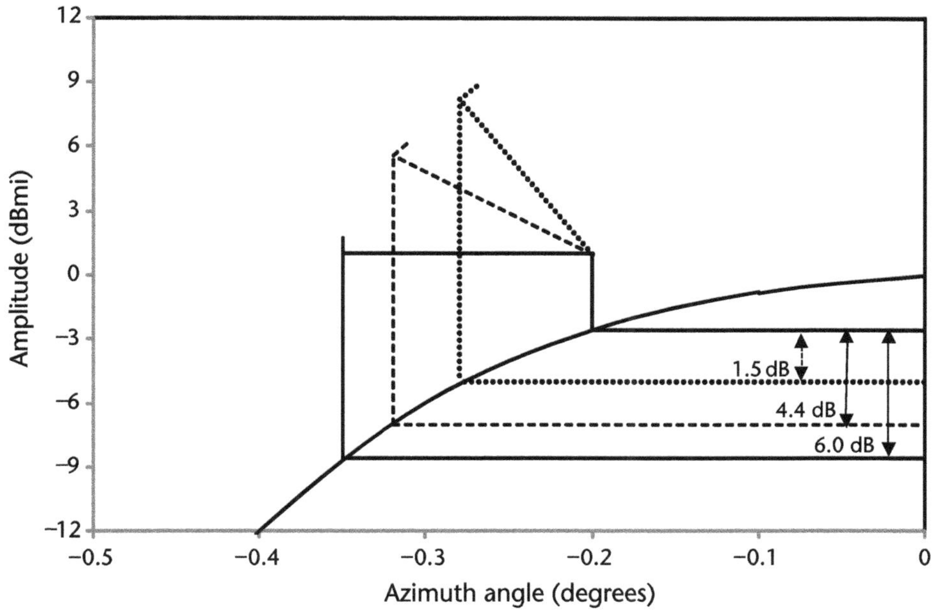

Figure 5.12 Effect of aircraft orientation with respect to the radar on the DOA error.

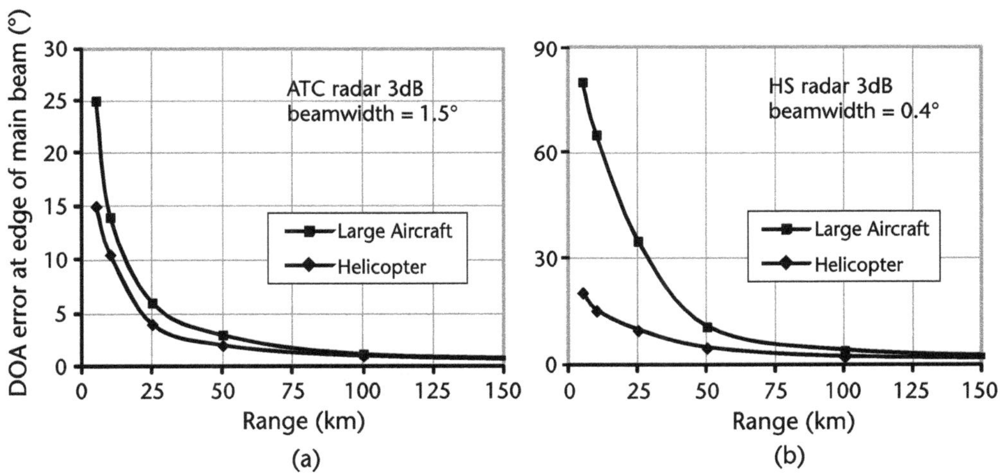

Figure 5.13 DOA errors at the edge of the main beam for ATC and HS radars.

ter ESM configuration has smaller DOA errors than the large aircraft installation at the edge of the main beam due to the smaller separation of the ESM antennas.

5.5 Effect of Radar Beamshape on DOA Error

Most scanning radars have narrow azimuth beamwidths, with typical ATC radars having 3-dB beamwidths of 1.5°. HS radars have 3-dB beamwidths as small as 0.3° meaning that there is likely to be a significant difference between the input pulse amplitude seen at separated antennas for this type of radar.

Figures 5.14 and 5.15 contrast the reception of pulses from an ATC radar and a HS radar at a range of 8 km from an aircraft with 30-m ESM antenna separation. The aircraft subtends an angle of 0.2° to a circle of radius 8 km, and is drawn to scale on the figures.

For both cases, only the pulses at the peak of the radar beam scan are seen by the ESM system at the correct DOA. Pulse A, which is next to the peak pulse on the leading edge of the radar bean is seen with a difference of 0.6 dB in ESM

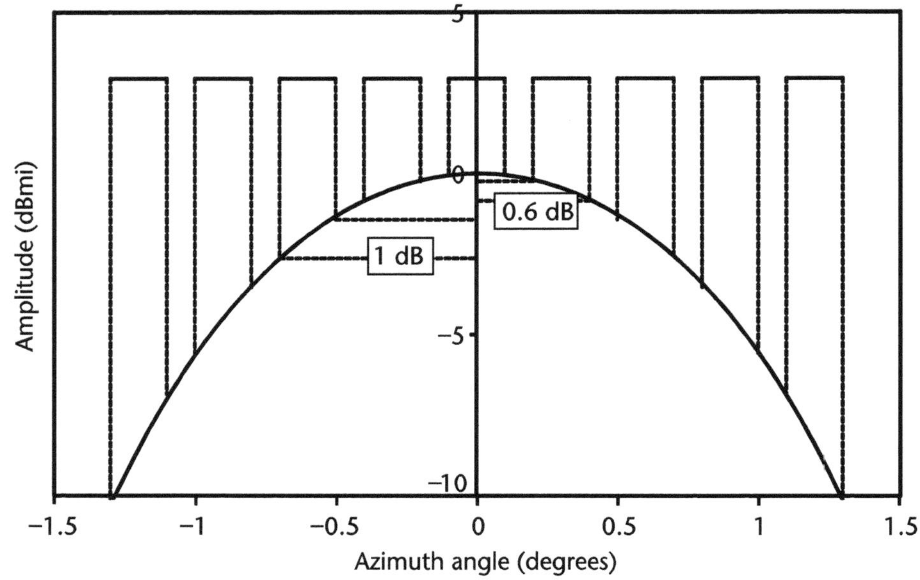

Figure 5.14 ATC antenna radiation pattern relative to an ESM at 8 km.

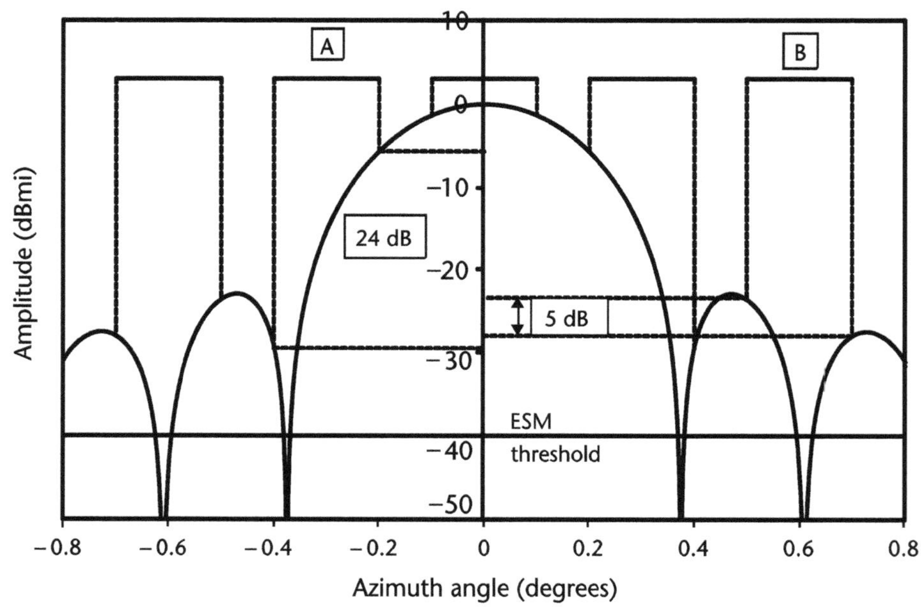

Figure 5.15 HS antenna radiation pattern relative to an ESM at 8 km.

antenna activation levels, for the ATC radar, leading to a DOA error typically of 2°. However, the same pulse emitted by the HS radar has a much larger difference in activation levels for the relevant antennas, of 24 dB, with a consequent increase in DOA error.

For the HS radar, it is only the peak pulse that is seen on both forward and aft antennas; for all other pulses, the ESM sees the main beam on only one antenna and the sidelobe on the other. Therefore, at short range, a radar with a narrow 3-dB beamwidth is very difficult for the ESM system to detect as most pulses have large DOA errors. Intercept creation and track correlation are compromised as a result.

5.6 Effect of Elevation on DOA Error

The second effect changing the measured DOA in amplitude comparison systems is the elevation error effect. For a given range and elevation, the DOA error increases near the boresight of either receiving antenna, and it is only at zero elevation that there is no elevation error effect. The elevation error effect decreases to the point where directly between two antennas the effect does not exist at all, for any elevation angle.

Figure 5.16 illustrates this point by showing the amplitude difference between two antennas at various elevation angles.

An example of the elevation error effect is seen in Figure 5.17 showing DOA profiles for a single scan of a typical ATC radar at a high constant elevation angle, 30°. In the first plot, the ESM antennas are colocated. To produce the second plot, a 20-m ESM antenna configuration has been assumed. The correct DOA of the emitted signal is at the boresight of one of the ESM antennas in both cases.

When the antennas are colocated, there is no variation in DOA across the scan, and the effect of the elevation is to shift the DOA by about 5°. The same effect is

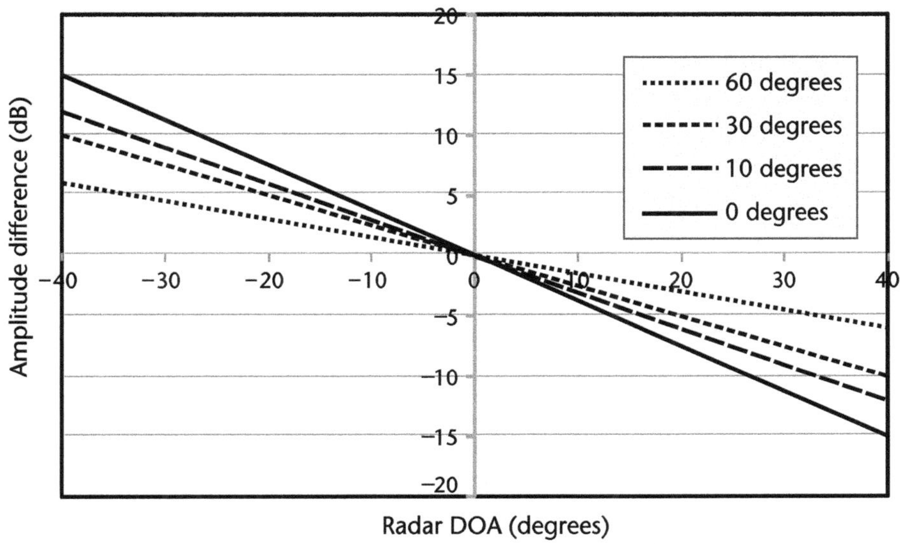

Figure 5.16 Amplitude difference for two ESM antennas.

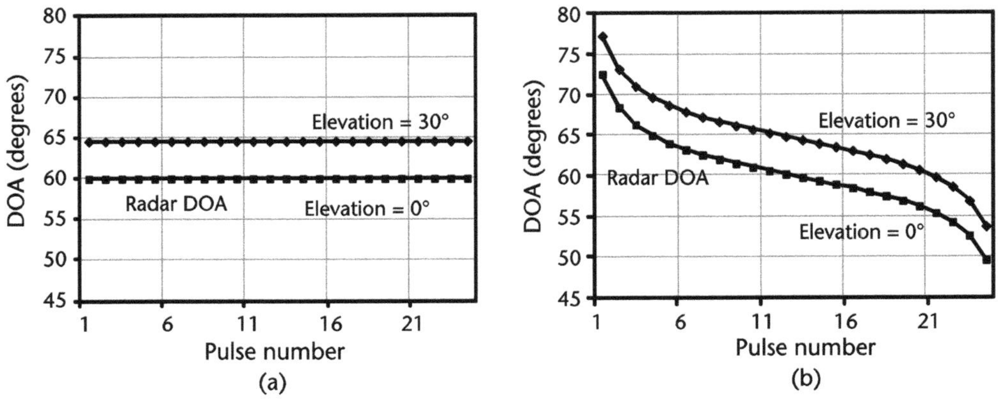

Figure 5.17 Effect of elevation on DOA calculation for colocated and for separated antennas configurations.

seen for the separated ESM antenna configuration with a shift of 5° in the entire DOA profile.

5.7 Solutions to ESM Antenna Separation Problem

The separation of ESM antennas causes DOA errors in amplitude comparison ESM systems, so it is important to address the issue in the design of the ESM system or to take steps to correct DOA errors in post-flight analysis.

The effect of the antenna separation in amplitude comparison systems is to see a slope in the DOA for pulses received from the main beam of a radar as the aircraft travels past the radar. Extra tracks are created due to the DOA spread leading to confusion about the correct DOA of the radar and to multitrack clutter on the ESM operator's screen.

Pulses may be missing from the pulse train seen from a radar as they are not seen on enough antennas. PRI calculation is affected as a result. Poor PRI calculation leads to extra track creation as intercepts cannot be correlated to existing tracks. Poor PRI calculation also leads to misidentification. Some radars may be missed altogether from the ESM picture due to the lack of pulses at reasonable DOA being available for intercept creation.

Possible solutions to the problem of DOA errors in amplitude comparison ESM systems are to colocate the ESM antennas, to select pulses from the peak of the beam for DOA calculation, and to adjust the activation levels of the receiver antennas.

5.7.1 Colocate the ESM Antennas

When the ESM antennas are separated, there is a slope in DOA across the main beam. However, when the antennas are colocated, the pulse is seen at the same activation level on all antennas and there is no DOA error. This solution may not be feasible as it may not be possible to find a location where all the ESM antennas are free from obscuration by platform surfaces.

5.7.2 Use Pulses from the Peak of the Radar Beam

By selecting pulses near the peak of the radar main beam, the errors due to antenna separation are minimized. If pulse detection is limited to these pulses, the larger DOA errors seen in the left graphs in Figure 5.18 are not taken into account in further ESM processing. When only the pulses near the peak of the beam are used, the ESM system should be able to average the DOAs from pulses either side of the peak and arrive at the correct DOA for the radar.

5.7.3 Adjust Antenna Activation Levels to Give an Indication of DOA Error

When there is no difference between the input amplitude to all three ESM antennas, the activation levels lie on a straight line so that the correct DOA is determined. In Figure 5.19 the activation levels are for a correct DOA of 70°.

When the input amplitudes to the three antennas are different, due to the separation of the ESM antennas and the shape of the radar beam, the activation levels no longer lie on a straight line, as shown in Figure 5.20.

The activation levels do not lie on a straight vertical line in Figure 5.20, and the DOA is calculated as 56°, an error of 14°. Table 5.1 shows the input amplitude and antenna activation levels for each of the three antennas for the cases in Figures 5.19 and 5.20.

By using the measured activation levels and the calculated (albeit incorrect DOA) to estimate the input amplitude to each antenna, it is possible to determine whether the calculated DOA is in error or not.

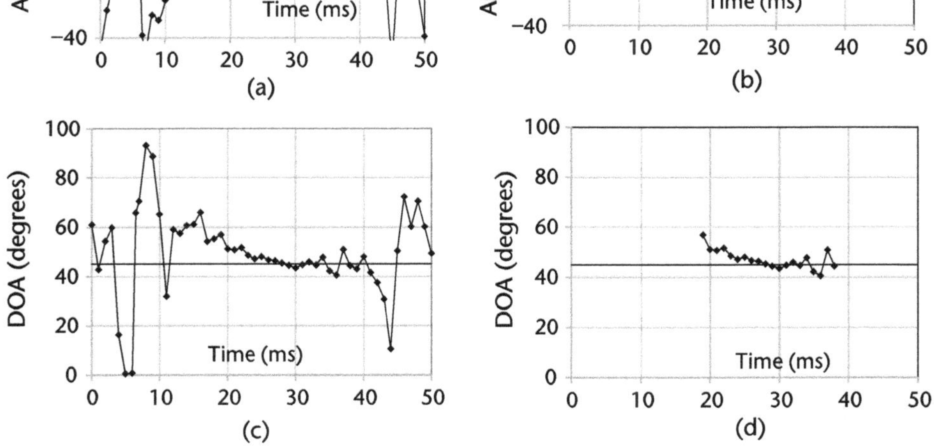

Figure 5.18 Single radar scans showing the reduction in DOA error for peak pulses.

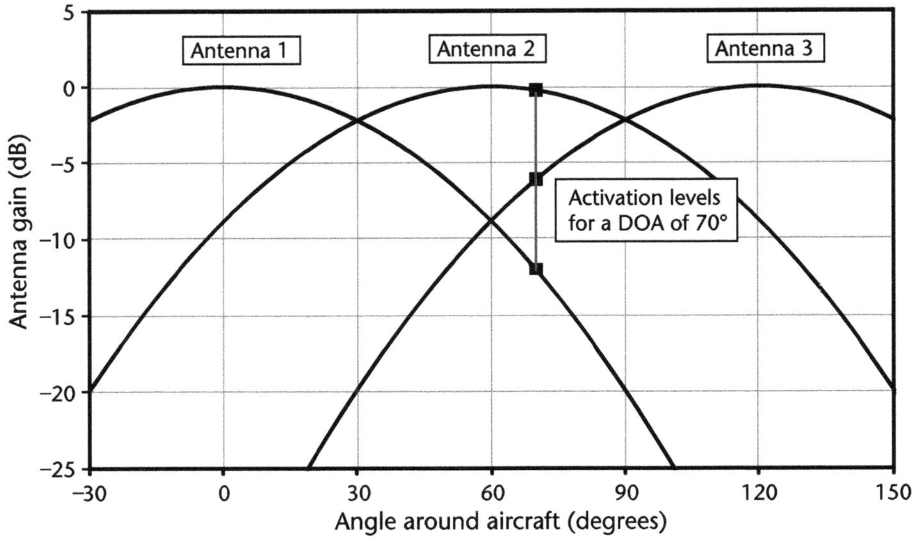

Figure 5.19 ESM antenna beamshapes with activation levels for a correct DOA of 70°.

Table 5.1 Input Amplitude and ESM Antenna Activation Levels for Correct DOA Calculation and for a DOA Error of 14°

	Input Amplitude (dB)	*ESM Activation Level (dB)*
DOA = 70°		
Antenna 1	0	–12.0
Antenna 2	0	–0.2
Antenna 3	0	–6.1
DOA = 56° (14° error)		
Antenna 1	–7	–19.6
Antenna 2	–5	–5.0
Antenna 3	–17	–23.5

It looks as if it should be easy to fix amplitude comparison systems. What is needed are the relative values of the input amplitudes for each ESM antenna. Unfortunately, this information is not available. All that is known are the activation levels of the individual ESM antennas and the calculated DOA.

The DOA verification process begins by estimating the input amplitudes to each antenna by reversing the calculation of the amplitude comparison algorithm.

For antenna 1 in Figure 5.20, an estimate of the incident amplitude (A_{in} = –11.3 dB) is made using the equation:

$$A_{in} = k\left(\theta_c - \theta_o\right)^2 + A_{out} \tag{5.12}$$

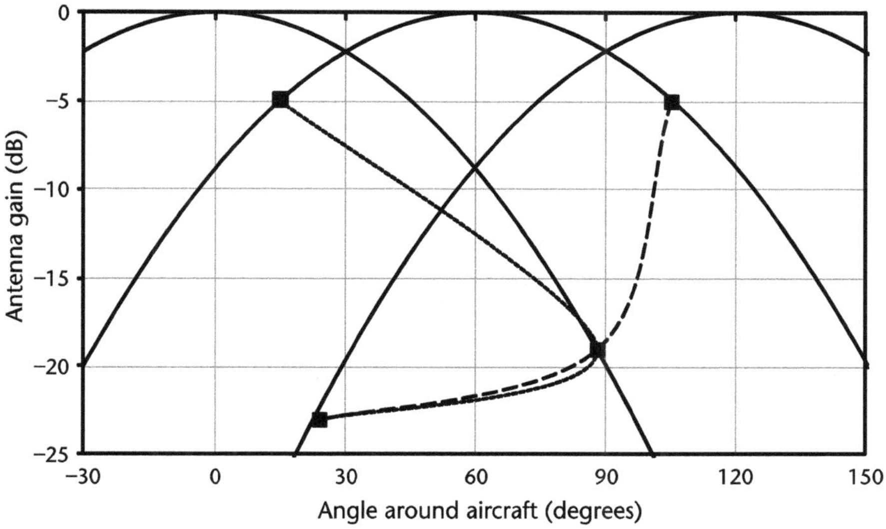

Figure 5.20 Activation levels for ESM antennas when input amplitudes are unequal.

where $k = 12/(\theta_b)^2$, θ_b = ESM antenna 3-dB beamwidth (70°), θ_c = calculated value of DOA (56°), θ_o = antenna boresight (0° in this case), and A_{out} = ESM antenna activation level (−19.6 dB).

Similar calculations lead to values of −5.2 dB and −13.1 dB as estimates of the input amplitudes for antennas 2 and 3, respectively. These are not the correct values, as an incorrect DOA value was used in their calculation. It is the only one available at this stage, so there is no choice but to use it. What this set of calculations does indicate is that the pulse amplitudes incident at the ESM antennas were not equal, so it is likely that the DOA is in error.

A method of mapping the incorrect DOA and the individual antenna activation levels to the correct DOA can be devised. There are several issues to be considered, the speed of processing, the amount of data that would need to be stored, and the ambiguities in the ESM antenna beamshapes that must all be considered.

It is possible that a form of pattern matching such as a neural network could be used to map the calculated DOA and ESM antenna activation levels to the correct DOA, but at the time of this writing, this method has yet to be devised.

References

[1] Stott, G. F., "DF Algorithms for ESM," *Proc. of 6th Conference on Military Microwaves*, 1988.

[2 Wiley, R. G., *Electronic Intelligence: The Analysis of Radar Signals*, Norwood, MA: Artech House, 1993.

[3] Ipek, A. V., "Implementation of a Direction Finding Algorithm on an FPGA Platform," Thesis submitted and accepted to the Middle East Technical University, October 2006.

CHAPTER 6

Time Difference of Arrival ESM

TDOA systems make use of the measurement of the time of arrival (TOA) of pulses at each of the ESM antennas to determine the DOA of a radar. This type of system is often implemented on large aircraft where the distance between antennas, located at the extremities of the platform, is large enough to allow a DOA calculation with good resolution to be carried out. TDOA systems typically operate at RF between 2 and 18 GHz.

6.1 DOA Determination in TDOA ESM Systems

The antenna configuration of a typical TDOA system and the baselines between the antennas used to find the difference in TOA at each pair of antennas are shown in Figure 6.1.

There are six baselines shown on Figure 6.1: Forward-Aft (FA), Starboard-Port (SP), Forward-Port (FP), Starboard-Forward (SF), Port-Aft (PA), and Starboard-Aft (SA).

The TDOA ESM system relies on the difference in the time it takes a pulse to reach each antenna for the determination of the DOA of the pulse. For example, a pulse at 90° with respect to the aircraft nose is seen first by the starboard antenna and then at the same slightly later time by both the forward and aft antennas.

TDOA (ΔT) can be calculated for a given DOA (θ) using the following equation:

$$D_T = (d \sin\theta) / c \tag{6.1}$$

where d is the distance separating the two antennas and c is the speed of light (3×10^8 m/s).

The geometry of the calculation of TDOA is shown in Figure 6.2.

If a single baseline is used ambiguity arises in the determination of DOA, so measurements must be taken on more than one baseline to resolve this issue. Figure 6.3 shows a plot of TOA difference against DOA for all six baselines in a system with forward and aft antennas separated by 30m and port and starboard antennas 35m apart.

The overlap in TDOA versus DOA profiles show how using multiple baselines can resolve DOA ambiguity. For example, a TDOA of 100 ns on the SP baseline

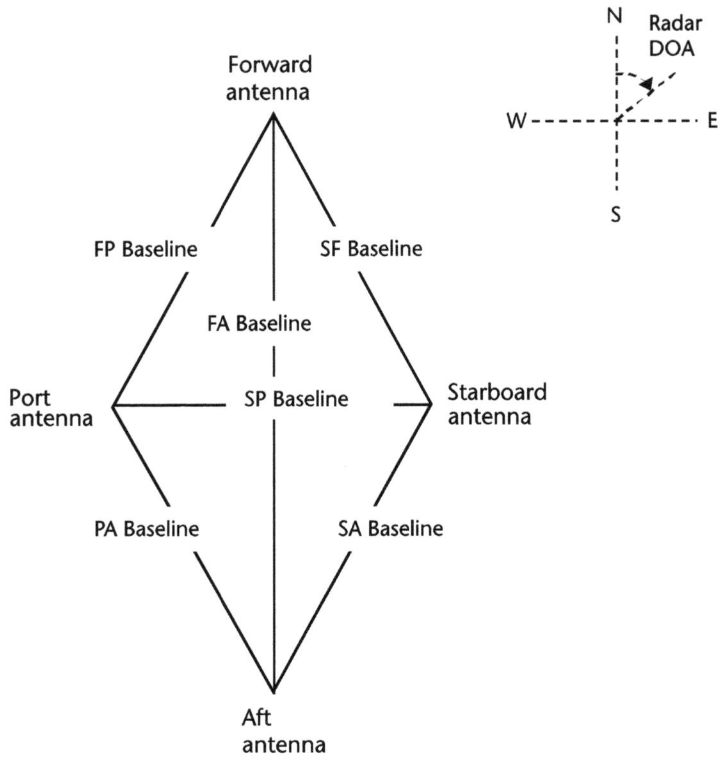

Figure 6.1 Antenna configuration for typical TDOA system.

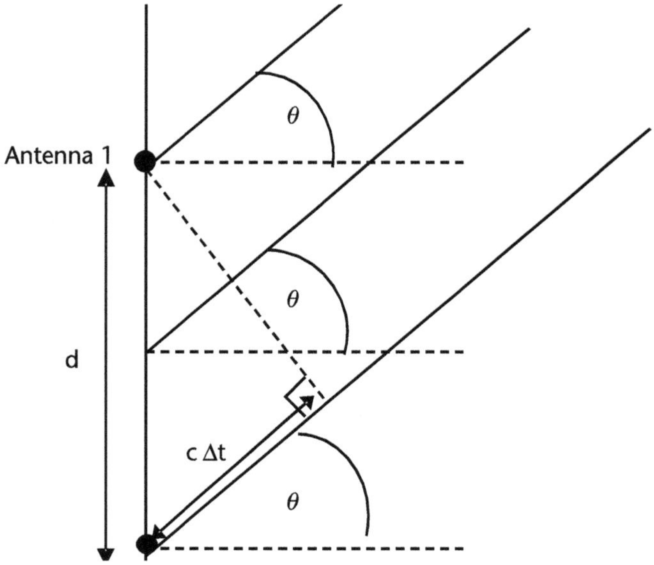

Figure 6.2 Geometry for TDOA calculation.

could equate to a DOA of 60° or 120°. A TDOA measurement of 58 ns on the SF baseline confirms that the DOA is 120°.

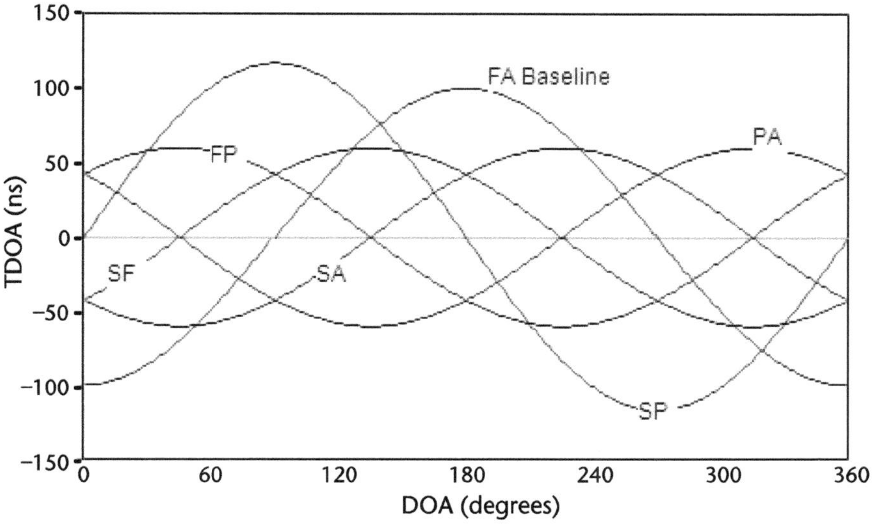

Figure 6.3 Time difference against DOA for each baseline in a TDOA ESM.

6.2 TOA Measurement

Pulse rise time is defined as the time taken for the pulse amplitude to increase from 10% to 90% of peak pulse amplitude. The number of watts per nanosecond is linear throughout the rise time, but the rise in dBW/ns is not linear. Figures 6.4 and 6.5 illustrate this point as it is important in the discussions on pulse measurement time that follow.

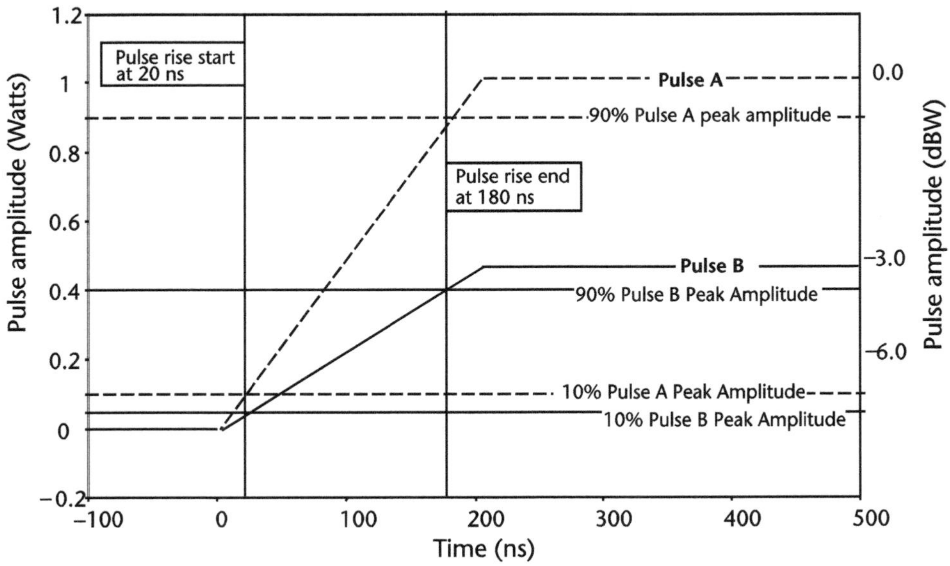

Figure 6.4 Rise time of pulses in watts/nanosecond.

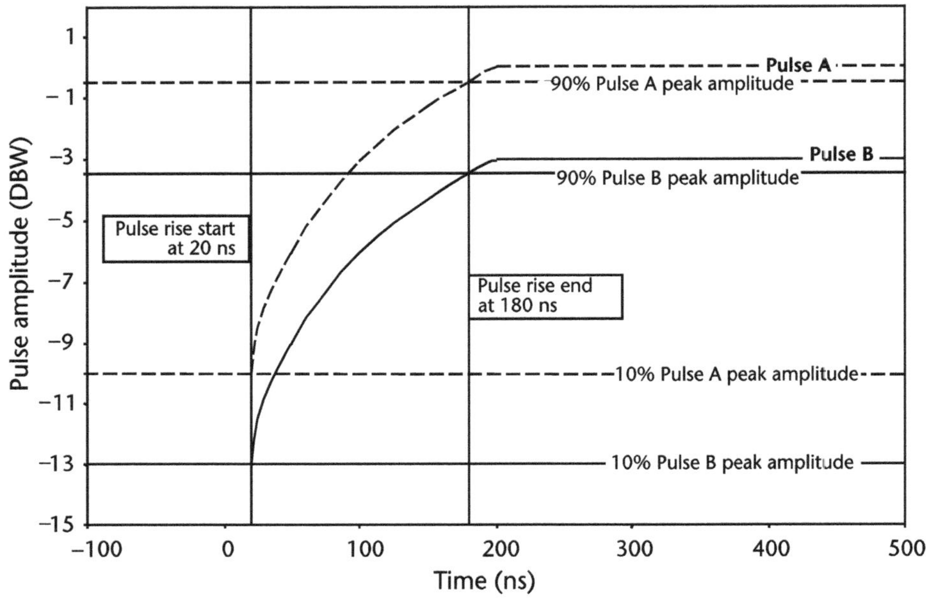

Figure 6.5 Rise time of pulses in dBW/nanosecond.

Figure 6.4 shows the linear slope in rise time seen for two pulses, one with a peak amplitude half that of the other, but both with the same rise time of 160 ns. The rise time starts at 10% of the peak amplitude at 20 ns and ends at 90% of the peak amplitude at 180 ns. On the right *y* axis, the amplitude in dBW is marked.

Although the number of watts per nanosecond is linear for both pulses shown on Figure 6.4, the slope in rise time (i.e., the number of watts per nanosecond is not the same for the two pulses).

The profile of the rise of the pulse in dBW/ns is not linear, as shown in Figure 6.5.

There are two possible methods of measuring the TOA of a pulse. In the first method, the TOA of a pulse at each antenna is recorded when the amplitude of the pulse reaches a threshold. In Figure 6.6, a pulse of width 1 μs has a rise time of 80 ns. (A pulse with such a long rise time relative to PW is unlikely in reality, but is used here for clearer illustration of pulse TOA measurement.)

In Figure 6.6, the pulse TOA is measured 85 ns after the start of the pulse using the amplitude threshold method of time measurement for an ESM with a sensitivity of −50 dBmi.

A second method of measuring TOA records the intrapulse amplitude profile and moves back in time down the profile from the peak by a set number of decibels. Figure 6.7 shows the measurement of the TOA of a pulse where the time measurement is made at a point 3 dB lower than the peak of the pulse amplitude. In this case, the pulse TOA is measured 100 ns after the start of the pulse.

Both methods of time measurement, the amplitude threshold and the amplitude reduction techniques, are compared in the following sections to show their effect on DOA accuracy.

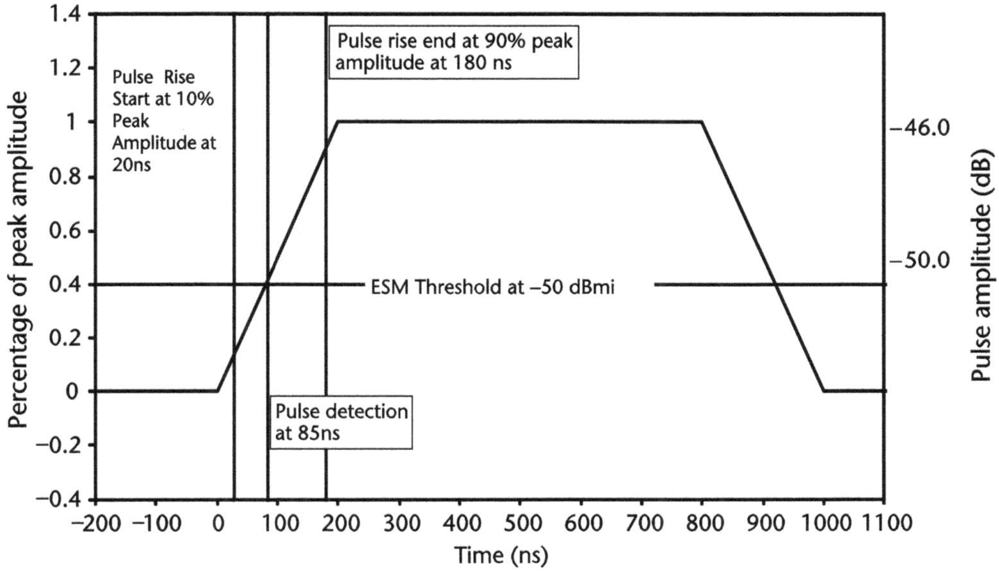

Figure 6.6 Measurement of TOA of a pulse using amplitude threshold method.

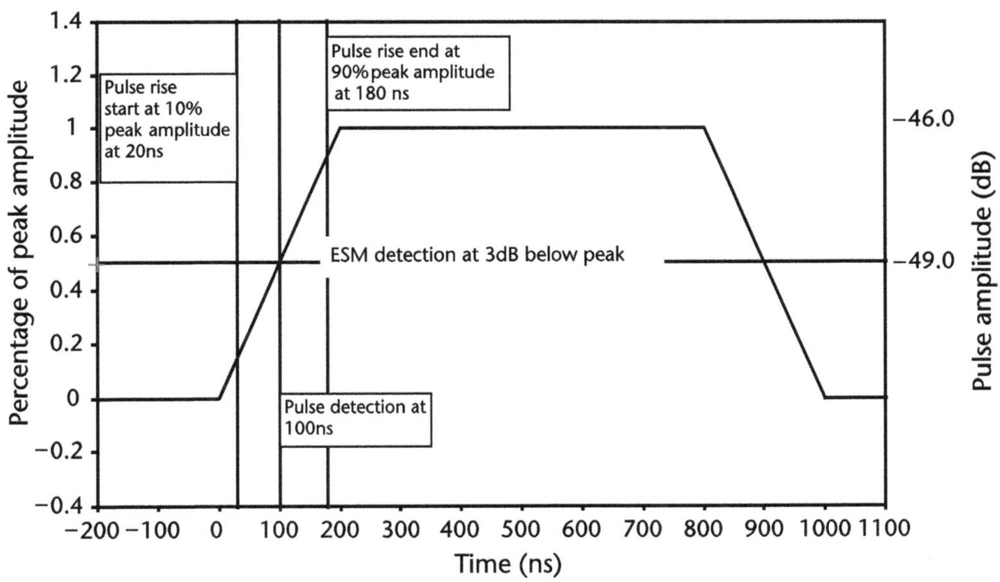

Figure 6.7 Measurement of TOA of a pulse using the amplitude reduction method.

6.3 ESM Antenna Separation Effect

The separation of the ESM antennas, while essential for the operation of TDOA ESM systems, actually contributes to DOA errors for scanning radars. Pulses from scanning radars are not received at equal amplitude at all ESM antennas due to the shape of the radar beam and the separation of the ESM antennas. Pulses seen at different amplitudes at separated antennas cross the pulse detection threshold at different times for each antenna, due to the rise times of the pulses.

It is usually assumed in the design of TDOA systems that the pulse arrives with equal amplitude at each of the ESM antennas. If this were the case, then a simple correlation could be made between the TDOA at a pair of antennas and the DOA of the pulse.

In reality, the shape of the radar beam means that the same amplitude is only seen at all relevant receiver antennas at the peak of the beam. The ESM platform subtends a discernible azimuth angle to the radar beam (see Chapter 5) as shown in the example in Figure 6.8 where the range from the ESM platform to the radar is 25 km and the separation of the ESM antennas is 30m. Figure 6.8 illustrates the effect of the antenna separation on the arrival of three pulses from a single scan of a radar operating at 3 GHz with a 3-dB beamwidth of 0.9°, a PRI of 0.8 ms and a rise time of 160 ns. (All the radars considered in this chapter are assumed to operate at 3 GHz as at this RF the power received in dBmi is the same as that received in dBW/m^2.)

The peak amplitude at which each pulse is received by each antenna is shown in Figure 6.8. These amplitudes are the same for either of the two methods of time measurement, which are discussed below. It is assumed that the FA baseline is being used for the measurement of TDOA in this case.

First, the amplitude threshold method of measuring TOA is considered. Figure 6.9 shows the intrapulse amplitude profiles of pulse A, at the peak of the beam, for both forward and aft antennas. Both antennas see this pulse at an amplitude of about –29.5 dBmi and the time at which the ESM amplitude threshold of –50

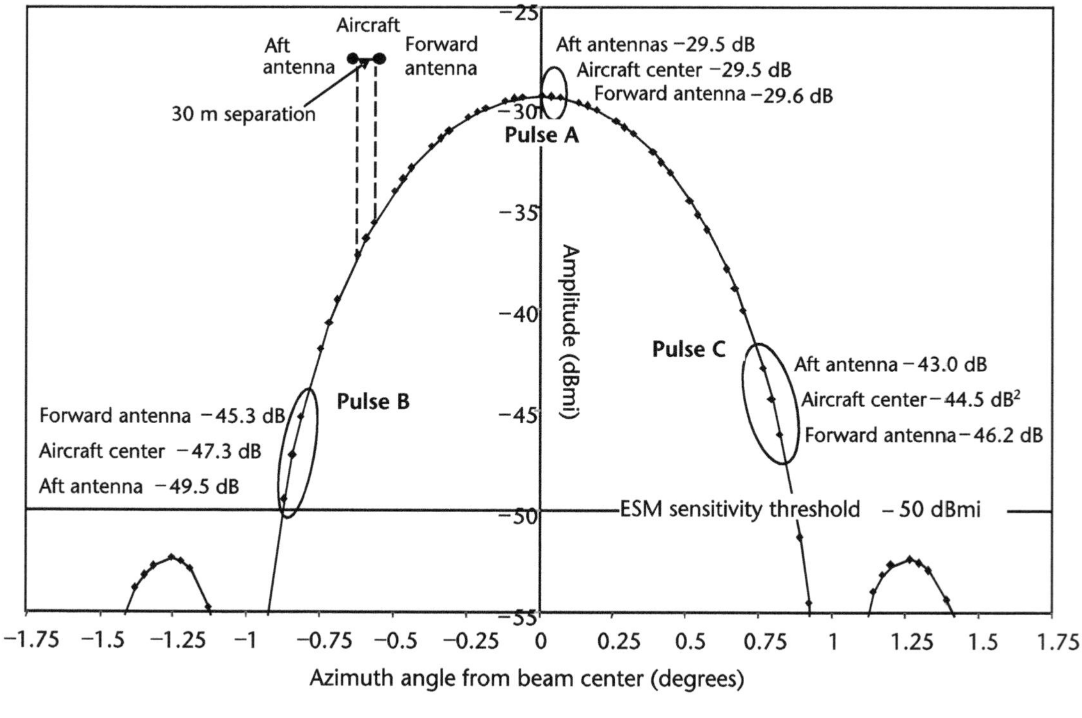

Figure 6.8 Effect of antenna separation on the amplitude received at two ESM antennas.

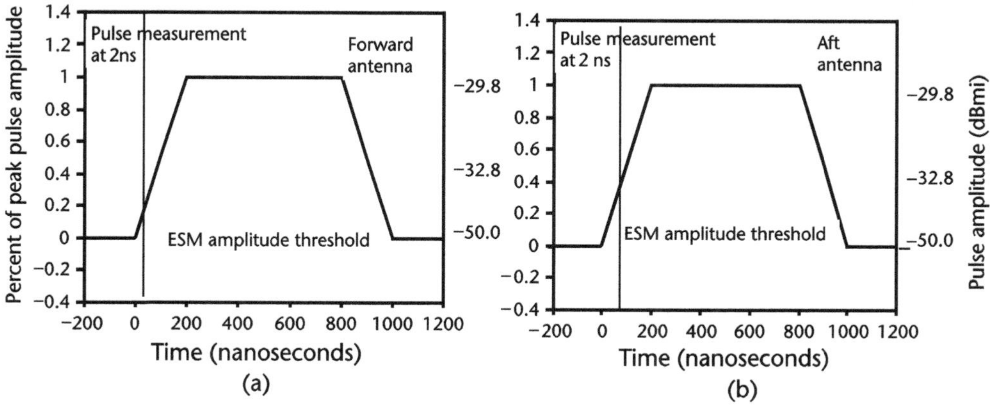

Figure 6.9 Intrapulse amplitude profiles for pulse A received at forward and aft antennas detected using the amplitude threshold method.

dBmi is crossed is almost identical for both antennas at 2 ns after the pulse starts to rise. The detection at 2 ns after the pulse starts takes place just above the ESM sensitivity level, which is 20.5 dBmi below the pulse peak amplitude. Any difference in TOA between the two antennas should be solely due to DOA and not due to the radar beamshape in this case.

The intrapulse amplitude profiles of pulse B, which is on the trailing edge of the main beam are shown in Figure 6.10. The aft antenna sees the pulse with amplitude –49.5 dBmi and the pulse has risen for 180 ns before the ESM amplitude threshold of –50 dBmi has been reached.

However, the forward antenna sees the same pulse at the higher amplitude of –43.5 dBmi, four times higher than the amplitude seen at the aft antenna. The ESM amplitude threshold is crossed at the much shorter time of 70 ns for the aft antenna. The TDOA for the FA baseline for this pulse is therefore in error by 110 ns, which translates into a DOA error of about 33°.

Pulse C on the leading edge of the main beam is detected by the aft antenna 40 ns after it begins to rise, but the forward antenna does not see it until 80 ns after it

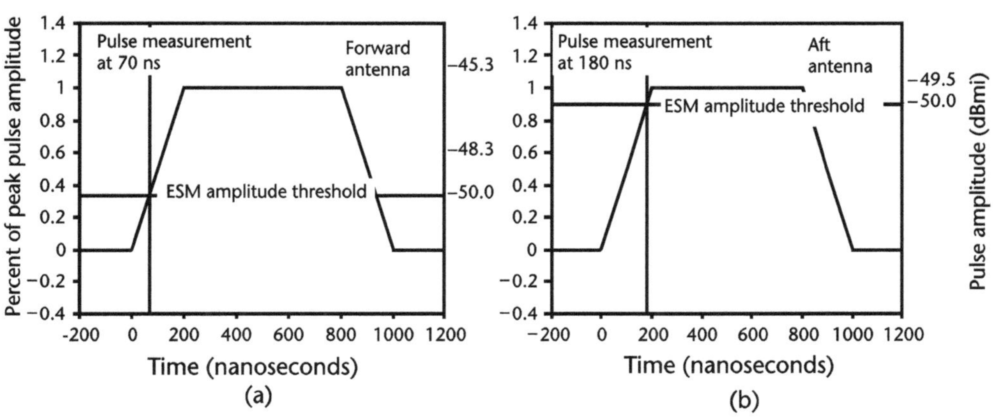

Figure 6.10 Intrapulse amplitude profiles for pulse B received at forward and aft antennas detected using the amplitude threshold method.

has begun to rise. The 40-ns error in the TDOA between the two antennas leads to a DOA error of about 13° for this pulse.

When the amplitude reduction method of time measurement is used, pulse A is seen by both forward and aft antennas at 100 ns after it has begun to rise, as shown in Figure 6.11. Although this is very different from the time of arrival of 2 ns using the amplitude threshold method, the fact that the pulse is seen at the same time after its start by both antennas means that there is no DOA error for the amplitude reduction method of TOA measurement.

Pulse B is seen at 100 ns after the start of the pulse at the forward antenna, as shown in Figure 6.12. However, it is not detected at all by the aft antenna, as the peak amplitude seen at the aft antenna, of –49.5 dBmi is not more than 3 dB higher than the ESM detection threshold of –50 dBmi in this case.

Pulse C is seen at 100 ns by both the forward and aft antennas using the amplitude reduction method of time measurement. There is therefore no DOA error due to ESM antenna separation and the radar beamshape for pulse C, whereas an error of 13° is seen when the amplitude threshold method is used.

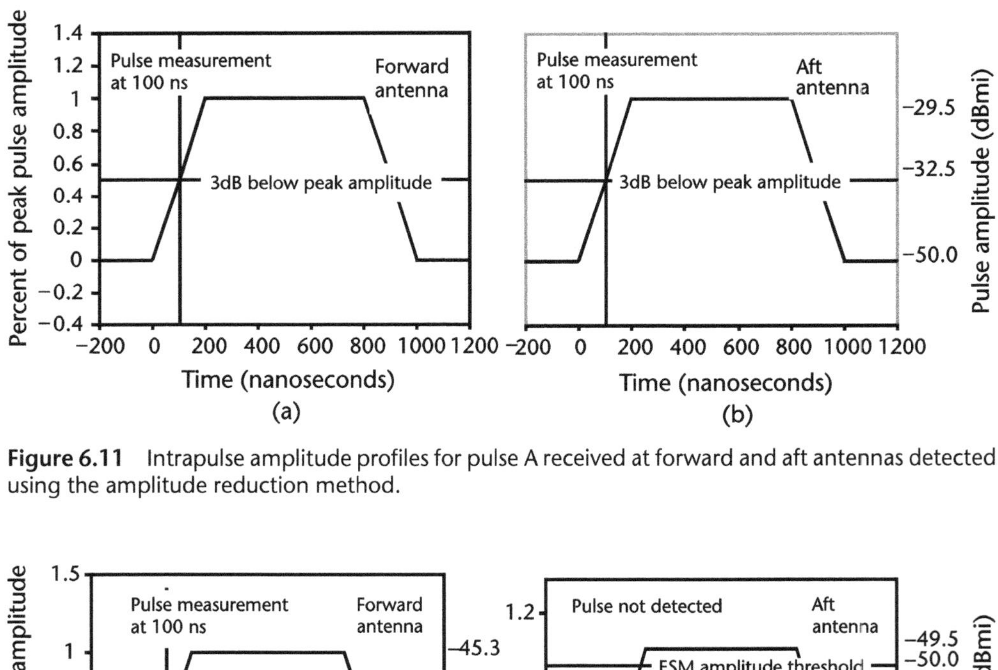

Figure 6.11 Intrapulse amplitude profiles for pulse A received at forward and aft antennas detected using the amplitude reduction method.

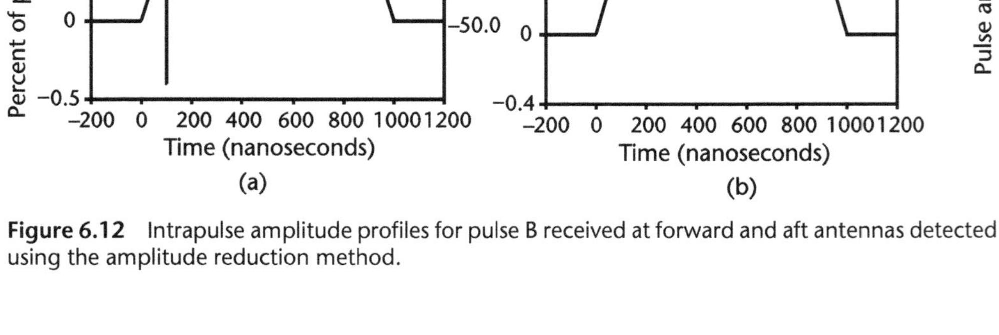

Figure 6.12 Intrapulse amplitude profiles for pulse B received at forward and aft antennas detected using the amplitude reduction method.

The two methods of measuring the time of arrival of the three pulses in Figure 6.8 are compared in Table 6.1.

Although the DOA errors are removed, for those pulses that can be detected, by using the amplitude reduction method, the effect of using amplitude reduction is to reduce the sensitivity of the ESM system. An amplitude reduction ESM system using a 3-dB amplitude reduction can only see radars at half the range of an amplitude threshold system with the same quoted sensitivity.

The reduction in the number of pulses that can be seen by the amplitude reduction ESM may mean that track formation is not carried out for some radars and that identification is delayed due to fewer pulses being available for the determination of PRI sequences.

In amplitude threshold systems that suffer large DOA errors for pulses near the ESM detection threshold, pulses that are likely to have large DOA errors can be used for PRI determination, even if they are not used in the calculation of the DOA of an ESM track.

6.4 Effect of ESM Sensitivity on DOA Errors

For scanning radars, a TDOA system using an amplitude threshold to measure TOA with widely spaced ESM antennas suffers sufficient differences in TOA for pulses that are not near to the peak of the beam to cause errors in DOA. The effect of antenna separation is worse for pulses whose peak amplitude is near the detection threshold or sensitivity of the receiver.

Figure 6.13 shows an example of how a single pulse of 3 GHz on the side of a main radar beam is received by three different ESM sensitivity levels, of –50, –55, and –60 dBmi. The forward and aft antennas are separated by 30m in the example on Figure 6.13 and the range to the radar is 25 km. The bearing of the radar with respect to the aircraft is only significant in that the angle subtended by the aircraft to the radar beam is smaller when the FA axis of the aircraft is not perpendicular to the ground range from the radar to the center of the aircraft.

For the lowest level of sensitivity, –50 dBmi, the aft antenna receives the pulse at only 0.5 dB above the threshold.

When the amplitude threshold TOA measurement method is used, TOA measurements for the pulse differ by 110 ns for the forward and aft antennas, translating into a DOA error of 33°. The DOA error is significantly improved for the reception of the same pulse by an ESM system of sensitivity, –55 dBmi. In this case the TDOA error is reduced to 25 ns, leading to a DOA error of 7°. Finally, an ESM

Table 6.1 Summary of TOA Results for the Two Methods of TOA Measurement

	DOA Error (degrees)	
	Amplitude Threshold	*Amplitude Reduction*
Pulse A	0	0
Pulse B	33	Not Seen
Pulse C	13	0

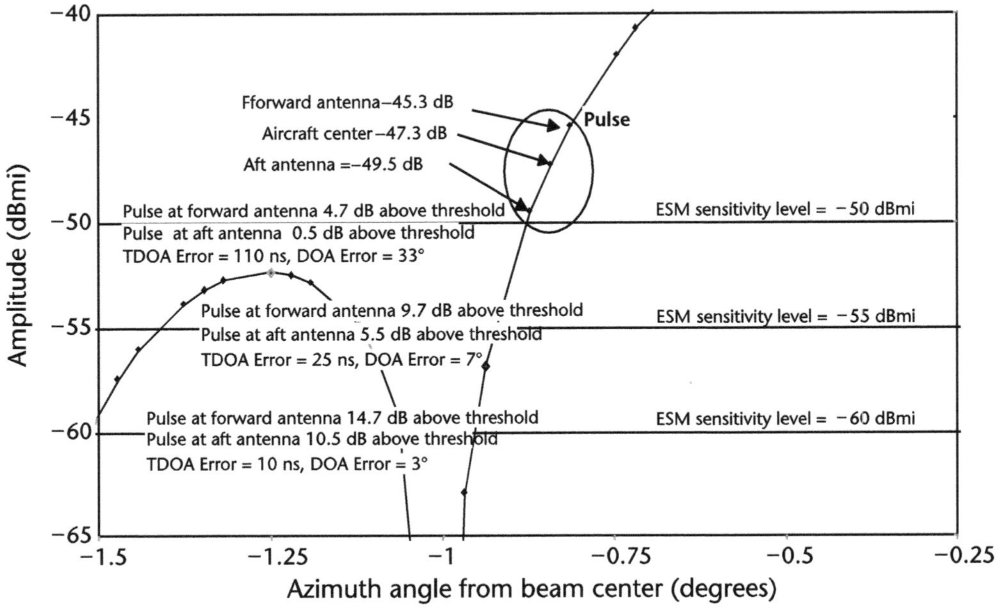

Figure 6.13 Reception of a single pulse at three levels of ESM sensitivity.

system of sensitivity –60 dBmi sees a TDOA difference of just 10 ns with a DOA error of 3° for the same pulse.

When the amplitude reduction method is used to measure TOA, the pulses are either subject to no TOA measurement error due to ESM sensitivity or they are not seen at all on one of the antennas. The pulse is not seen at all on the aft antenna when the ESM sensitivity level is –50 dBmi when the amplitude reduction method of measuring TOA is used. It is seen with a DOA error of 33° using the amplitude threshold method.

However, with ESM sensitivity levels at –55 and –60 dBmi, the pulse is seen with no DOA error using the amplitude reduction method. This compares with 7° and 3° errors for the amplitude threshold method.

Although the DOA errors due to ESM sensitivity are removed by using the amplitude reduction method, this technique results in a 3-dB reduction in ESM sensitivity so the ESM system can see only half as far as one with the same quoted sensitivity level using the amplitude threshold method of TOA measurement.

6.5 Effect of ESM Antenna Beamshape on DOA Calculation

The sensitivity quoted for an ESM is usually that at the peak of the ESM antenna beam. Even with the wide beamwidths of 70°, which are typical of ESM receiver antennas, there is considerable variation in sensitivity from the peak of the ESM antenna beam to the point at which a single ESM antenna beamshape overlaps that of the adjacent antenna beam peak.

Figure 6.14 shows two overlapping ESM antenna beamshapes which are separated by 90°, each having 70° 3-dB beamwidths. There is a difference in sensitivity between the two antenna beams of 20 dB at either beam peak.

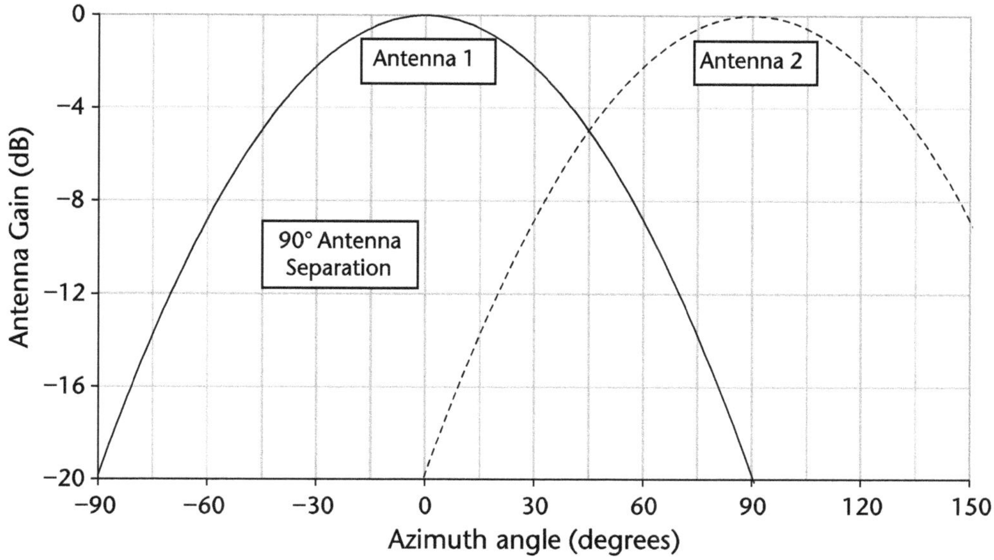

Figure 6.14 ESM receiver antenna beams separated by 90°.

The effect that the ESM antenna beamshape has on the DOA calculation of a pulse is shown in the following graphs (Figures 6.15 and 6.16) where pulses at DOA 0° and 30° DOA from the same radar at the same amplitude level are considered. The ESM antennas are separated by 90°, with 20m between the location of the forward and starboard antennas being used to calculate the TDOA and then the DOA of the pulse. The range to the radar is 15 km in this case so the azimuth angle of the aircraft subtended between the forward and starboard antennas is about 0.1°, leading to discernible difference in amplitude at each antenna for a single pulse.

For a DOA of 0°, the pulse is on the boresight of the forward antenna at 0° and 90° away from the boresight of the starboard antenna. The sensitivity of the forward antenna is –60 dBmi and that of the starboard antenna is –40 dBmi. Figure 6.15 shows the amplitude levels at which the pulse is received, taking account of the radar beamshape and the amplitude above the threshold at each antenna.

The ESM receives 10 pulses from a single radar scan that are above the lower of the two ESM antenna sensitivity levels (i.e., above –40 dBmi). Half of the main beam is shown in Figure 6.15, so the amplitudes of 5 pulses above the ESM sensitivity level are shown.

For the single pulse highlighted on Figure 6.15, the forward antenna measures the TOA within 1 ns of the start of the pulse, whereas the starboard antenna does not detect the pulse until 150 ns after the pulse start, leading to a DOA error of 50°.

If the TOA measurement is carried out using the amplitude reduction method, the pulse is not seen on the starboard antenna, so no DOA estimation is carried out. The number of pulses for which DOA calculations are carried out using this method is reduced from 10 when the amplitude threshold is used to 8 when the amplitude reduction method is used.

When a pulse of DOA of 90° arrives at the ESM platform, the situation is reversed in that the forward antenna has a detection threshold of –40 dBmi and the

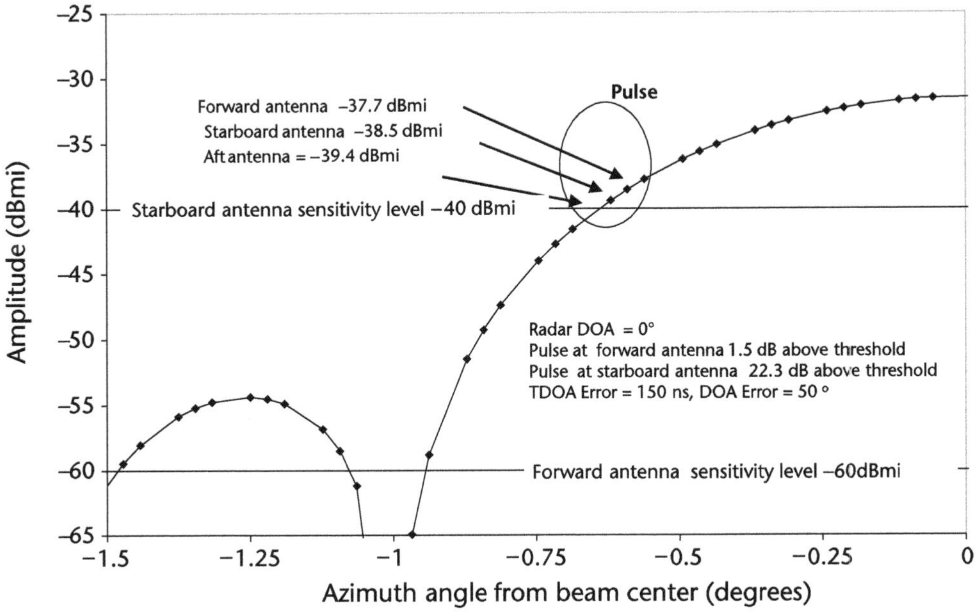

Figure 6.15 Reception of a single pulse at DOA 0° by an ESM with antennas separated by 90°.

starboard antenna has 20 dB more sensitivity, as the radar is on the boresight of the starboard antenna.

In this case using the amplitude threshold method, the starboard antenna measures the TOA of the pulse within 1 ns of its start, whereas the forward antenna does not see it until 120 ns after it has begun, leading to a DOA error of 36°. If the TOA measurement is carried out using the amplitude reduction method, the pulse is not seen at all on the forward antenna. This is because the effects of radar beamshape and ESM antenna beamshape combine to allow a drop of only 2.3 dB before the ESM detection threshold is reached and it is therefore unable to detect the pulse.

Figure 6.16 shows the reception of two pulses from a single radar scan at DOA 30° where the starboard and forward antennas have sensitivity levels of –51.2 and –58.0 dBmi, respectively.

Pulse 1 in Figure 6.16 is seen 1 ns after its start by the forward antenna and at 10 ns by the starboard antenna, leading to a DOA error of 3°. Pulse 2, near the sensitivity level of the starboard antenna, is seen at 130 ns by the starboard antenna. It is seen at 25 ns by the forward antenna leading to a TDOA error of 105 ns and a subsequent DOA error of 31°.

When using the amplitude reduction method, pulse 1 is detected with no time difference between the forward and starboard antennas and hence has no DOA error. However, pulse 2 is not detected on the starboard antenna using the amplitude reduction method as the pulse amplitude is only 2 dB above the starboard antenna sensitivity level, only –51.2 dBmi at 30° DOA.

Although DOA errors due to the ESM antenna beamshape are removed when the amplitude reduction method is used, the number of pulses seen in the main beam scan is reduced from 14 for the amplitude threshold method to 12 for the amplitude reduction method for the 30° DOA case.

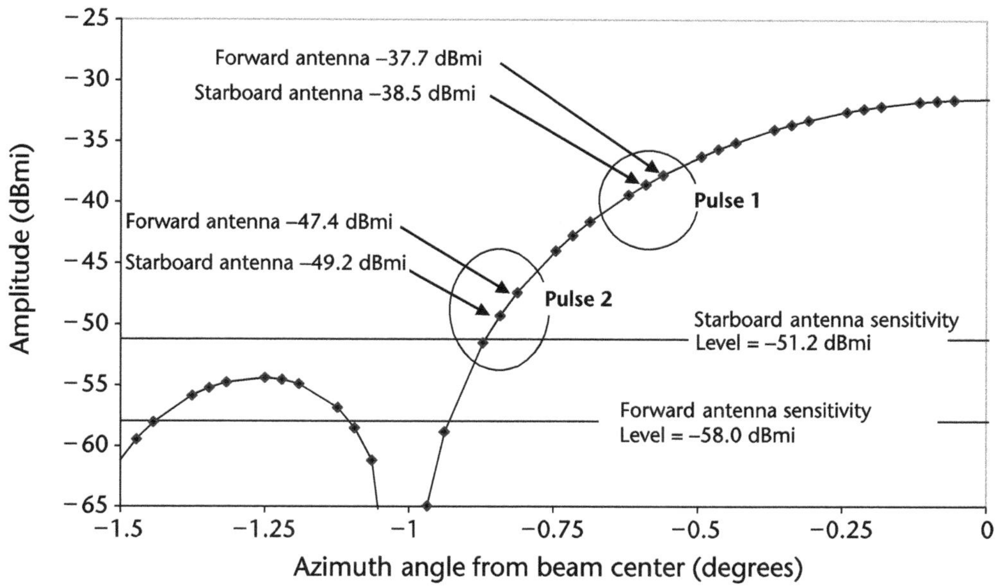

Pulse 1
Radar DOA =30°
pulse at forward antenna 20.2 dB above threshold
pulse at starboard antenna 12.7 dB above threshold
TDOA Error = 10 ns, DOA Error = 3°

Pulse 2
Radar DOA =30°
pulse at forward antenna 10.6 dB above threshold
pulse at starboard antenna 2.0 dB above threshold
TDOA Error = 105 ns, DOA Error = 31°

Figure 6.16 Reception of two pulses from a single scan at DOA 30° by an ESM with antennas separated by 90°.

6.6 Effect of ESM Antenna Configuration on DOA Calculation and Resolution

To determine the effect of the magnitude of ESM antenna separation on the calculation of DOA, two different ESM antenna configurations are considered. They correspond to the installations likely to be found on a large aircraft, such as a surveillance platform, and on a smaller fighter aircraft. Figure 6.17 shows a diagram of the two configurations.

A simulation of the reception of two pulses, A and B, by a large aircraft TDOA ESM installation is shown in Figure 6.18. Two levels of ESM sensitivity (at –50 dBmi and –60 dBmi) are shown and the radar parameters are those of a typical surveillance radar (with a 3-dB beamwidth of 0.3°) at a range of 80 km to the ESM aircraft.

At an ESM sensitivity of the –50 dBmi pulse A is seen with a TDOA error of 20 ns and a subsequent DOA error of 6°, due to the difference in amplitude seen at the forward and aft antennas. Pulse B is not seen at this sensitivity level. When the sensitivity is increased to –60 dBmi, the DOA error in pulse A due to the antenna separation effect is reduced to 2°, but the DOA error for pulse B is 16°. If the range to the radar is increased, then pulse B may not be seen at –60 dBmi and the error in DOA for pulse A is increased as pulse A is nearer to the sensitivity threshold of the receiver.

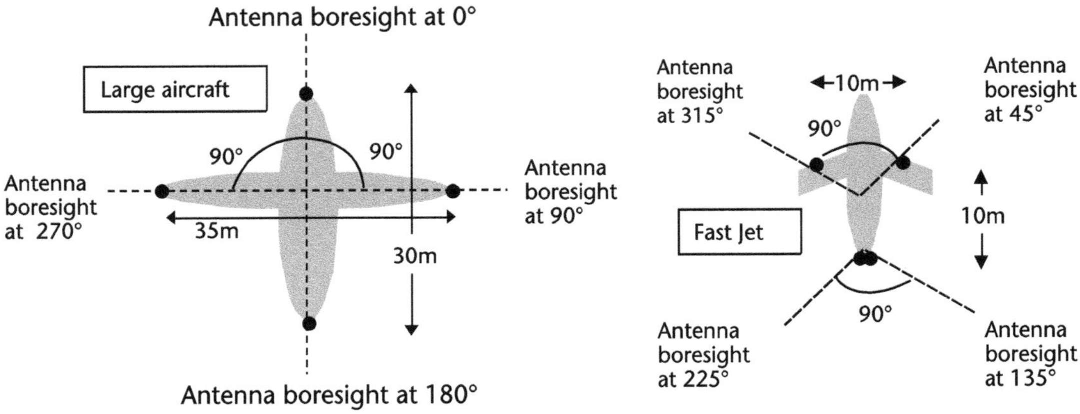

Figure 6.17 Typical ESM antenna configurations.

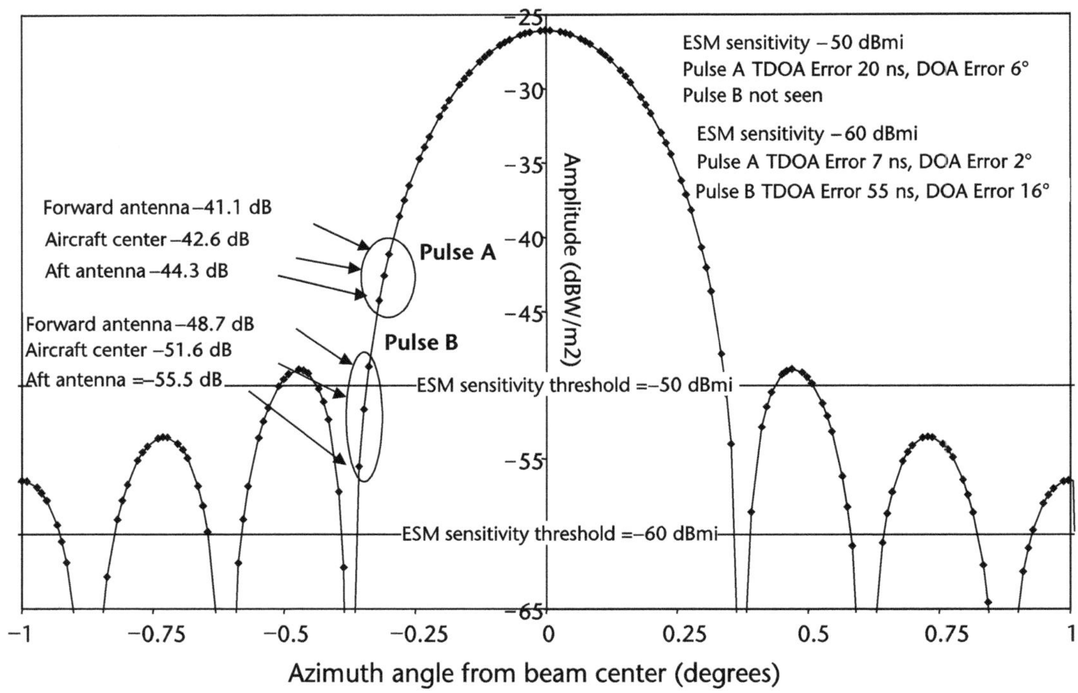

Figure 6.18 Reception of pulses by a large aircraft TDOA ESM system.

Figure 6.19 shows a simulation of the reception of exactly the same scan of pulses, this time by a fast jet ESM antenna configuration.

The difference in received amplitudes at the forward and aft antennas is reduced for the closer ESM antenna spacing in the case of a Fast Jet TDOA ESM system, and the DOA errors seen are also reduced. For example, at the –60-dBmi sensitivity level, pulse A is seen with a TDOA error of 55 ns for the large aircraft ESM, whereas the same pulse is seen with a TDOA error of only 20 ns for the Fast Jet installation.

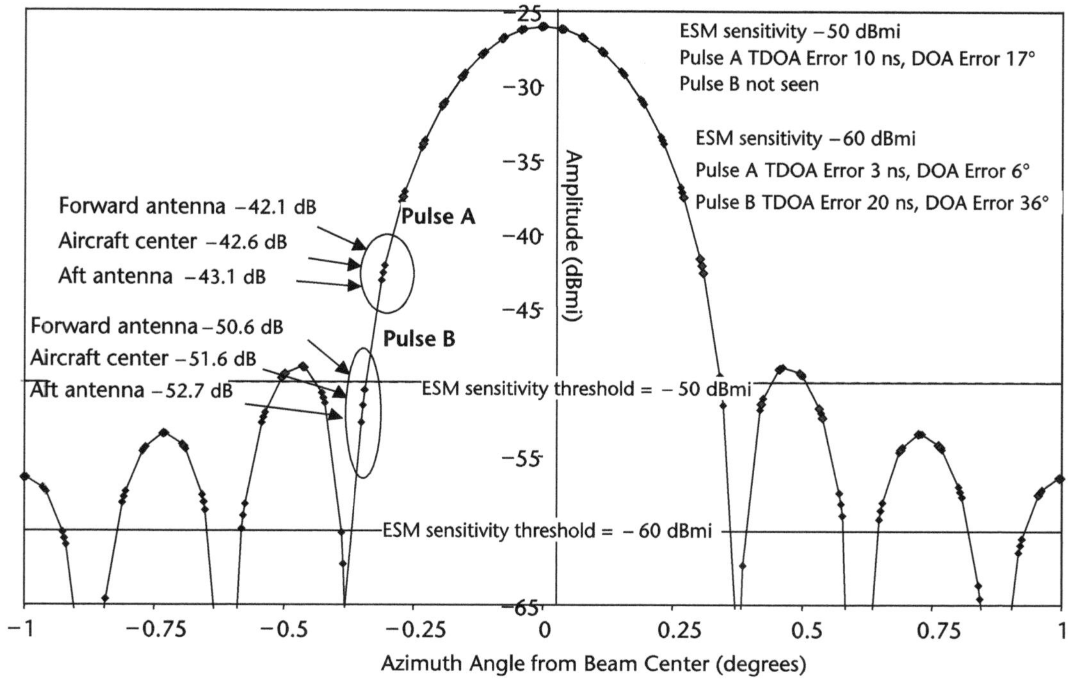

Figure 6.19 Reception of pulses by a Fast Jet TDOA ESM system.

Even though the TDOA errors are smaller for all pulses shown for the Fast Jet installation in Figure 6.19 than for the large aircraft case in Figure 6.18, the DOA resolution is worse. Figure 6.20 shows the translation from TDOA to DOA for four different ESM antenna separations. This graph shows that at the point of crossover of two antennas the DOA can (in theory) be determined exactly, but as the DOA moves away from being at an antenna crossover point, the resolution degrades, following different profiles according to the antenna spacing.

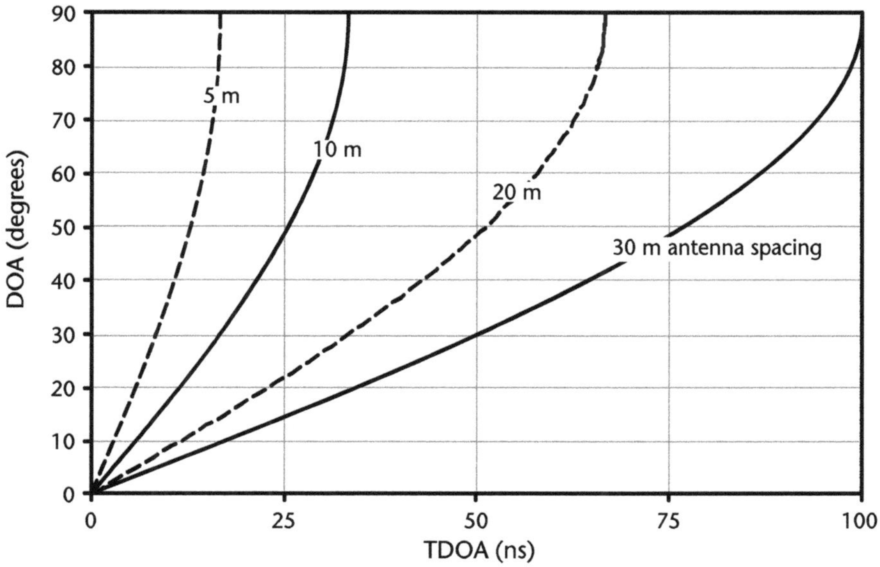

Figure 6.20 Translation of TDOA into DOA for various ESM antenna separations.

Figure 6.21 shows four extracts from the TDOA to DOA translation plot, one for each of the antenna separations shown. The purpose of these extracts is to compare the resolution at about 55° from one of the antenna boresights for each antenna separation. The 30-m separation gives a DOA resolution of about 1°/ns, compared to 3°/ns for a 10-m separation and only 6°/ns for a 5-m separation.

The DOA resolution for various ESM antenna separations can be calculated assuming that a pulse is received from a radar at range r and DOA θ relative to the boresight of an array comprising two antennas separated by d, as shown in Figure 6.22. From Figure 6.22 and rearranging (6.1):

$$\sin\theta = (c\Delta t) / d \tag{6.2}$$

$$\Delta t = (d\sin\theta) / c \tag{6.3}$$

$$\delta\Delta t / \delta\theta = (d\cos\theta) / c \tag{6.4}$$

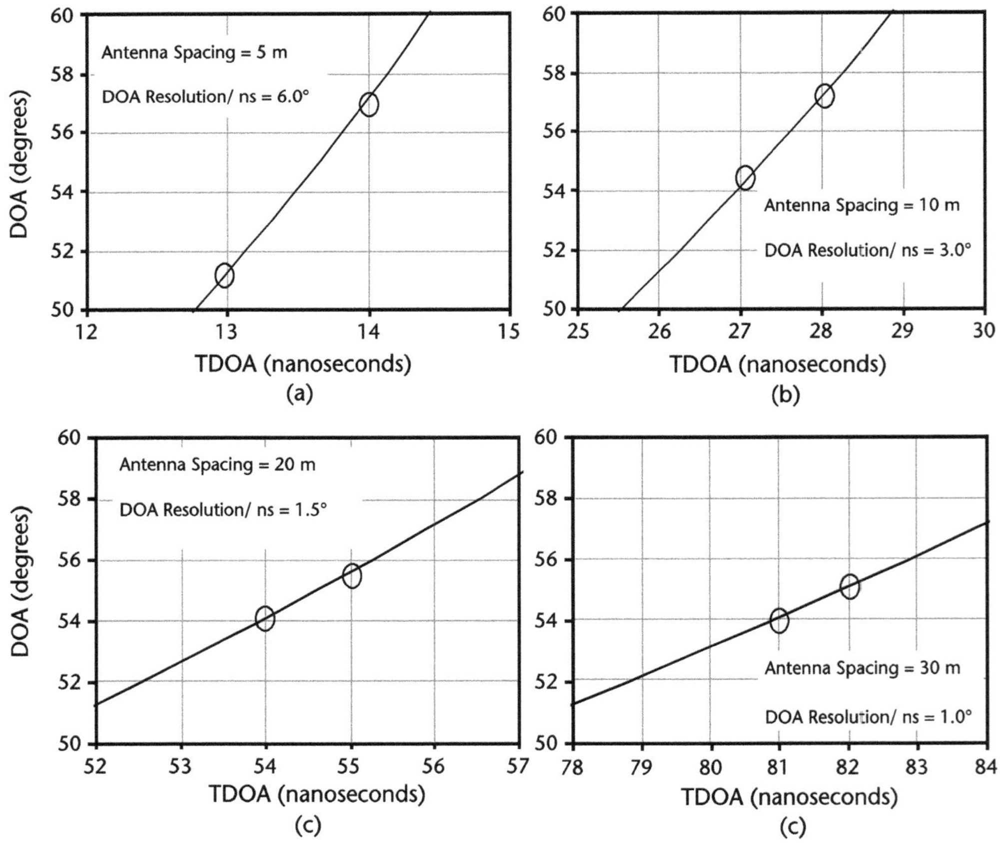

Figure 6.21 DOA resolution of various ESM antenna separations at DOA 55°.

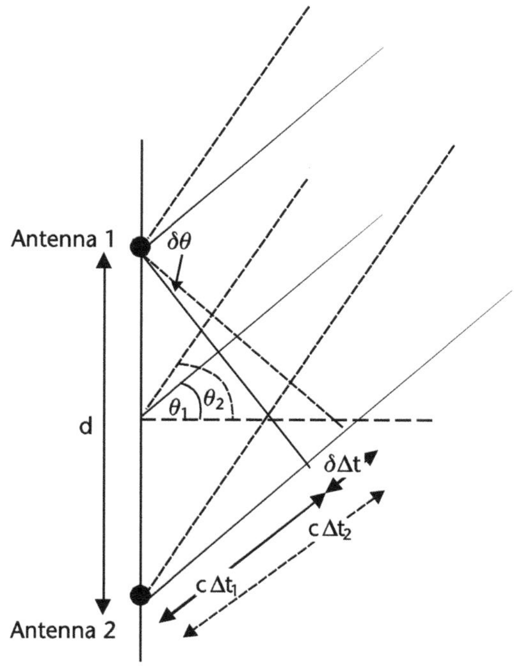

Figure 6.22 Geometry of calculation of DOA resolution.

$$\delta\theta = (\delta\Delta t c) / (d \cos\theta) \tag{6.5}$$

where $\delta\Delta t$ is the time resolution and $\delta\theta$ is the DOA resolution.

The DOA resolution at each DOA is given in Figure 6.23 for ESM antenna separations from 1m to 30m assuming a 1-ns time resolution. As expected, the DOA resolution increases with increasing antenna separation. Figure 6.23 has been generated assuming that there are two ESM antennas separated by 180° (boresights at 0° and 180°).

Figure 6.24 shows a smaller DOA range from 0° to 5° to make the detail of the DOA resolution available for larger antenna separations more obvious.

Directly between the antennas a DOA resolution of ~0.5° can be achieved for a 30-m antenna separation. This reduces to 1.75° for a 10-m separation. However, within 20° of the boresight of either antenna, the DOA resolution has degraded significantly for the 30-m separation and is much worse for smaller antenna separations.

The best DOA resolution achievable for a 5-m antenna separation is 3.5° with a 1-ns time resolution. However, using a 0.1-ns time resolution improves the DOA resolution for all antenna separations by a factor of 10 over the 1-ns cases.

Figures 6.25 shows the DOA resolutions for 0.1-ns time resolution for various antenna separations. The 5-m antenna separation has a best DOA resolution of 0.35° directly between the two antennas, but a DOA resolution worse than 5° at each antenna boresight for a 0.1-ns time resolution.

Figures 6.23 to 6.25 show that using just two antennas for DOA estimation leads to poor DOA resolution near the boresight of each antenna, with no DOA

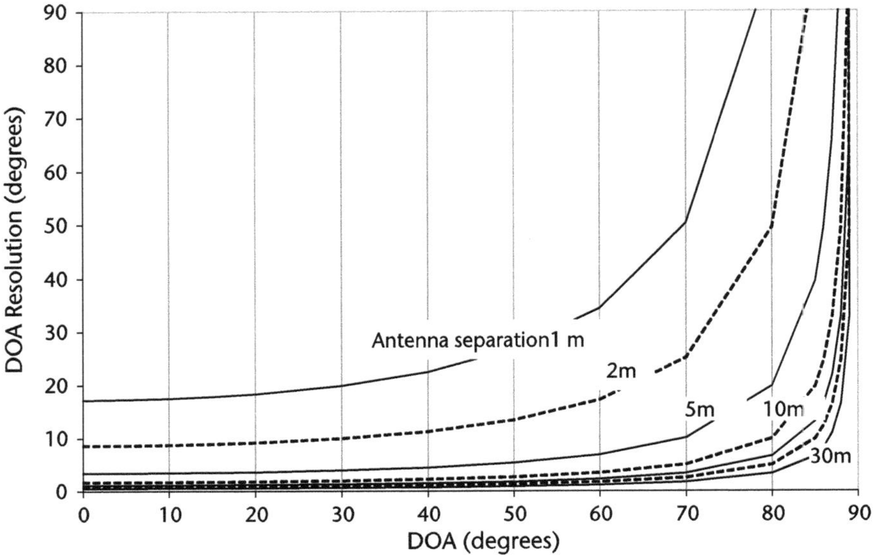

Figure 6.23 DOA resolution for various ESM antenna separations for 1-ns time resolution.

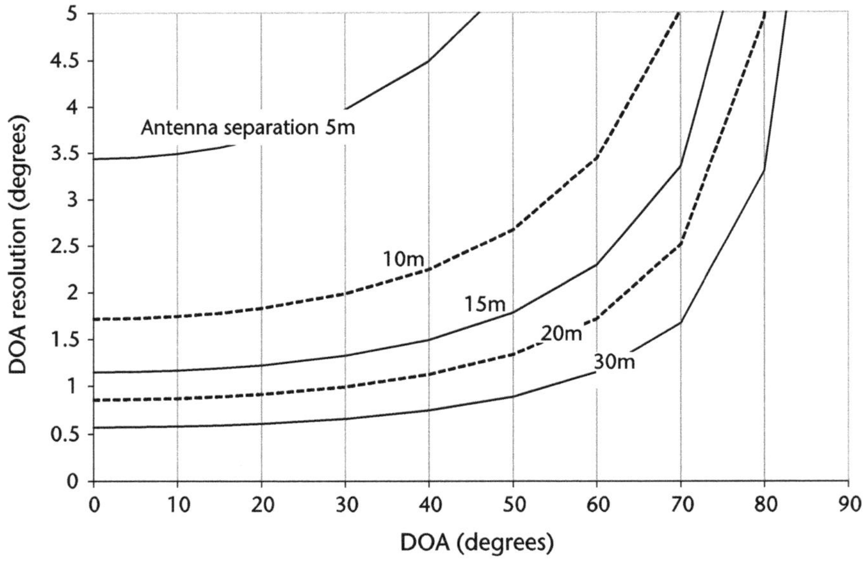

Figure 6.24 DOA resolution (0° → 5°) for various ESM antenna separations for 1-ns time resolution.

calculation possible exactly at the antenna boresight. The effect of the ESM antenna beamwidth also reduces the capability for DOA measurement away from the crossover point of the coverage of two antennas.

6.7 TDOA Histograms and DOA Uncertainty

In an ideal ESM system, a set of pulses from a single radar is received with the same TDOA for a given baseline. Each baseline should have a single peak of the same width as the time resolution of the ESM, at different TDOA for each baseline, as

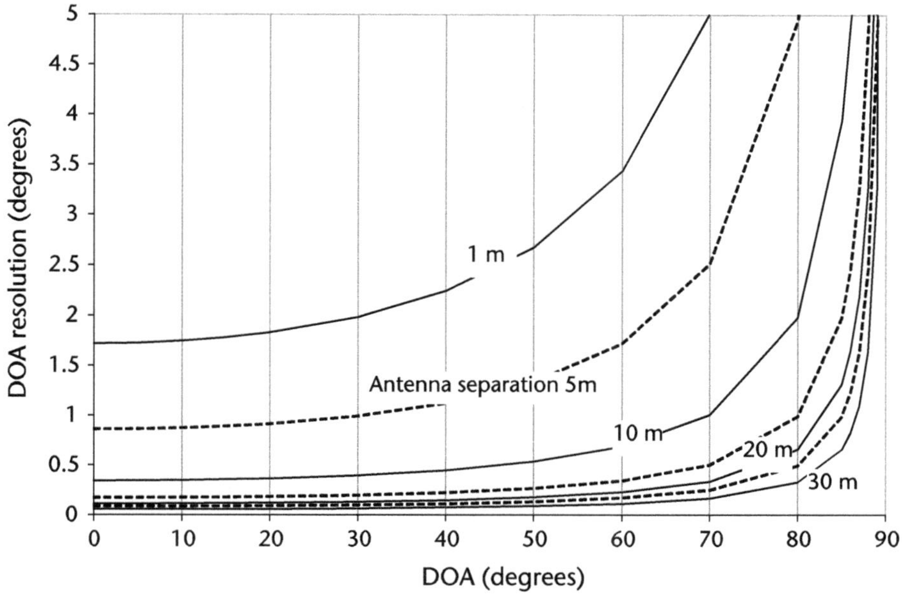

Figure 6.25 DOA resolution (0° → 5°) for various ESM antenna separations for 0.1-ns time resolution.

shown in Figure 6.26. This histogram is for pulses received from a radar of DOA 300° on the FA, PA, and PF baselines of an ESM system. The pulses are not seen on the starboard antenna as the aircraft fuselage prevents the pulse from arriving at this antenna; hence, there are no peaks seen for the pulses from this radar on the SP, SF, and SA baselines.

In the real world, TDOA histograms always have a spread in TDOA, sometimes to the extent that the TDOA peak is several tens of nanoseconds wide, as in the example in Figure 6.27.

The spread in the TDOA histogram is due to errors in TOA measurement for the reasons given in this chapter. However, TOA errors also arise because of intrapulse multipath interference, described in Chapter 15. For a given TOA measurement resolution, the resolution in DOA is proportional to the number of nanoseconds difference in TOA for each baseline. It is independent of radar range (R) and depends only on the separation of the ESM antennas (d), assuming $R >> d$.

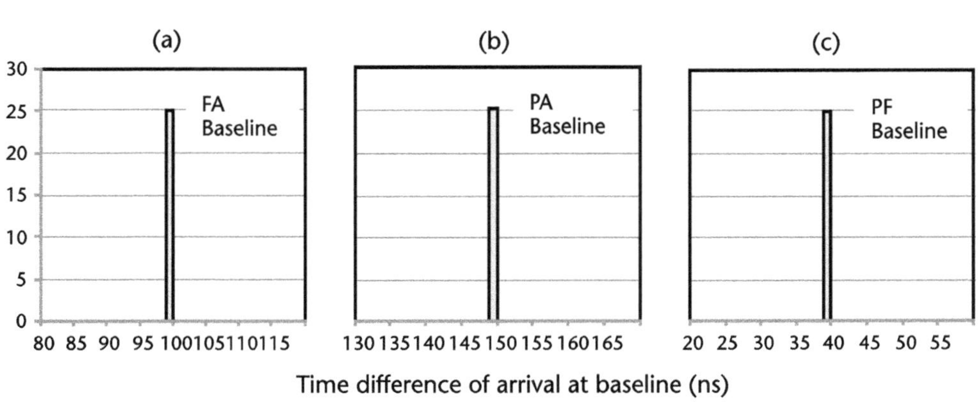

Figure 6.26 Ideal TDOA histograms for a pulse at DOA 300°.

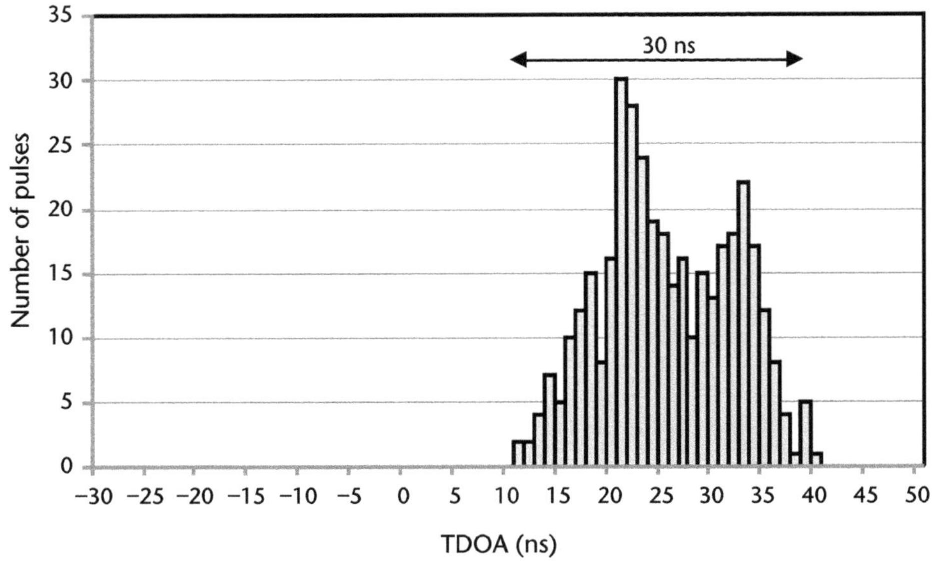

Figure 6.27 TDOA histogram for DOA determination in practice.

A plot of the TDOA against DOA can be used to determine the uncertainty in DOA for a spread in TDOA. A plot of TDOA against DOA for the FA baseline of length 30m is shown in Figure 6.28.

A TDOA spread of 20 ns leads to different DOA uncertainty according to where the TDOA values lie on the curve in Figure 6.28. For example, in the left plot on Figure 6.29, a TDOA spread of 20 ns for an ESM system receiving pulses from a radar at approximately 30° with respect to the aircraft nose leads to a DOA uncertainty of 24° for the FA baseline.

In Figure 6.29, where pulses are received from a radar located at DOA approximately 95° with respect to the aircraft nose (nearer the crossover point between

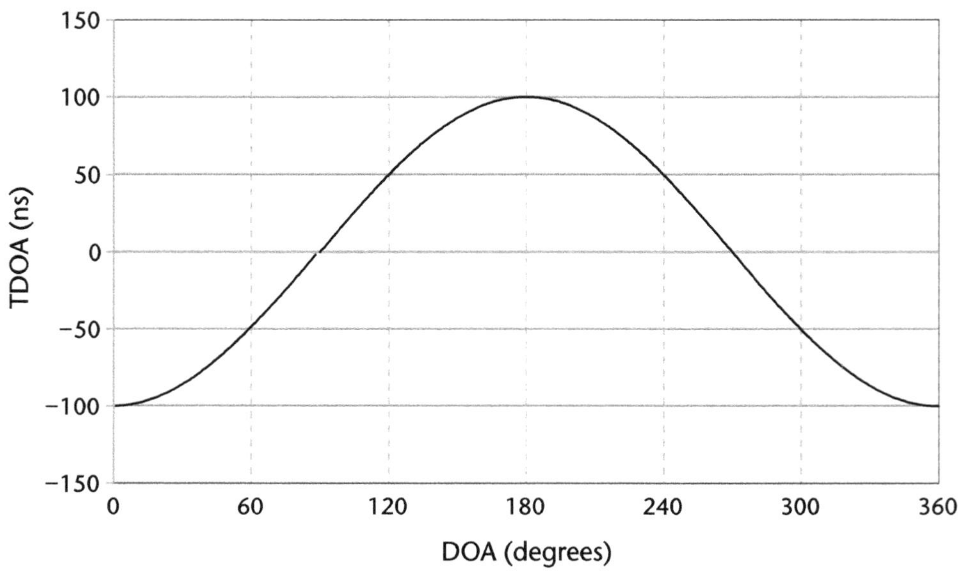

Figure 6.28 TDOA against DOA for FA baseline.

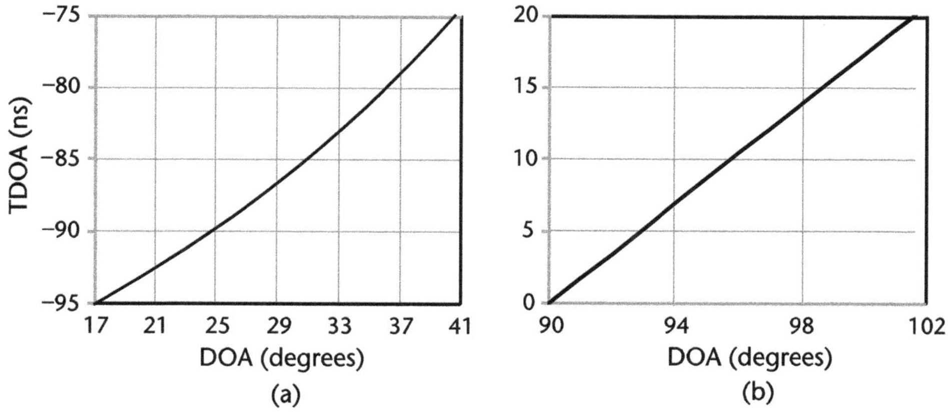

Figure 6.29 TDOA against DOA for FA baseline.

the forward and aft antennas), a spread of 20 ns in TDOA leads to a DOA uncertainty of 12°.

The use of TDOA histograms as the first stage of the pulse deinterleaving process is described in more detail in Chapter 8.

CHAPTER 7

Phase Comparison/Interferometer ESM

Phase comparison ESM systems make use of the measurement of the phase difference of pulses received by antennas along a linear or circular array to calculate DOA. It is an instantaneous measurement technique with monopulse capability (i.e., DOA can be calculated for a single pulse), as is the case for amplitude comparison systems.

7.1 DOA Calculation in Phase Comparison Systems

An interferometer consists of a set of antenna elements in a straight line with different spacings between the elements to obtain TOA relationships that can be transposed into a measurable phase difference for the determination of DOA. The concept can also be extended to a circular array, where interferometer techniques can be applied to cover 360° of azimuth from antennas mounted at one point.

The outputs of the antennas differ in phase from each other in proportion to the extra time that it takes for a plane wave to travel a greater distance to the further antenna. A signal is detected at the same time if it is equally distant from both antennas, so the phase difference is zero, creating a phase difference null along the antenna array boresight. The measurement of the time difference or phase difference, either greater than or less than zero, gives the angle of arrival (AOA) of the incoming signal with respect to the antenna array. Figure 7.1 shows the geometry of a single baseline two-antenna interferometer.

The distance traveled by the pulse to antenna 1 is R in Figure 7.1 and the distance to antenna 2 is $(R + s)$. The path difference (s) is given by:

$$s = d\sin(\theta) \tag{7.1}$$

where d is the antenna separation and θ is the angle of arrival at the antenna array $(-90° \leq \theta \geq +90°)$.

Translating the path difference into a phase difference involves the use of the radar wavelength, λ. The phase change for each wavelength is 2π so the phase difference, $\Delta\phi$, for the path difference (s) is given by:

$$\Delta\phi = (2\pi d\sin\theta)/\lambda \tag{7.2}$$

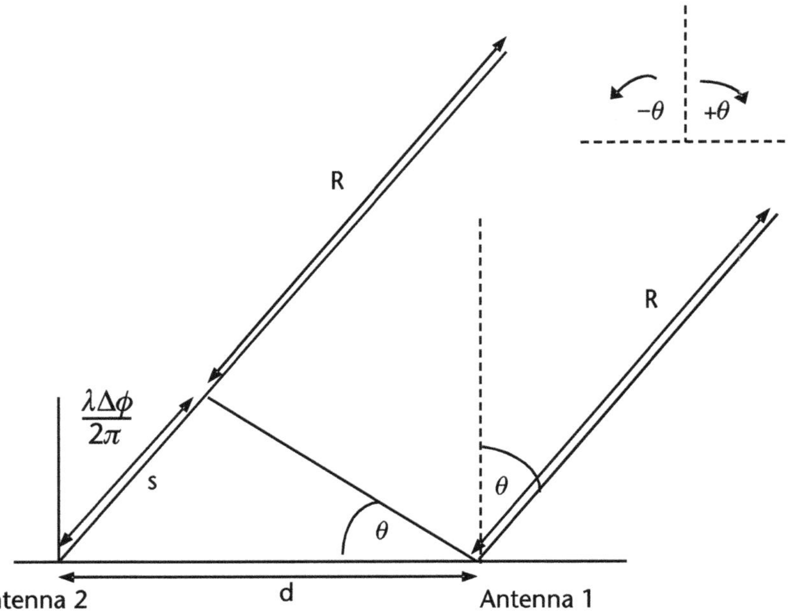

Figure 7.1 Geometry of phase comparison ESM AOA measurement.

where d = antenna separation in meters, θ = AOA with respect to the planar array, λ = radar wavelength in meters, and $2\pi = 360°$.

The phase comparison system measures phase difference, so angle (θ) can be found by rearranging (7.2) as follows:

$$\theta = \sin^{-1}\left(\lambda\Delta\phi / 2\pi d\right) \tag{7.3}$$

Consider the example of the time and associated phase differences for a radar operating at 3 GHz. The pulse travels at the speed of light (3×10^8) m/s and takes 3.33 ns to travel a distance of 1m. A typical pair of phase comparison antennas is separated by 10 cm so the maximum time difference of arrival of this radar wave (with AOA 90°) at each antenna is 0.33 ns. The radar operating at 3 GHz with a wavelength of 0.1m has a phase cycle time of 0.33 ns. The wavelength of this radar is the same as the antenna separation as the antennas are 10 cm apart. A pulse from this radar at an AOA of 90° arrives at antenna 2 at 0.33 ns later than at antenna 1. This is equal to the cycle time for this radar so the phase difference between the two antennas is 360°, calculated by the ESM system to be 0°.

A pulse at AOA 60° from the same radar arrives at antenna two 0.288 ns later than the pulse arrival time at antenna one. The phase difference ($\Delta\phi$) in this case is calculated to be ≈ 312°. A similar calculation for a pulse with an AOA of 30° gives the time difference between the pulse arrival at the two antennas as 0.1667 ns and a phase difference of 180°.

Figure 7.2 shows a graph of phase difference for a 3-GHz pulse for the case where the two antennas are separated by 10 cm. The AOA can be read from the graph for all phase difference between −360° and +360° differences, although the resolution changes as the AOA approaches ±90°.

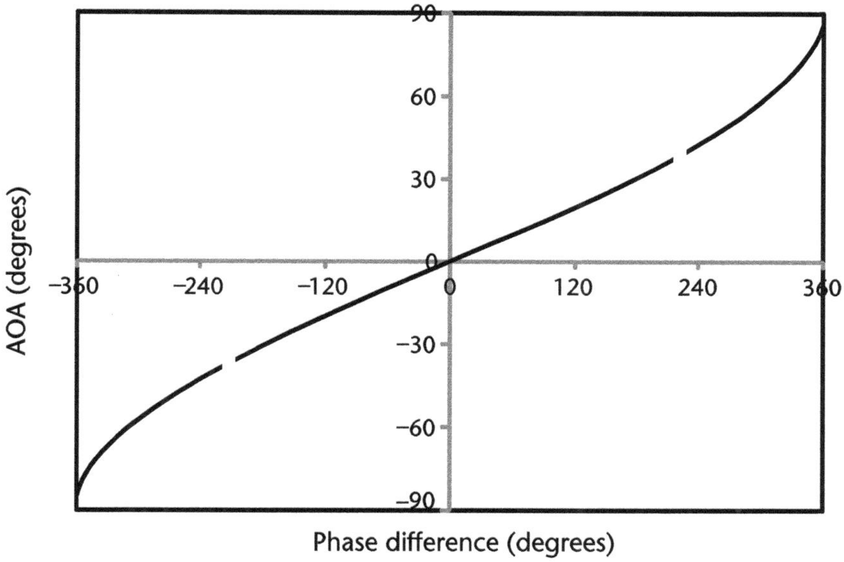

Figure 7.2 AOA versus phase difference for a 3-GHz pulse.

The example shown in Figure 7.2 is for a pulse with a wavelength that exactly matches the separation of the two antennas (i.e., 10 cm). When a pulse from a radar of a different wavelength, for example, 9 GHz, arrives at the same antenna array, the plot of AOA against phase difference looks like the one in Figure 7.3.

For the 9-GHz radar with an AOA of 90°, there is a phase difference of 1,080° between the two antennas as read from Figure 7.3. In reality, the phase difference that is measured can only be in the range ±360°. A phase difference of 1,080° is recorded as 1,080 − (3 × 360°) = 0° by the phase measurement system. This leads

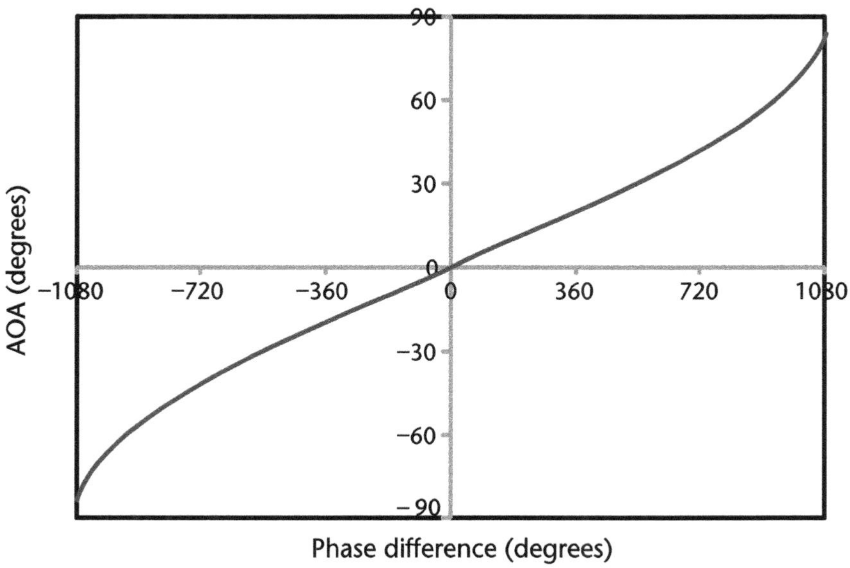

Figure 7.3 AOA versus phase difference for a 9-GHz pulse.

to ambiguity in the determination of the AOA, as shown in Figure 7.4 where the values of phase difference that fall outside the range of ±360° are corrected to provide only those phase differences that are in the measurable range.

The phase differences shown in Figure 7.4 give three possible values for AOA. To resolve this ambiguous situation, multiple antennas are used so that different antenna spacings can be used in the calculation of AOA.

For example, Figure 7.5 shows the phase differences seen at two sets of antennas with the separation of one pair of antennas being 5 cm and the second pair having a 10-cm separation.

A phase difference of 120° is recorded for the antenna pair with a separation of 10 cm. This value of phase difference equates to three different AOAs, at 32°, −7°, and −52°. To resolve the ambiguity in AOA, the phase difference (285°) from the second antenna pair with a spacing of 5 cm gives an AOA of 32°, confiming this as the correct AOA of the pulse.

The fine angular measurement is accomplished by using the antenna pair with the widest space because it has the greatest phase change per degree of DOA. However, unambiguous measurement is found using the smallest baseline possible.

The correct operation of a phase comparison system makes the assumption that there is no phase error at the individual receiver antenna. However, there are factors that affect the phase of the pulse, as discussed below.

7.2 Effect of Antenna Separation on DOA Calculation

The separation of antennas is the cause of DOA errors in both amplitude comparison and TDOA ESM systems. In these types of systems, there is usually a separation of several meters between antennas. However, in phase comparison systems where

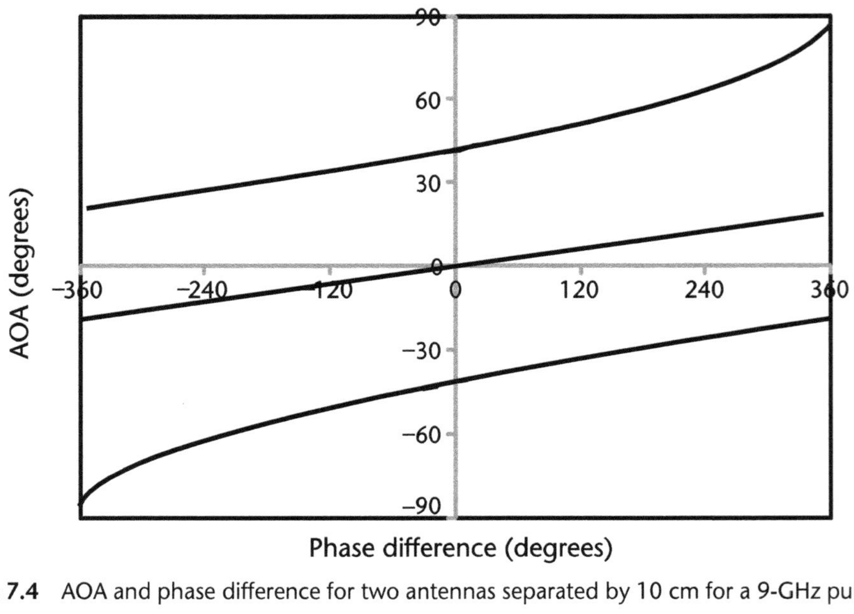

Figure 7.4 AOA and phase difference for two antennas separated by 10 cm for a 9-GHz pulse.

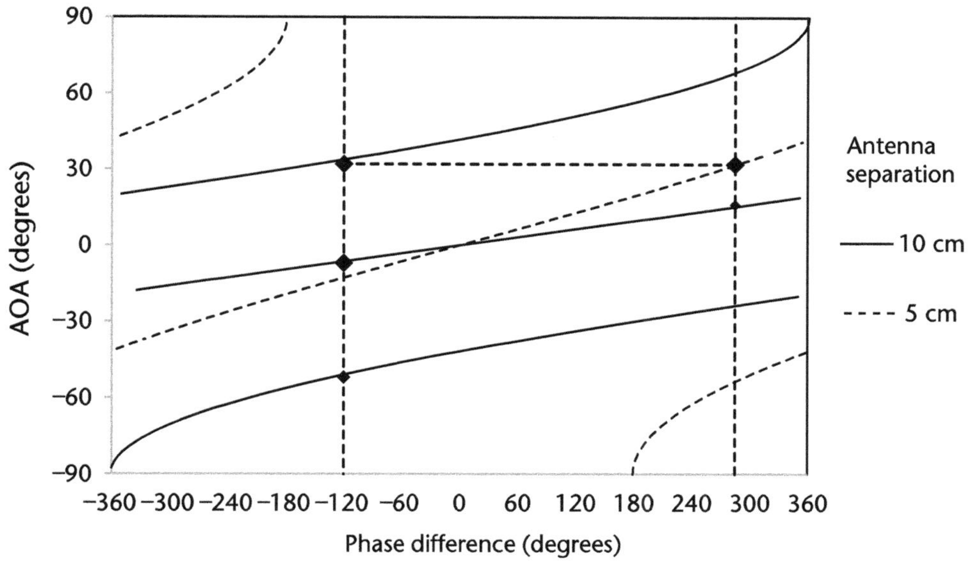

Figure 7.5 AOA versus phase difference for 2 sets of antennas for a 9-GHz pulse.

the antennas are only separated by several centimeters, the effects of a slope in the DOA as the radar beam sweeps past the ESM platform are not seen.

In phase comparison systems, the effect of antenna separation is that different antenna separations lead to different phase differences, as shown in Figure 7.6 where a phase difference of 180° is seen for two antennas separated by 10 cm for an AOA of 30°. Antenna separations of 5 cm and 15 cm lead to phase differences of 80° and 270° for the same pulse at an AOA of 30°.

The phase differences between the two antennas are different for each antenna separation as the slope in the AOA versus phase difference is different, leading to differences in the resolution to which the DOA can be calculated.

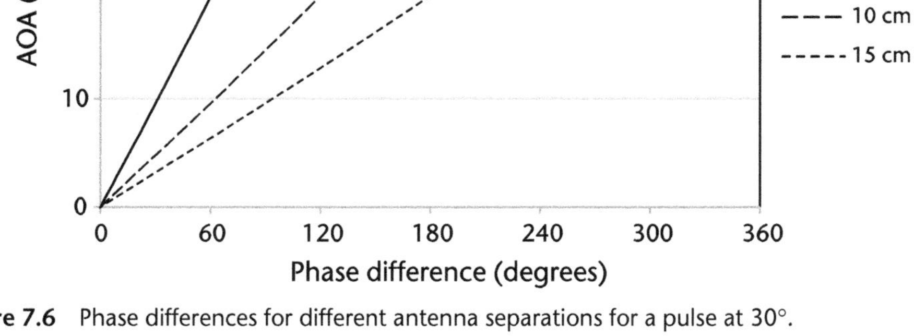

Figure 7.6 Phase differences for different antenna separations for a pulse at 30°.

7.3 Resolution of DOA

The resolution in the AOA depends on the AOA and on the antenna separation as shown in Figure 7.7.

The lines showing AOA versus phase difference are curved, indicating that there is a change in resolution in AOA as AOA changes. Figures 7.8 and 7.9 show more detail of the resolution in the AOA, first for an AOA of 30° and second for an AOA of 85°.

For the 10-cm antenna separation, the AOA resolution is about 1° for a 1° phase difference for AOA near 85°. For the 5-cm antenna separation the resolution in AOA is reduced to 2° for a 1° change in phase difference.

Table 7.1 shows the resolution in AOA available for each 1° phase change for AOA values taken from different parts of the AOA versus phase difference curve.

For AOA values between 0° and 60°, the resolution in AOA is better than 1° for a 1° change in measured phase difference, with finer resolution for larger antenna separations. For an AOA of 85°, the resolution to which AOA can be measured is degraded significantly for all antenna separations, and for a 5-cm antenna spacing, the AOA at 85° can only be measured to a resolution of 3.5 for a 1° phase difference measurement.

7.4 Effect of Radar Beamshape

The small spacing between the antennas in phase comparison systems means that the effects of radar parameters such as the beamshape do not affect the measurement of the DOA. The pulses are detected at the same amplitude at all the antennas, but multipath effects occurring very near to the start of the pulse (within a couple of nanoseconds) can cause amplitude reduction on one or some of antennas and a change in the phase profile.

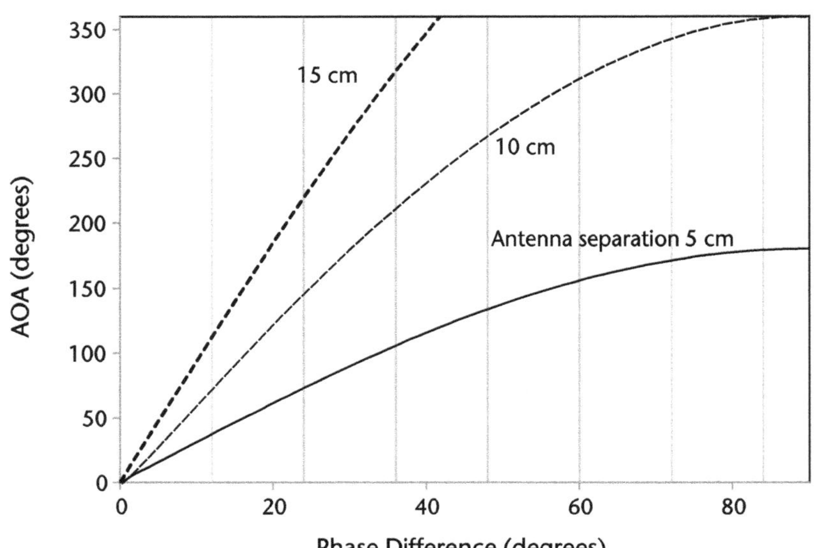

Figure 7.7 AOA versus phase difference for different antenna separations.

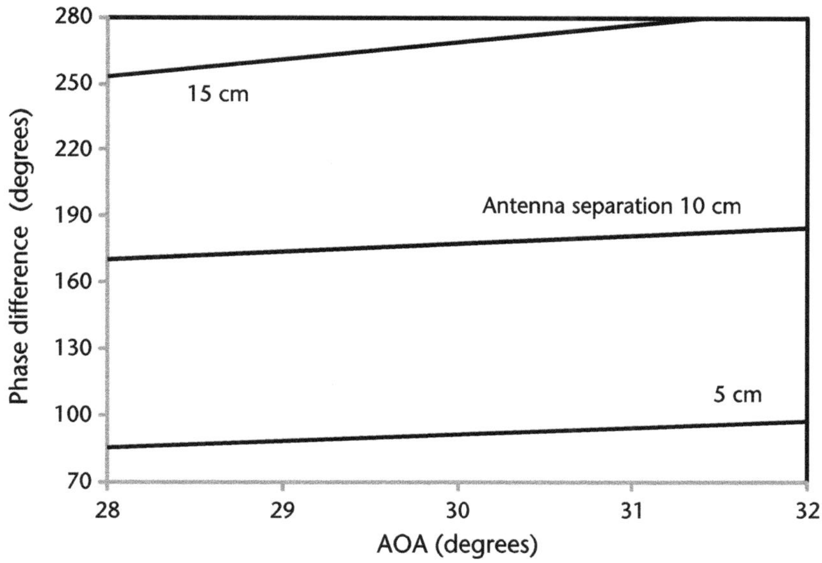

Figure 7.8 Resolution in AOA for different antenna separations for an AOA of 30°.

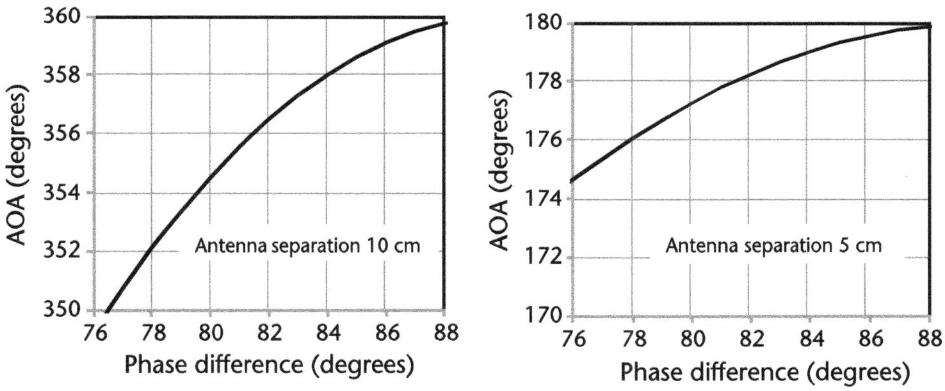

Figure 7.9 Resolution in AOA for different antenna separations for an AOA of 85°.

Table 7.1 Resolution of AOA for Various Antenna Separations

	AOA Resolution in Degrees for 1° Phase Resolution			
Antenna separation (cm)	*AOA 1°*	*AOA 30°*	*AOA 60°*	*AOA 85°*
5	0.334	0.364	0.645	3.44
10	0.167	0.200	0.328	2.22
15	0.111	0.125	0.219	1.53

However, if the antenna spacing is wide (e.g., ~30m for a large aircraft installation), the radar beamshape may affect the calculation of phase difference. All ESM receivers rely on a noise threshold to determine that a pulse is present. For radars that have narrow beamwidths (~1.5° or less), the pulse TOA is affected by the radar beamshape (Chapter 6). In theory, the phase difference should not be affected even though there is a significant TOA difference between the two antennas. There is considerable scope for multipath interference and, as the pulse is sampled throughout its lifetime, this is likely to cause a significant error in the calculation of the DOA. Section 15.5 shows more detail on multipath interference in phase comparison systems.

7.5 Effect of Pulse Modulation

Frequency modulation on a pulse (FMOP), called chirp, affects the phase profile of the pulse throughout its lifetime. In considering how chirp affects phase comparison ESM systems, a start is made by looking at the effect of a linear chirp of 100 MHz/μs on the frequency profile of a pulse, as shown in Figure 7.10. Such a large chirp is not usual for current radars but is chosen here to illustrate the effects of chirp on phase measurement.

Figure 7.11 shows the effect of the chirp of 100 MHz/μs on part of the chirped pulse phase profile. The unmodulated pulse phase profile is also shown on the graph.

There is a beating effect seen on the superimposed profiles of the two pulses (unmodulated and chirped) because of the cyclical nature of the phase profiles (i.e., they repeat every 360°). The beating effect means that the phase difference between the unmodulated and chirped pulses changes through the lifetime of the pulses.

Shorter sections of the pulses appear in Figure 7.12 to show how the chirp changes the phase profile at different stages during the pulse lifetime.

The chirped pulse phase profile effectively "walks" past the unmodulated profile as seen in Figure 7.12 where there are four very different phase profiles seen within the space of less than 100 ns. The phase difference between the modulated and chirped pulses does not depend on the time at which it is measured with respect to the pulse lifetime.

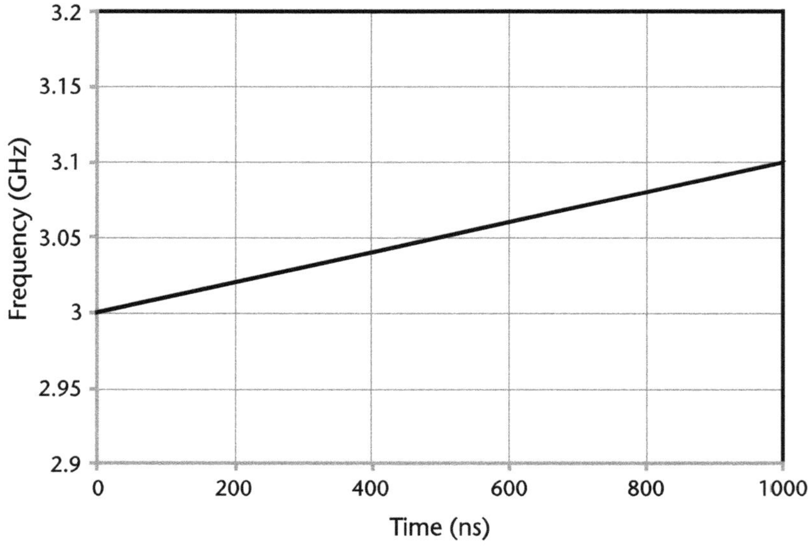

Figure 7.10 Frequency profile for a chirped pulse.

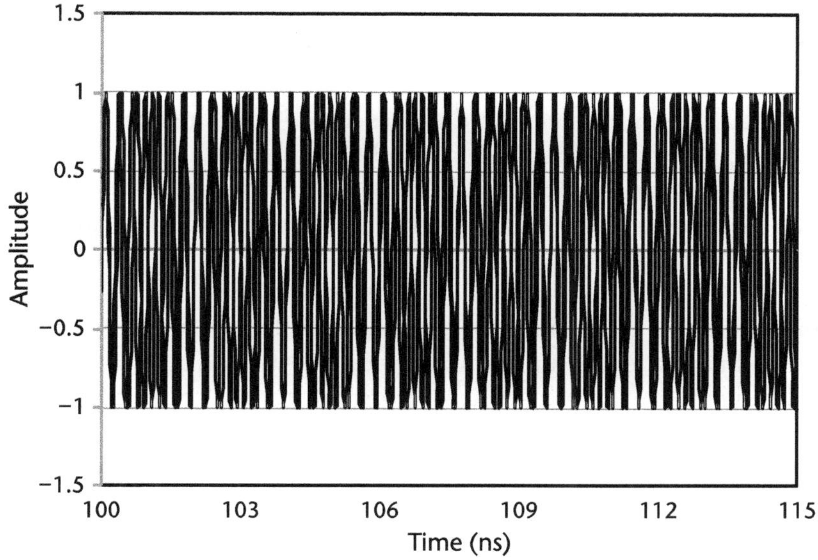

Figure 7.11 Superimposed unmodulated and chirped phase profiles.

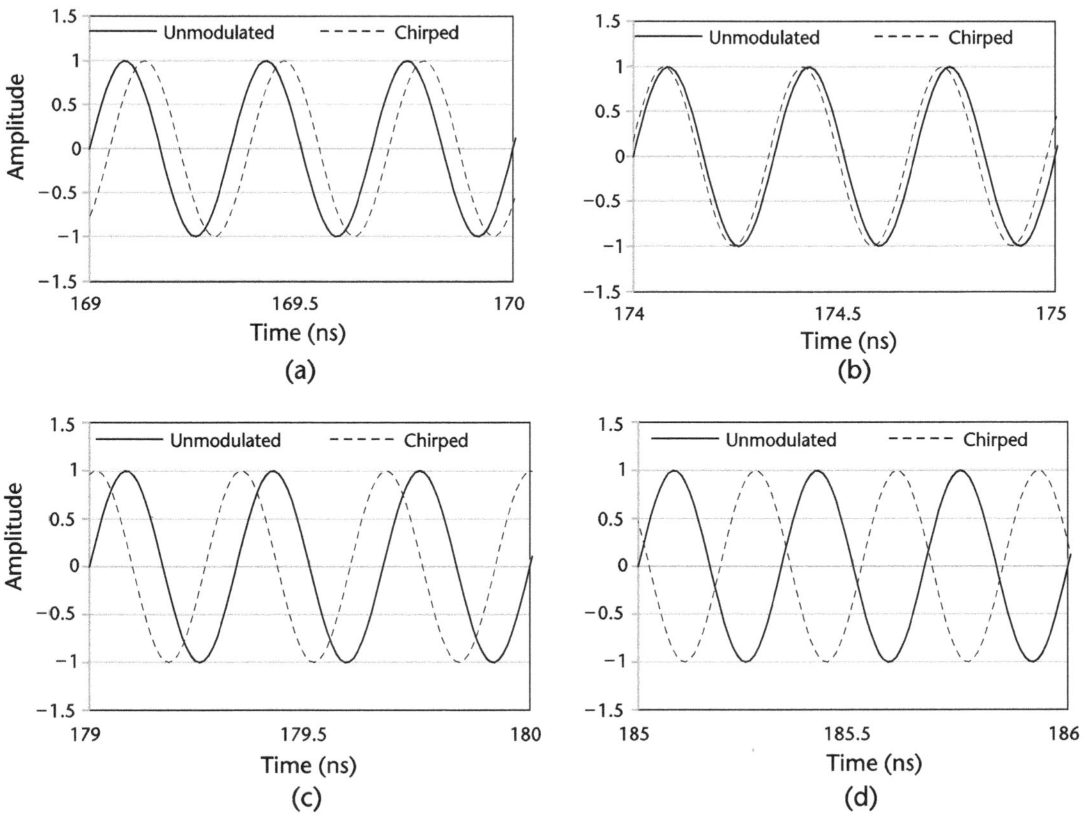

Figure 7.12 Unmodulated and chirped pulse phase profiles.

To find the effect of the phase walk on the DOA calculation, the unmodulated pulse is considered first. The phase profiles of a portion of the unmodulated pulse are shown on Figure 7.13 for the two antennas used to find the phase difference. The phase seen at each antenna at the sample time of 211.4 ns is marked on the graphs.

At 211.4 ns, the measured phases are 85° for antenna 1 and 125° for antenna 2. The phase difference is –40° and the AOA is –6°.

For comparison, Figure 7.14 shows the phase profiles on the two antennas of a pulse starting at the same time as the unmodulated one but this time with the chirp of 100 MHz/μs.

The phase measurements are 160° at antenna 1 and 220° at antenna 2, obviously different from the phases measured at the two antennas for the unmodulated pulse, but still giving a phase difference of –40° and a DOA of –6°. The chirp considered in this example is 100 MHz/μs, larger than chirps typically used by current radars but it does not affect the calculation of the DOA. It is only when the chirp becomes large enough for the frequency of the pulse to have changed between the measurements of phase at each antenna that chirp contributes to the AOA errors.

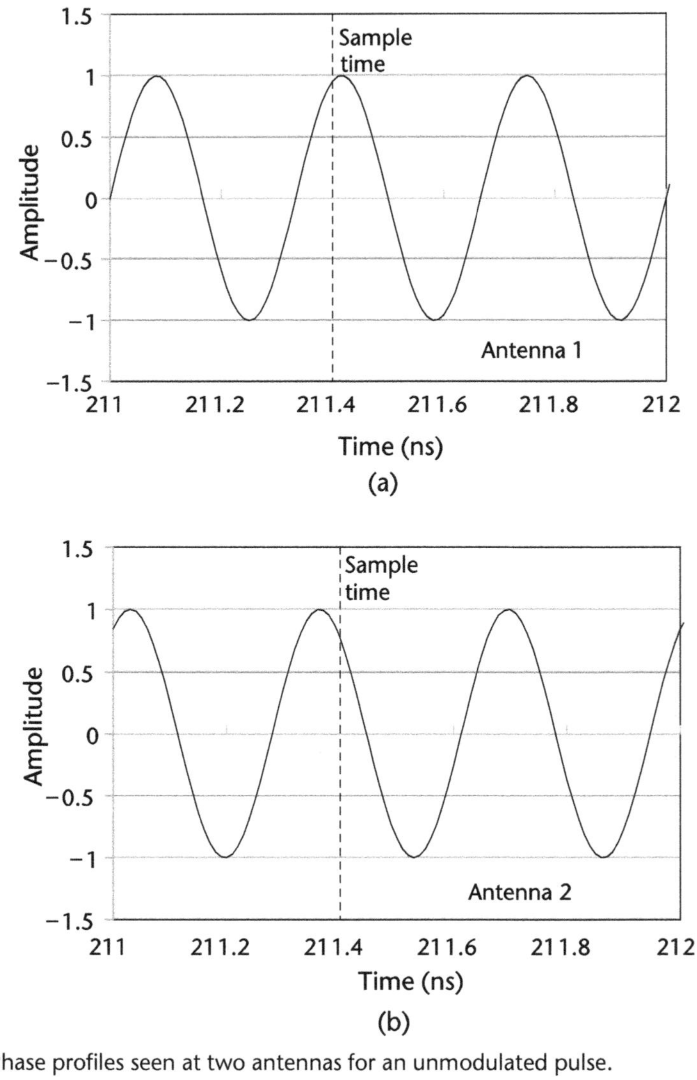

Figure 7.13 Phase profiles seen at two antennas for an unmodulated pulse.

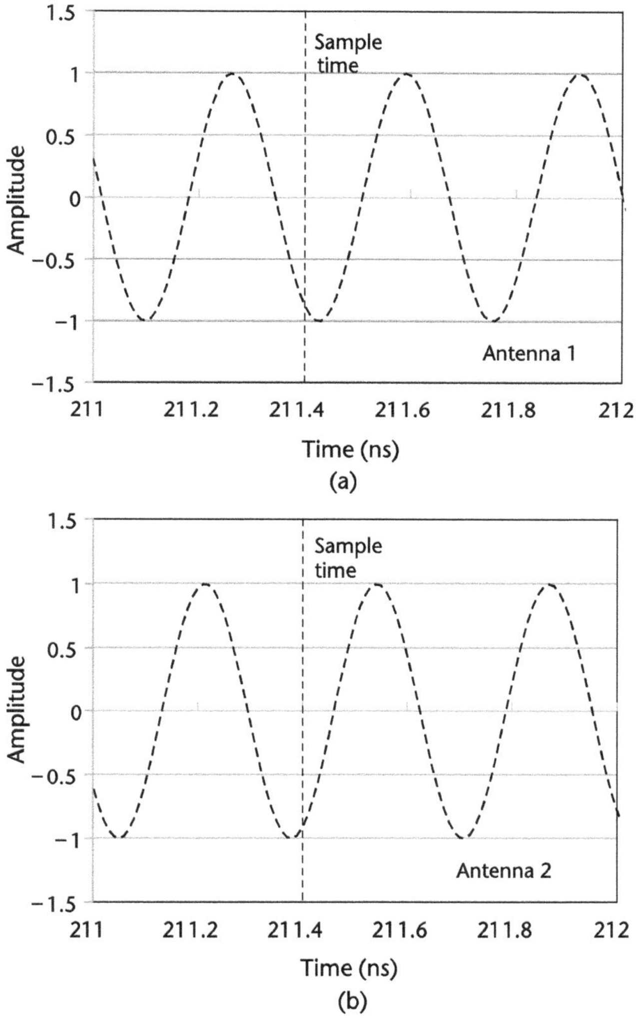

Figure 7.14 Phase profiles seen at 2 antennas for a chirped pulse.

7.6 Effect of Pulse Shape

The phase profile of a chirped pulse that is affected by multipath (i.e., where the pulse phase and amplitude are altered by a reflection of itself during the pulse lifetime) is shown in Figure 7.15.

The effect of multipath is to double the peak amplitude of the pulse but also to introduce nulls in the amplitude profile where the phases of the direct path pulse and the reflected path pulse are 180° out of phase, leading to total destructive interference.

The troughs in the amplitude profile may lead the ESM to truncate the measurement of the pulse width and therefore reduce the number of samples that are available for phase comparison. Another outcome of multipath is the random changes in phase that are introduced throughout the pulse lifetime. This has very serious consequences for the operation of phase comparison systems. A description

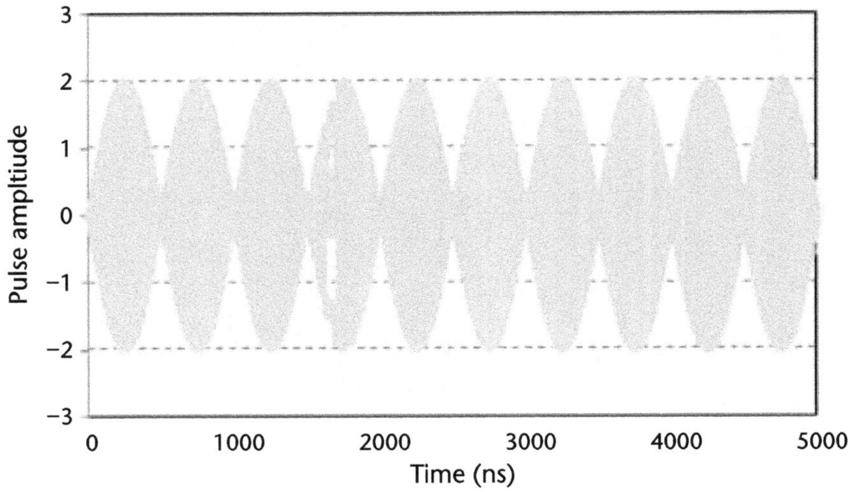

Figure 7.15 Phase profile of a chirped pulse affected by multipath.

of the effect of multipath on phase comparison systems is given in Chapter 15 and some ideas for reducing the damaging effects of multipath are given in Chapter 17.

7.7 Effect of Radar RF

Unambiguous AOA measurement is possible as long as the antenna separation is equal to or less than the radar wavelength. For pulses of RF of 3 GHz, this means that antenna separations of up to 10 cm can be used to determine the AOA without the need for confirmation using other antenna separations. When considering higher frequencies, it is not practical to reduce the antenna spacing to be within a wavelength of the radar, especially as small antenna spacings lead to poor angle resolution.

Figures 7.16 and 7.17 show the phase difference versus AOA for RFs of 9 GHz and 18 GHz for an antenna spacing of 10 cm.

There are four ambiguous values of the AOA for each phase difference for the 9-GHz pulse and six ambiguous values for the 18-GHz pulse. As discussed in Section 7.1, the use of a second set of antenna separations solves the ambiguities.

7.8 Phase Comparison Systems in Practice

Taking a single sample from each antenna pair to calculate the phase difference at a particular time is likely to cause error in the measurement of the AOA. All pulses are subject to multipath, as discussed at length in this chapter and elsewhere in the book. Each pulse is subjected to many reflections and, each time a reflection affects a pulse, a phase change takes place due to the phase difference between the existing composite pulse and the new reflection path. If one of a pair of antennas is affected by multipath and the other is not, then a phase error results. The phase error can have any value so the derivation of a solution for this is not possible. Chapter 15

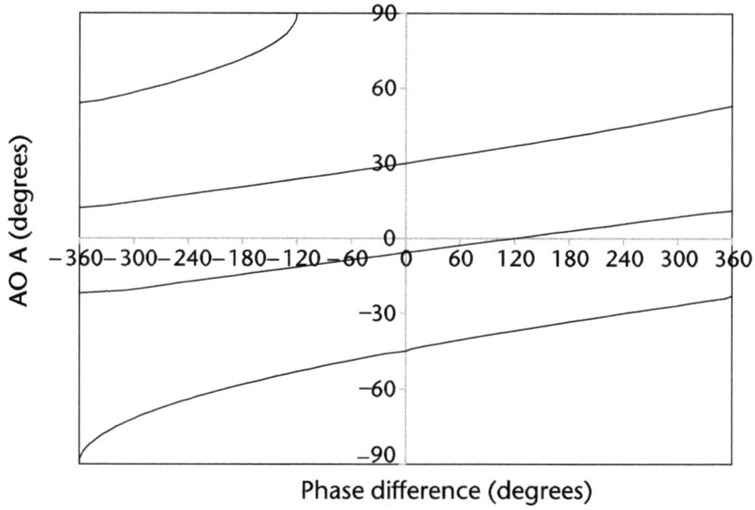

Figure 7.16 AOA versus phase difference for a 9-GHz pulse for antenna spacing of 10 cm.

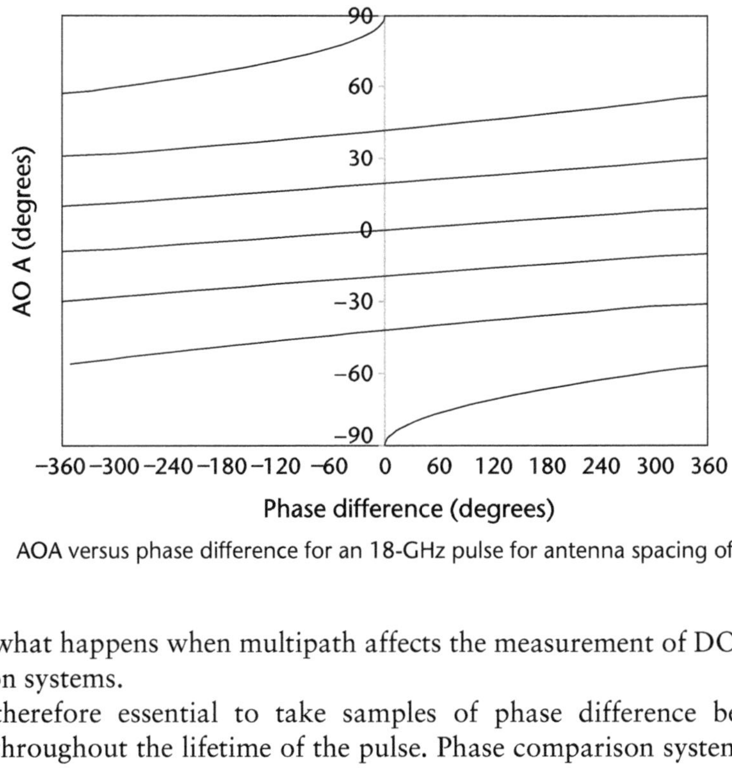

Figure 7.17 AOA versus phase difference for an 18-GHz pulse for antenna spacing of 10 cm.

describes what happens when multipath affects the measurement of DOA in phase comparison systems.

It is therefore essential to take samples of phase difference between the antennas throughout the lifetime of the pulse. Phase comparison systems typically sample every nanosecond, meaning that there are approximately 1,000 samples available in a pulse of width 1 μs.

It is to be hoped that the average AOA calculated from the samples will lead to a correct value for the AOA, but as seen in the example pulse data set in Figure 7.18, this is not usually the case. The spread in the AOA for each scan of the pulses in Figure 7.18 is up to 40°. In reality, it is very difficult to counter problems caused by multipath, and, in practice, pulse data sets like the one shown in

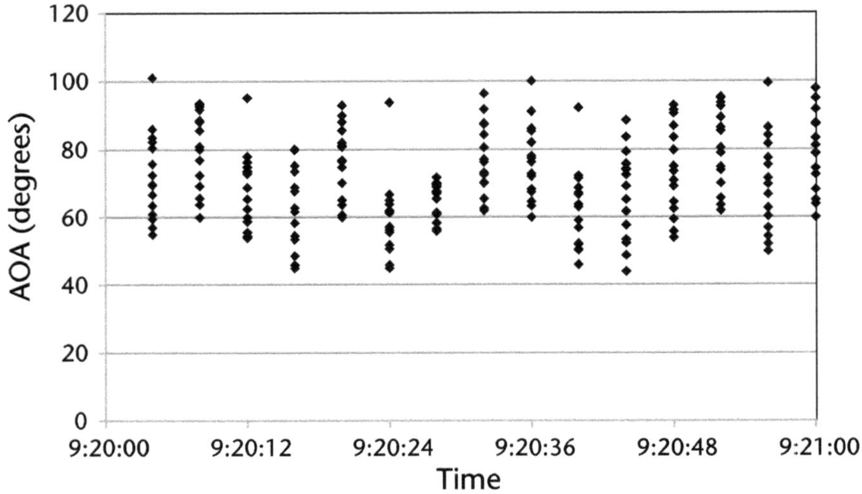

Figure 7.18 An example of a typical pulse data set from a phase comparison ESM system.

Figure 7.18 are commonly seen in current systems. The most successful ESM systems are those that combine techniques for DOA measurement.

7.9 Long Baseline Phase Comparison ESM Systems

The discussion in this chapter has mostly been about short baseline interferometry, where the separation between antennas is of the order of centimeters. Phase comparison systems may also be configured with several meters separation between antennas. In these cases, there is increased scope for phase difference errors due to multipath. The time at which the phases are sampled must be the same for each antenna for the phase comparison method to work. As discussed in Chapter 15, the problem with multipath affecting long baseline phase comparison systems is that, for at least some of the phase difference samples, the phase is measured on one antenna before multipath has taken effect and on the other after the multipath interference has taken place.

CHAPTER 8

Deinterleavers and ESM Processing

The output from an ESM receiver (antenna subsystem and parameter measurement unit) is a set of pulse descriptor words (PDW) containing the measured parameters of the pulses such as RF, PW, DOA, amplitude, and TOA. Table 8.1 shows an example of a set of PDWs recorded by an ESM system.

Most ESM systems then use conventional deinterleaving techniques to attempt to separate the pulses into clusters according to their parameters and to produce tracks, ideally one for each radar. All new pulses are grouped with pulses of similar parameters, usually RF and DOA in the case of amplitude comparison systems, RF and TDOA in the case of TDOA systems, and RF and phase difference for phase comparison systems.

PRI determination is carried out and the output of the deinterleaving process is a chain of pulses, an intercept, with similar parameters. These intercepts are used to create or update a track that is displayed to the ESM operator. A track typically consists of the maximum, minimum, and mean values for parameters such as RF, PRI, PW, and DOA. Additionally, for each parameter, a parameter agility code may be determined, giving information such as RF agility, PRI type, and scan type. Table 8.2 shows some examples of track descriptions.

A merging process attempts to identify if two or more tracks originate from the same source. If they are deemed to be from the same radar, the two tracks are merged.

In the final stage of ESM processing, a radar identification algorithm takes each track and matches it with entries in the ESM radar library. The radar library contains information about parameters for each radar type and its functional mode of operation. More details on this can be found in Chapter 10.

Table 8.1 Example of a Set of PDWs

RF (MHz)	*PW (μs)*	*DOA (°)*	*Amplitude (dBmi)*	*TOA*
9,084	0.16	223.59	–20.0	14:18:13.4683
3,011	26.32	230.63	–17.5	14:18:13.46843
9,082	0.12	230.63	–19.5	14:18:13.46864
9,084	0.08	232.03	–20.5	14:18:13.46889
3,012	26.36	229.22	–17.0	14:18:13.46981
9,082	0.08	275.63	–22.0	14:18:13.47041
9,084	0.16	277.03	–22.5	14:18:13.47065

Table 8.2 Examples of Track Descriptions

Track No.	*DOA (°)*	*RF mean*	*RF max.*	*RF min.*	*RF type*	*PRI mean*	*PRI max.*	*PRI min.*	*PRI type*	*PW mean*	*PW max.*	*PW min.*	*Amp. max.*	*Identity*
1	23.5	9,084	9,085	9,083	Fixed	223	223	222	Fixed	0.6	0.7	0.5	−19	XXX
2	145.2	3,011	3,015	3,008	Agile	599	605	585	Jittered	0.7	0.7	0.6	−23	YYY
3	256.73	2,960	2,955	2,965	Hopper	1,120	1,340	1,040	Stagger	1.3	1.5	1.2	−22	ZZZ
4	58.2	3,280	3,282	3,279	Fixed	1,501	1,502	1,500	Fixed	2.1	2.2	2.0	−37	Unknown

8.1 Deinterleaving Techniques

Almost all ESM systems begin the process of deinterleaving by clustering pulses in terms of RF and DOA, or DTOA. This is followed by PRI extraction to arrive at separate intercepts for each radar that the ESM has deemed to be in the cluster. Attempts are made to correlate each intercept to an existing track or to create a new track if there are no suitable tracks already in existence.

Most ESM systems work with buffers of pulses, typically not more than a couple of thousand pulses received in a specific time-frame. The pulse buffer time may be fixed in the case of an ESM that is wide open in RF or set up to be different for each narrow RF band for narrower-band receivers. Chapter 4 discusses RF scan strategies.

The number of pulses to form an intercept is often chosen to be a minimum of six pulses to allow for parameter extraction while allowing most radars at long range to be detected. Advantages and disadvantages for choosing to use six pulses are described in Chapter 3.

Most ESM systems use clustering techniques followed by parameter histograms for the deinterleaving process, but there are some alternative deinterleaver techniques available. Brief descriptions of Graph Theory and Predictive Gate [1] methods of deinterleaving are described below, but other methods including the use of Kalman filters [2], Incremental Clustering for Evolving Data (ICED) [3], neural networks [4], and geometrical deinterleaving methods [5] have also been devised. The use of the Radar Clock Period estimation as a means of extracting pulses from specific radars is a technique that is currently under development.

8.2 DOA/Frequency or DTOA/Frequency Cluster Algorithms

This type of deinterleaver works by assigning each pulse in a buffer to a cell in a two-dimensional (2-D) histogram of DOA or DTOA and RF. If more than a minimum number of pulses are found in a cell, those pulses are processed further to determine the average parameters for a cluster. Typical cell sizes are 1 MHz by 5° (for DOA deinterleavers) or 1 ns (for TDOA deinterleavers) with a minimum pulse requirement of six pulses for a cell to be deemed to be populated. Clusters are formed from individual cells or from more than one cell if adjacent cells are occupied. It should be noted that, for TDOA systems, the pulses that are seen on the most populous baseline (see Chapter 6) are processed first, followed by the next most populous baseline, and so forth, until all baselines have been considered.

Figure 8.1 shows an example of a 2-D histogram forming the first part of the signal processing in a TDOA system. In this diagram two clusters of pulses are seen, one from a radar with fixed RF and the other from an RF agile radar.

8.3 Time of Arrival Difference Histogram

In the next stage of deinterleaving, each DTOA/RF or DOA/RF pulse cluster is processed through the time arrival difference histogram (TOADH) to extract the PRI of the radars in the pulse cluster. Figure 8.2 shows diagrammatically how the TOA differences are calculated.

The TOADH process is very important in understanding how an ESM system works; to make this process clear, Table 8.3 shows an example of the calculation of a TOADH table for two radars with interleaved pulses. The two radars have fixed PRIs of 880 μs and 1,000 μs, respectively.

In Table 8.3 the TOADH is carried out to a depth of 4. A histogram is produced for each TOAD level using bin sizes of 10 μs in the example shown here. The final TOADH is generated by summing the results for each bin from all the individual histograms.

A histogram of TOAD values is plotted as shown in Figure 8.3 where there are peaks at the correct PRI for the two radars and also peaks at the first harmonics of the PRI for each radar (i.e., peaks at 880 μs and 1,760 μs for radar 1 and 1,000 μs and 2,000 μs for radar 2). The fact that there are at least half as many pulses in the second peak as there are in the first peak is the criterion for validating the PRI for each radar.

The simple example shown above is for radars with fixed PRI. Several passes are made though the DTOA/RF or DOA/RF clusters to extract the various types of PRIs. TOADHs are calculated to different depths and with different numbers of pulses for each category (i.e., for fixed PRI and for staggered and jittered PRI types). The process at each pass is the same as that for fixed PRI, except that the TOADH is calculated to a greater depth for the more complicated PRI types. Once

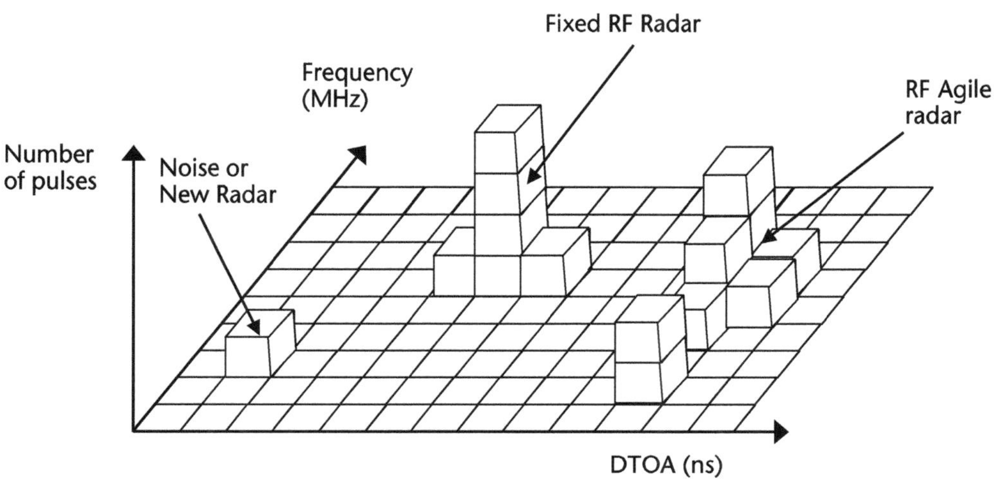

Figure 8.1 A 2-D frequency and DTOA histogram.

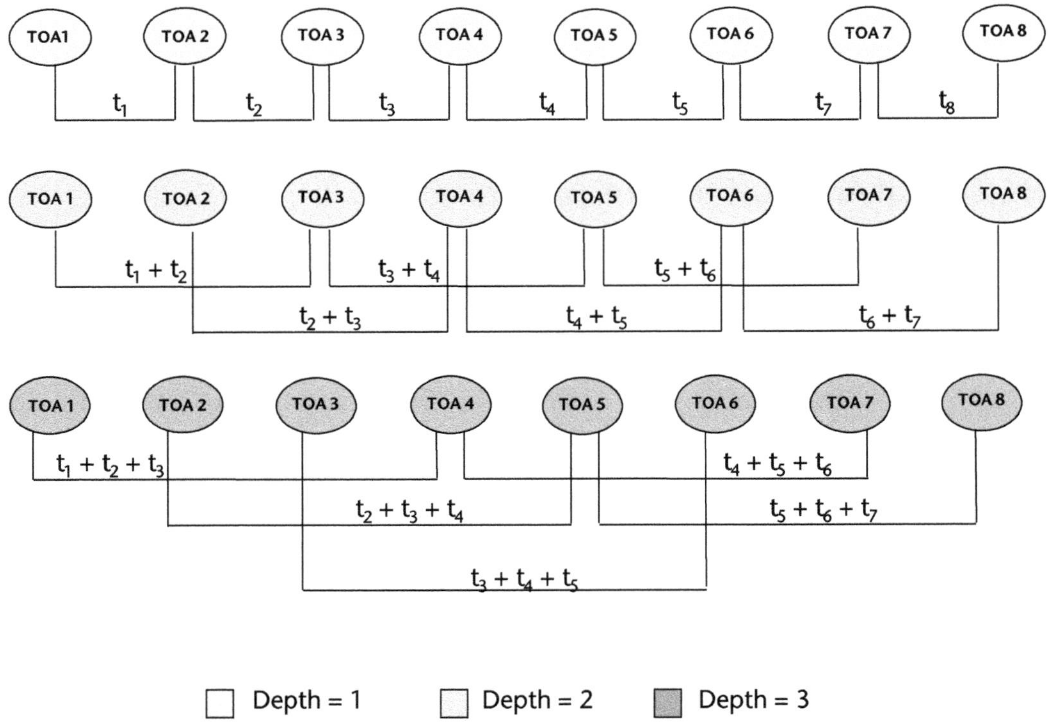

Figure 8.2 Calculation of time differences for a TOADH.

each pass is complete, the pulses that have contributed to the main peak and the secondary peak are deemed to come from the same radar and are extracted for further processing.

For the fixed PRI pass, the result from the TOADH is the actual PRI. For staggered PRI radars, the result from the TOADH process is the Pulse Group Repetition Interval (PGRI). Figure 8.4 shows an example of the calculation of a TOADH for three radars with simple two-position stagger sequences: Radar 1 has PRI elements of 600 μs and 650 μs, PGRI = 1,250 μs, Radar 2 has PRI elements of 720 *ms* and 680 *ms*, PGRI = 1,400 *ms*, and Radar 3 has PRI elements of 920 μs and 900 μs, PGRI = 1,820 μs.

There are four TOADHs shown on Figure 8.4, each to a different TOAD depth. When the depth is 1, there are no peaks seen as the TOAD between each pulse is less than any of the PGRI values for the three radars. A depth of 8 extracts the PGRI for all 3 radars but there are no validation peaks seen in this histogram. It is not until a depth of 16 is used that the three PGRI peaks and the three validation peaks are seen.

The actual depth for each pass through the TOAD process should depend on the maximum PRI stagger expected and a realistic estimate of the number of pulses expected from each radar in the RF environment. Ideally the density of the pulse environment should also be taken into account when deciding on the depth to which the TOADH is calculated, but, in reality, this adaptive processing is not used in current ESM systems.

Table 8.3 Example of a TOAD Process for Two Radars with Fixed PRI

Radar	*TOA*	*TOA1*	*TOA2*	*TOA3*	*TOA4*
1	880				
2	1,000	120			
1	1,760	760	880		
2	2,000	240	1,000	1,120	
1	2,640	640	880	1,640	1,760
2	3,000	360	1,000	1,240	2,000
1	3,520	520	880	1,520	1,760
2	4,000	480	1,000	1,360	2,000
1	4,400	400	880	1,400	1,760
2	5,000	600	1,000	1,480	2,000
1	5,280	280	880	1,280	1,760
2	6,000	720	1,000	1,600	2,000
1	6,160	160	880	1,160	1,760
2	7,000	840	1,000	1,720	2,000
1	7,040	40	880	1,040	1,760
1	7,920	880	920	1,760	1,920
2	8,000	80	960	1,000	1,840
1	8,800	800	880	1,760	1,800
2	9,000	200	1,000	1,080	1,960
1	9,680	680	880	1,680	1,760
2	10,000	320	1,000	1,200	2,000

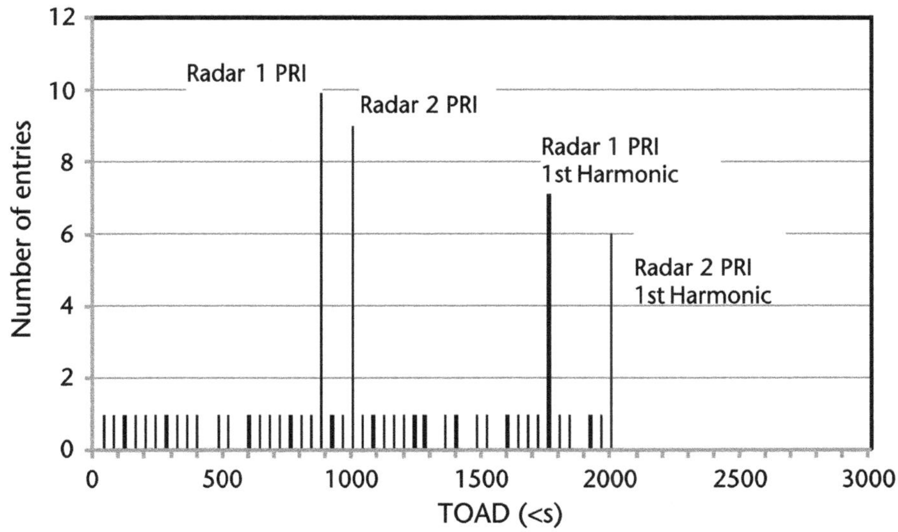

Figure 8.3 Example of a TOADH.

As described in Chapter 3, even in really dense RF environments, there are unlikely to be more than five interleaved radar pulse trains, so this is a starting point

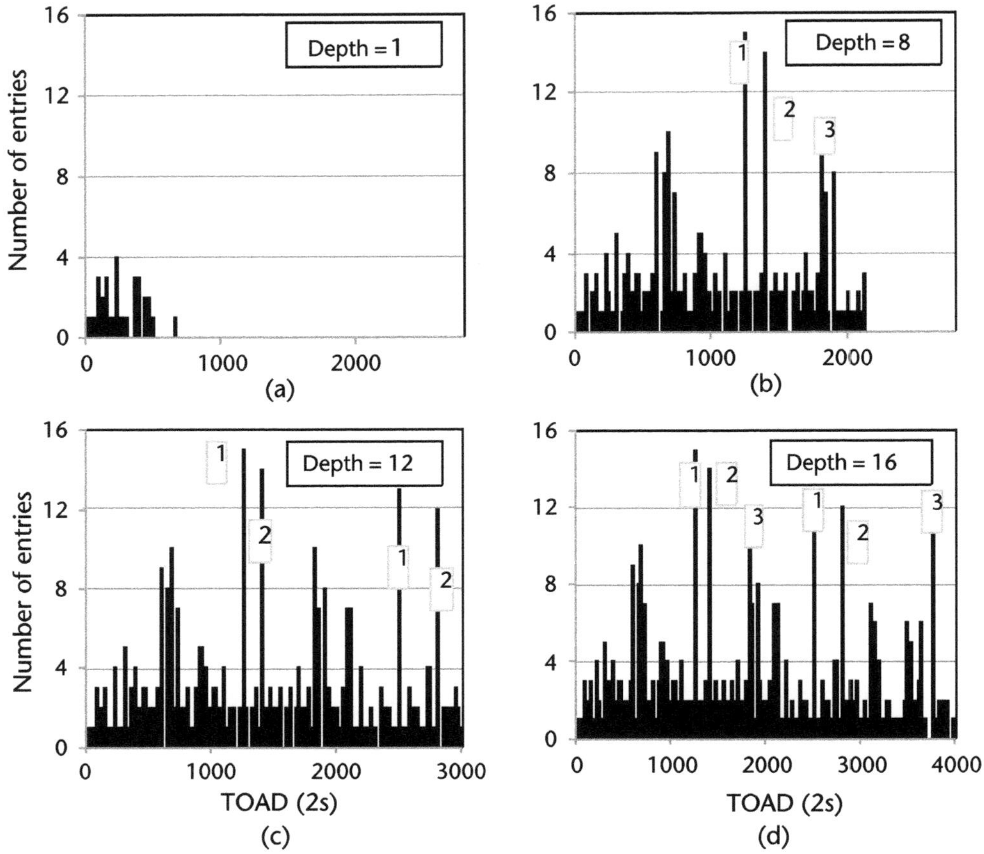

Figure 8.4 TOADHs for different depths for extraction of staggered PRIs.

for deciding the depth for the TOADH. In practice, the TOADHs are calculated to a depth set somewhere between 16 and 25. Some ESM systems have TOADH depths specified to take account of radars with very long and complicated stagger sequences. However, problems arise when there are missing pulses and when the ESM does not receive the full PRI sequence on a single scan of the radar.

If 30 or more pulses remain in an RF/DOA or RF/DTOA cluster, they are usually regrouped into pulses with a similar DTOA to a tolerance of perhaps 3 ns or a DOA of 5° and reported as complex intercepts. In reality, these complex intercepts are of little value to the ESM operator unless they have some feature, such as a distinctive PW, identifying them as coming from a particular radar type. The existence of such complex intercepts is very common as most ESM systems have difficulty in extracting PRI information for radars with multistagger PRI sequences.

Once all the pulses contributing to fixed RF intercepts have been extracted from a buffer of pulses, frequency agile signals are handled by grouping their pulses by DTOA or DOA alone and performing the PRI deinterleaving as above.

If a PRI element (or elements for staggered PRI radars) cannot be determined when the frequency criterion is removed from the TOADH processing, then the pulses are discarded. They are not allowed to form a complex PRI intercept without a fixed RF value.

8.4 Predictive Gates

Some conventional deinterleavers use predictive gate techniques to look for the next pulse in a chain, once PRI limits for a particular radar have been established. Figure 8.5 illustrates this process for a radar with fixed PRI.

In predictive gate systems [1], the limits that are set up to find the next pulse from the radar must be wide enough to cope with the potential PRI jitter, but narrow enough to prevent pulses from other radars from being included in the pulse chain. In a dense RF environment, where pulses are being received every few tens of microseconds, this causes problems for the PRI calculation process. This means that this type of deinterleaver is not appropriate for situations where there is a dense radar environment, such as the one described in Section 3.6.

8.5 Graph Theory Deinterleaver

This pulse association process is based on the use of the Graph Theory (GT) and nonlinear optimization. It attempts to determine if one or more pulse chains exist within a cluster. Parameter weights (PWT) are calculated and initially every pulse within a cluster is assumed to be linked to every other pulse with a look-ahead of typically 16 pulses. Figure 8.6 illustrates the look-ahead principle with an example with a look-ahead of 3.

The PWTs are calculated separately for each parameter (DOA, RF, PW, TOA, amplitude) according to the difference (*Diff*) between the parameter values for each pulse pair.

$$PWT_i = \left(W_i \times Diff_i\right) / Num\_pulses \tag{8.1}$$

where i is the parameter (DOA, RF, PW, TOA, or amplitude), W_i is the parameter weight, set according to the importance of the parameter in the deinterleaving process, $Diff_i$ is the difference in parameter value between the linked pulses, and *Num_pulses* is the number of pulses in a cluster.

Small differences are expected for simple radars, having high PWT. Large differences are seen for agile radars having low scores. The Link Weight (LWT) is the sum of all PWTs for the pulse pair. Large LWTs indicate that there is a greater chance that the given pair of pulses originates from the same radar.

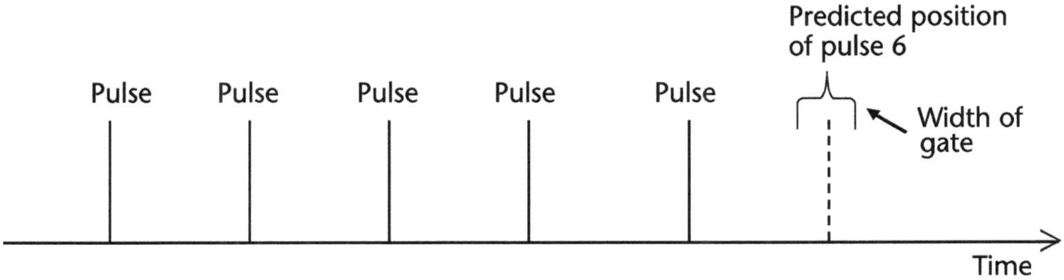

Figure 8.5 Predictive gate technique to group pulses according to the PRI.

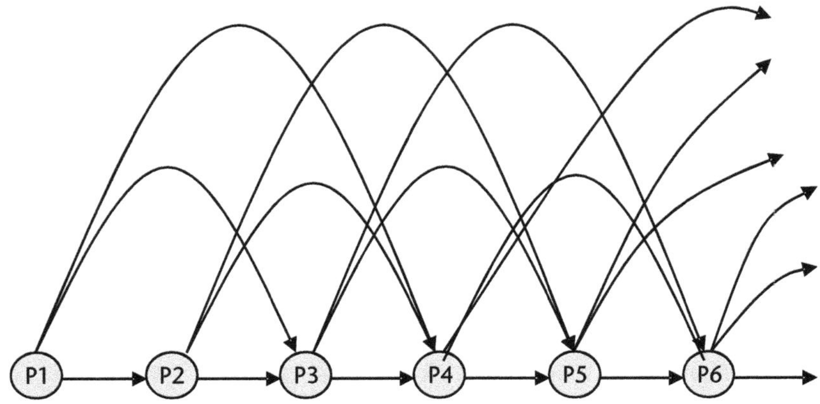

Figure 8.6 Generation of PWTs for look-ahead up to 3 for a pulse stream.

The next part of the process attempts to find a sequence of consecutive links such that the sum of link weights is a maximum. Each link has a LWT and a state. States start at 0.5 and go to either 0 (link invalid) or 1 (link valid). New states are calculated from the link weights and states of all neighboring links (other multiple links from the given pulse) in both forward and backward directions. Generally, links with high weight states force the states of neighboring links to 0. The result is that only one link from a pulse becomes active, giving a one-to-one pulse linking rather than one-to-many.

The final part of the process checks the chain for errors and splits them when they occur. For a given parameter, each pulse is checked to see if its parameter value differs by more than a tolerance from the previous pulse.

If it does, its position in the chain is marked as a possible place to split the chain. However, to prevent the splitting of batch hoppers, parameter differences of more than twice the tolerance are ignored. If the same position in the chain is marked for more than one parameter, the chain is split at the given position, as shown in Figure 8.7, and new chains are formed.

The entire GT process is then reapplied to the same buffer of pulses. However, this time in the weight calculation process, parameter weights are reinforced if links exist between pulses in the same chain and not otherwise. For optimum performance, the iteration is applied three times before the derived chains are sent to the merge and classification processes.

The merging process attempts to perform chain-to-chain merging and chain-to-track merging, updating existing track parameters in the process. This is a processing intensive operation and in most scenarios may not offer much advantage over the relatively simpler RF/DOA clustering and TOADH method. However, GT deinterleaving could have merit for radars that are so agile in RF that clustering is not a viable starting point for the deinterleaving process.

8.6 Radar Clock Period Deinterleavers

As described earlier in this chapter, conventional ESM deinterleavers take buffers of pulses and derive intercepts that are calculated to have come from the same radar.

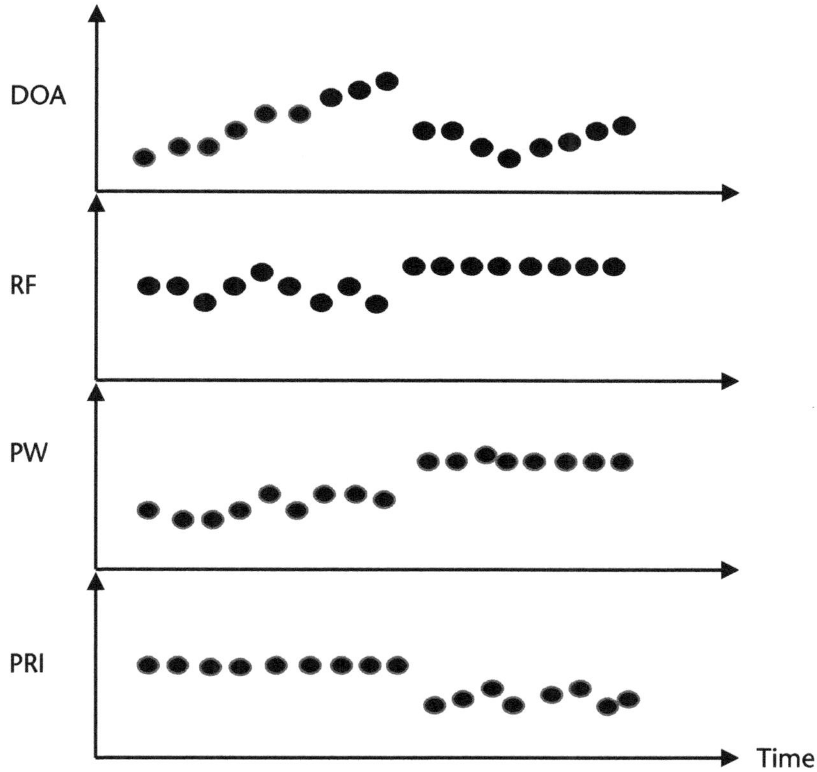

Figure 8.7 Determination of a split point in a pulse chain.

This process can fragment the pulse stream from a radar into several intercepts when the radar has agile parameters, and intercepts from a single agile radar can be difficult to correlate from scan to scan.

Radar clock period (RCP) deinterleaving is a new method of deinterleaving that does not rely on RF and PRI calculation. Many pulsed radars use a crystal controlled clock to generate stable PRI. On each cycle of the clock, the radar counts down from some initial value known as the down-count (DC), and, on reaching zero, a pulse is triggered. The DC may vary from pulse to pulse to generate a staggered PRI.

The primary RCP parameter is the CP, derived from the TOA of the pulses. The CP is fundamental to the operation of the RCP process as no other conventional parameter, such as DOA or RF, is used.

The RCP process analyzes an interleaved buffer of pulses to find CPs. It deinterleaves the pulses by sorting them into chains, where each PRI in a chain is an integer multiple of the same CP. The RCP process operates on buffers of pulse data, but does not correlate and track the results from buffer to buffer and does not characterize features such as agility, scan, or parameter levels. The output of the process is a table of pulses for each buffer, sorted into chains.

The CP is a parameter that may help to reduce ambiguities during the radar identification process. It may also lead to specific emitter identification (SEI) if both the radars and the RF sensors crystals are stable in the long term.

8.7 Parameter Classification Algorithms

Parameter classification algorithms attempt to describe the pulse-to-pulse behavior of RF, PW, and PRI for each pulse chain. The algorithms used to derive each parameter work by creating histograms of the respective parameters and determining if significant peaks exist and, if so, the number of peaks. The dispersion seen by the ESM for the parameter can be determined and this information can be used to decide the tolerance for the intercept-to-track correlation.

First, a histogram of all RF values in the chain is formed. If a single peak is seen containing more than about 80% of the pulses, the radar is deemed to have simple RF. Figure 8.8 shows an example of an RF histogram for a chain of pulses. In this case, a single peak is seen and the radar is declared to have fixed RF.

If several peaks are seen in the RF histogram, the radar is classed as agile and a determination of the type of agility is carried out. Figure 8.9 shows an example where there are several peaks in the RF histogram for an RF agile radar.

The final stage in the processing of the pulses appearing in a RF histogram with several peaks is to determine whether the radar is a batch hopper, in which case several consecutive pulses will appear at the same RF, or whether the radar is pulse-to-pulse agile in RF. The last two categories may be either regular, with the pulses appearing in an ordered pattern, or irregular where the pulses or groups of pulses change RF on a pseudo-random basis within the limits and at the levels of RF specified for the emitter.

The algorithms for the classification of RF and PW are similar, except in the case of PW where there is no testing for parameter regularity (i.e., regular pulse-to-pulse or regular batch hopper). Radars with PW agility usually only operate with a small number of different PWs. The purpose of the parameter histogram in this case is to determine the spread in measured PW so that the PW dispersion can be calculated and correlation tolerances (if used in the ESM track correlation algorithm) can be updated.

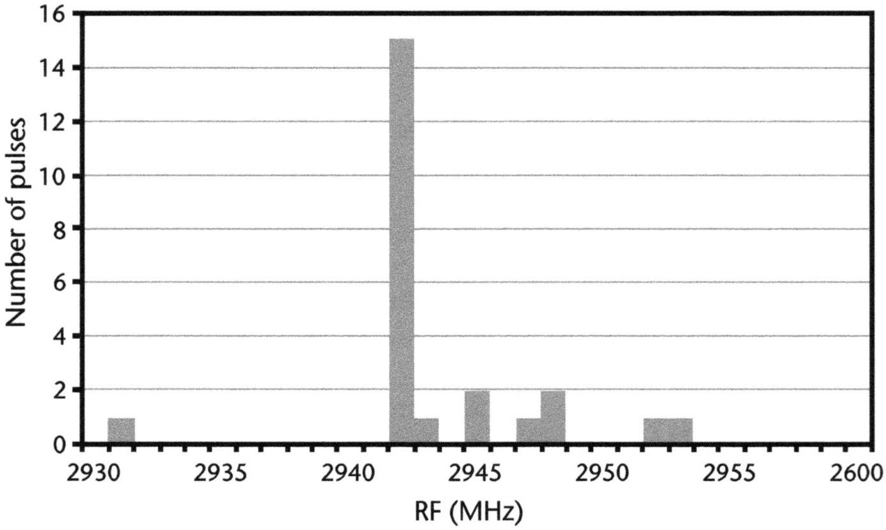

Figure 8.8 RF histogram for a fixed RF radar.

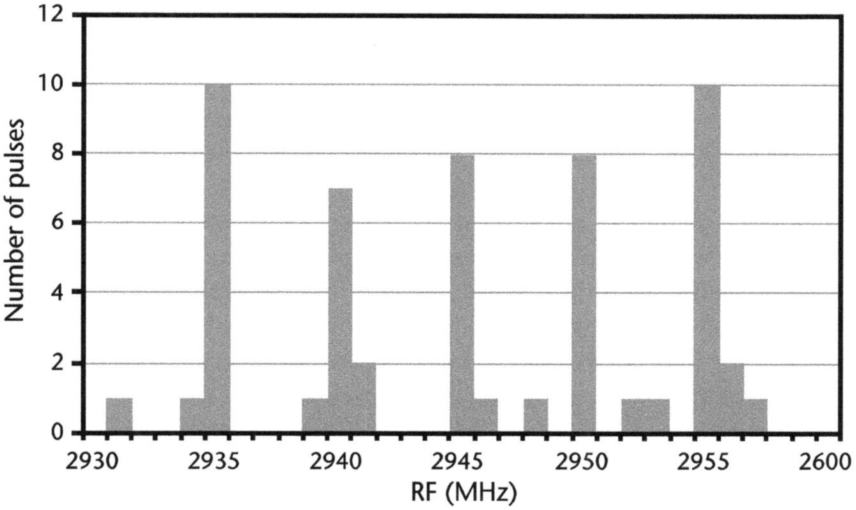

Figure 8.9 RF Histogram for an agile RF radar.

The PRI classification algorithm first tests for pulsed group behavior (doublets, triplets, and so forth with variable intragroup and intergroup intervals) within the chain. If grouped behavior is found, it generates two additional histograms for intragroup and intergroup. If pulse group behavior is not found, only the difference histogram is generated. In most PRI classification algorithms, if the largest peak in the PRI histogram contains at least 82% of the pulses, the emitter is labeled as having a simple PRI.

A proper determination of parameter agility can aid the identification of the radar. Problems arise where there are more levels of RF or PRI agility than can be seen in a chain of pulses extracted from a single pulse buffer or where there are only one or two pulses at each agility level in a single buffer. It may take some time before enough pulses are received to find the true nature of the agility and hence the identification of the radar. This is a common problem as most ESM systems are not able to process pulses from more than one buffer at a time, so any information on agility is not available from buffer to buffer. Correct determination of the radar parameters can be done in post-mission analysis so that useful information can be made available for the ELINT process, but using the ESM system in real time will result in misidentification for any but the most simple radars. This issue is described in more detail in Chapter 10.

8.8 Track Formation

When the intercepts have been formed and parameters have been determined, the ESM processor first attempts to correlate the intercept to an existing track containing intercepts from the same radar. Parameter tolerances are set up in ESM system to allow for the correlation process to take place. The tolerance in RF may be 1 or 2 MHz and that for the DOA may be a fixed small number of degrees. The DOA tolerance may be variable depending on the geometry of the ESM platform relative to the radar or the classification of platform hosting the radar (air, ship, land).

If the intercept can be correlated to an existing track, the parameters of the track, such as RF, PW, and PRI, are updated using the new information available in the intercept. The maximum and minimum values allowed for future track correlation are also updated, taking account of the parameter tolerances.

If the intercept cannot be correlated to an existing track, a new track is created. In a typical RF environment such as that described in Chapter 3, it is usual that only six pulses are needed for a valid intercept to be created. RF is usually measured well by ESM systems, but there are lots of reasons why PRI is miscalculated as described in Chapter 14. PW is often measured in error for pulses that have intentional modulation. Contrary to traditional thinking, DOA is actually the parameter that is most likely to be measured incorrectly. All these reasons mean that there are likely to be many more tracks created than there are radars in the RF environment.

Typically tracks are produced that are derived from the pulses of more than one radar (overmerging) or more than one track is produced from the pulses of a single radar (multitracking or fragmentation). The consequences of these limitations are that the ESM does not produce a correct picture of the RF environment. More details about track correlation and multitracking are given in Chapter 14.

References

[1] Hassan, H. E., "De-Interleaving of Radar Pulses in a Dense Emitter Environment," *Proc. of the Int. Conference on Radar*, 2003.

[2] Conroy, T., and J. B. Moore, "The Limits of Extended Kalman Filtering for Pulse Train De-Interleaving," *IEEE Transactions on Signal Processing*, January 1999.

[3] Bailie, S., and M. Leeser, "An FPGA Implementation of Incremental Clustering for Radar Pulse De-interleaving," https://www.ll.mit.edu//HPEC/agendas/proc10/Day1/PA02_Bailie_abstract.pdf.

[4] Ata'a, A. W., and S. N. Abdullah, "De-Interleaving of Radar Signals and PRF Identification Algorithms," *IET Journal of Radar, Sonar and Navigation*, November 2007.

[5] Keshavarsi, M., and A. M. Pezeshk, "A Simple Geometrical Approach for De-Interleaving Radar Pulse Trains," *UKSIM-AMSS 18th Int. Conference on Computer Modelling and Simulation*, 2016.

CHAPTER 9

Intercept-to-Track Correlation

Most ESM systems are only able to process pulses from one buffer of radar pulse data at a time, so the extraction of correct parameter values is difficult. This is especially true for radars that have parameter agility. The accuracy of measurement of RF and TOA of pulses may also contribute to errors in the determination of these values. Errors in pulse parameter values usually lead to multitracking, as described in Chapter 14.

However, when there are many radars with significantly overlapping parameter ranges, the opposite problem, overmerging of tracks, may occur. This leads to confusion in the ESM picture as there may be fewer tracks than there are radars in the environment, with each track being updated by intercepts from more than one radar. The most usual picture that the ESM system manages to extract contains a mixture of fragmented and overmerged tracks, giving a poor interpretation of the reality of the RF environment.

Sometimes pulses from more than one radar appear in a single intercept, but this is a rarer situation than the case where intercepts from several individual radars update a single track. Therefore, it is the intercept-to-track correlation process and the reasons for the failure of that process that are discussed in this chapter.

9.1 The Intercept-to-Track Correlation Process

Once an intercept has been created, the ESM attempts to correlate it into an existing track. It does this by considering each parameter in turn and comparing the intercept parameters with those of the tracks. The first parameter considered is usually the DOA. If the DOA of the intercept is within a few degrees of any of the existing tracks (typically 5°), then the track is considered as a candidate for further consideration.

RF is usually the next parameter to be compared, typically with a tolerance of ±2 MHz, followed by the PRI. PW may also be used a correlation parameter. Parameter types may also be used in the correlation process (e.g., RF type (fixed, hopper, agile) and PRI type (fixed, stagger, jitter)). A flowchart of a typical intercept-to-track correlation process is shown in Figure 9.1.

Figure 9.1 is a simplified version of the intercept-to-track correlation process. There are some steps in this process that require further explanation: RF matching (Figure 9.2) and PRI matching (Figure 9.3).

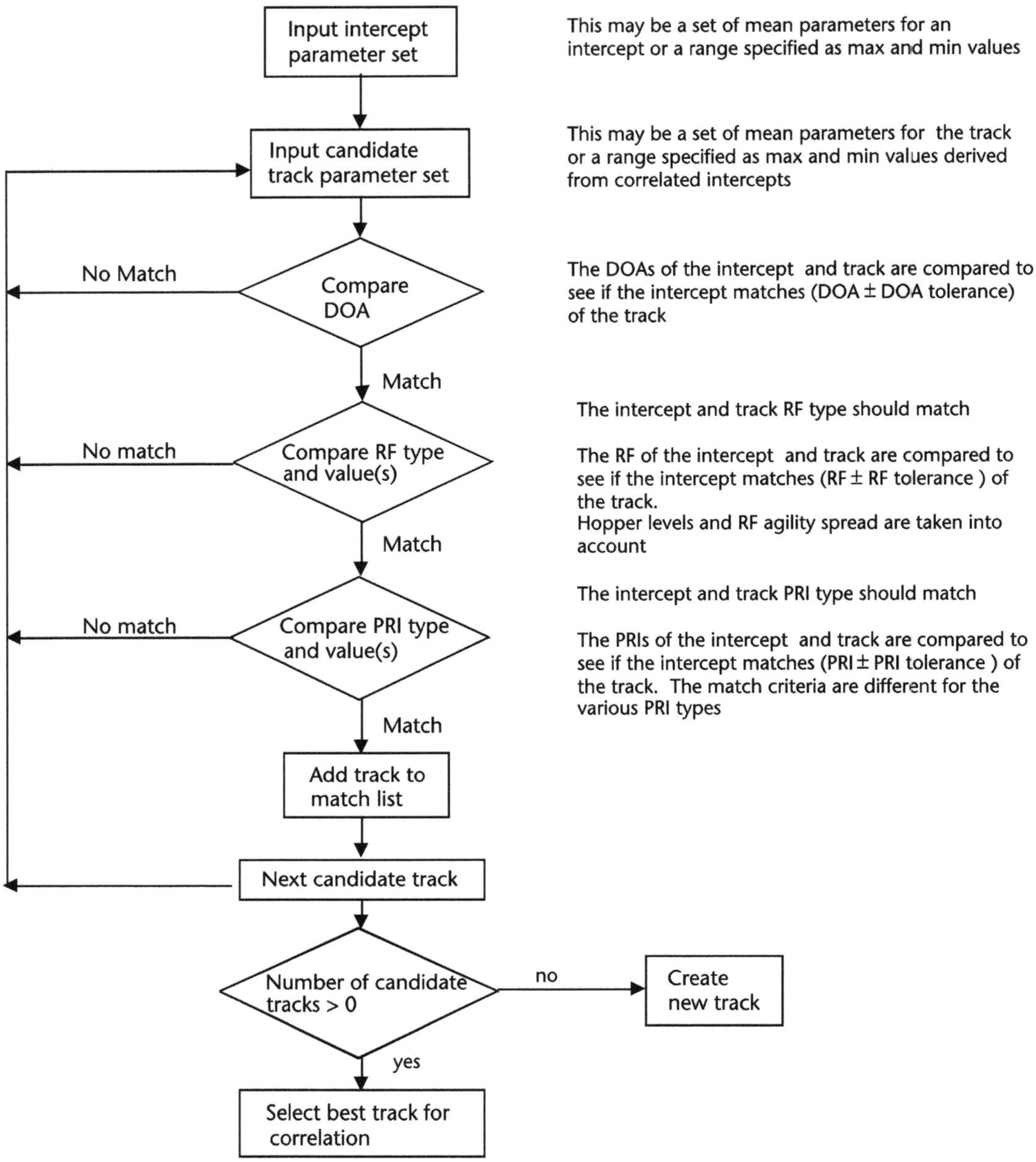

Figure 9.1 Intercept-to-track correlation process.

During intercept creation, the RF type should be classified as fixed, agile, or hopper. The RF matching process then proceeds as follows in Figure 9.2.

PRI matching is more complicated and the exact process is different for each ESM type. Figure 9.3 gives an indication of the steps needed to accomplish this part of the correlation process.

In many ESM systems, the classification of fixed PRI is encompassed in the staggered PRI definition (i.e., a fixed PRI signal is a one-position, one-element stagger). There are many reasons why pulses may be missed in the PRI sequence (see Chapter 14). To cater for this situation, the intercept PRI should be tested against

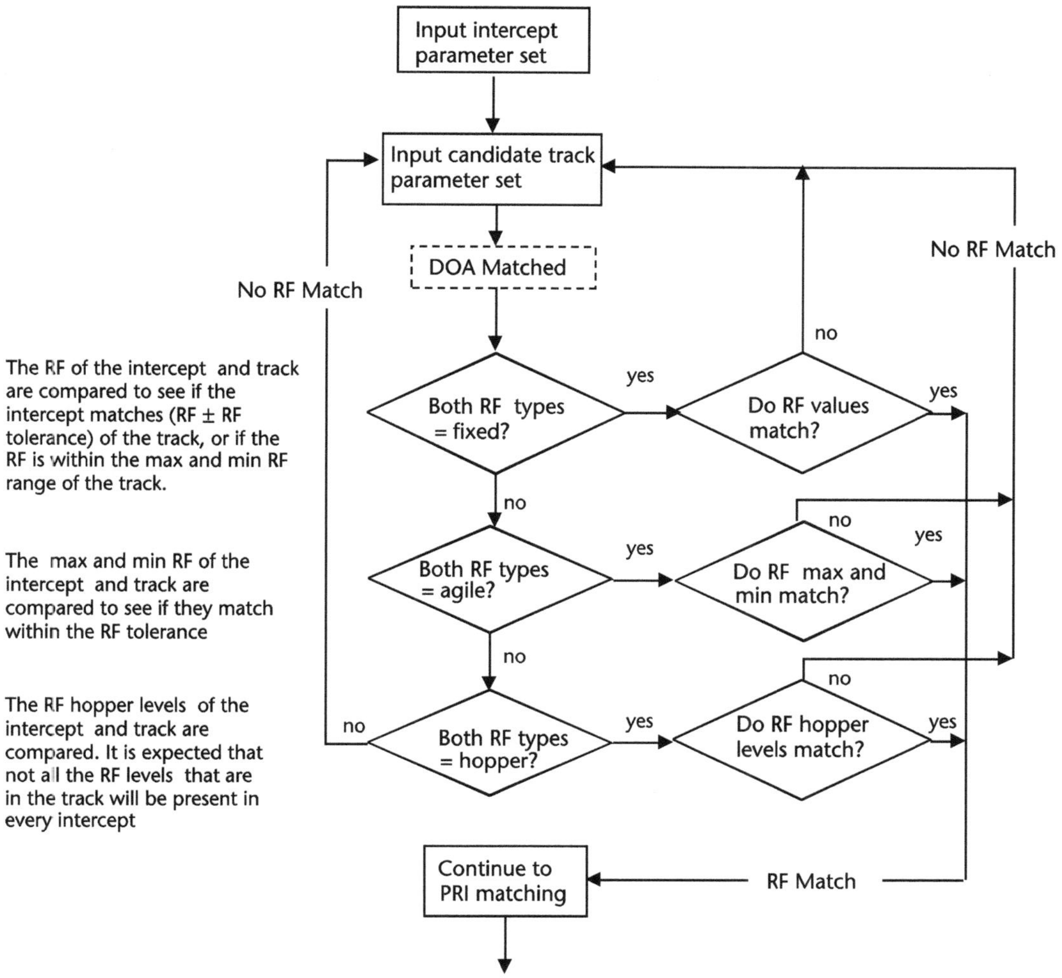

Figure 9.2 The RF matching process.

several harmonics of the track PRI to see if there is a match. For example, an intercept PRI of 2,550 μs should be allowed to correlate to a track with a PRI of 850 μs as the intercept PRI is the second harmonic of the track PRI.

For staggered PRI, the pulse group repetition interval (PGRI) may be calculated and stored in the track data. This allows intercepts where it has not been possible to distinguish all the elements in the PRI sequence and where only the PGRI has been calculated to be correlated into a relevant track.

An intercept with a complex PRI designation should be allowed to correlate into an existing track as the RF and DOA will have matched by the time the PRI processing takes place.

An alternative way for intercept-to-track correlation to take place starts by carrying out the identification of the intercept. The candidate tracks for the correlation process are then limited to those that have the same identification. A flowchart of the intercept-to-track correlation process when identity is used a correlation parameter is shown in Figure 9.4.

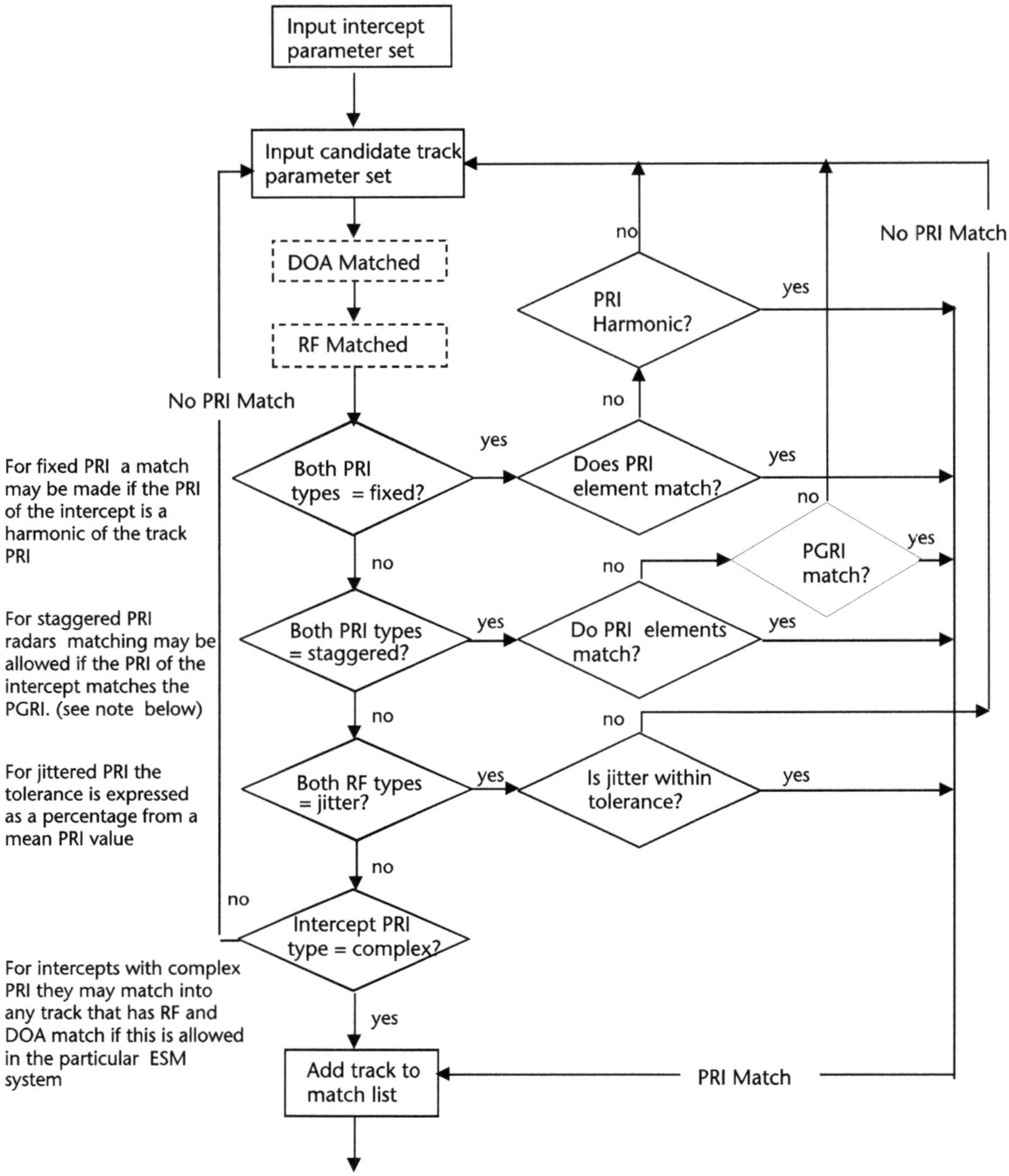

Figure 9.3 The PRI matching process.

Once the list of candidate tracks that match the identity of the intercept has been established, the parameter tests are carried out for each candidate. One advantage of using the identity in the correlation process is that different parameter tolerances can be set for each radar type.

In cases where more than one possible identity has been determined for the intercept (see Section 9.5), the list of available tracks for correlation is increased as those with all the possible intercept identities can be considered as candidates for correlation.

If an intercept cannot be identified and is labeled as unknown, then either all tracks are tested for correlation or only those tracks that have been created with the identity of Unknown are included in the correlation process.

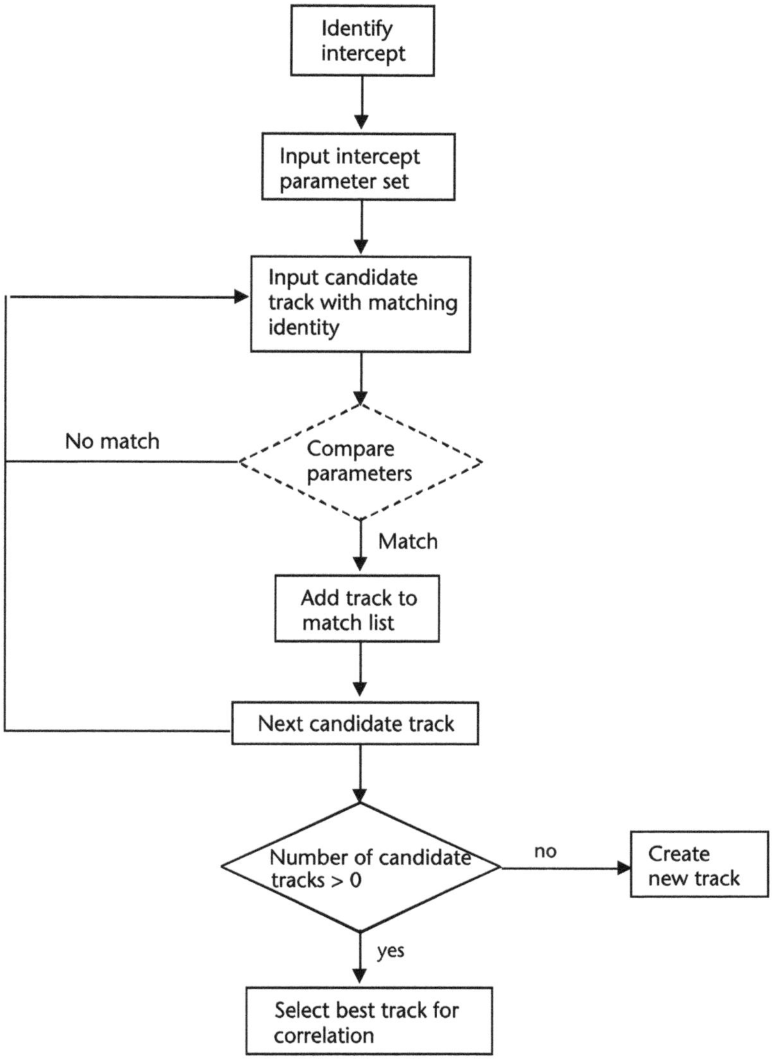

Figure 9.4 Intercept-to-track correlation process with identity as a correlation parameter.

If a track contains data on a radar location then instead of matching the DOAs of the intercept and track, a line-of-bearing (LOB) defined by the intercept DOA must lie within the error ellipse that has been calculated to encompass the radar location (see Chapter 11).

The hard part for the ESM system is that in some RF bands there are likely to be many types of radar with overlapping parameters. The use of identity as a correlation parameter therefore restricts the list of tracks available for correlation and may lead to the creation of extra tracks.

If the intercept can be matched to more than one track, it is correlated to the track with the smallest parameter differences. Once the best match is found, the parameters of the intercept are used to update the track parameters. If identification has not been used in the correlation process then at this stage the track identity is checked against the ESM radar library and the track identity updated, if necessary. For ESM systems that produce location fixes for radars, the last part of the correlation process, the geolocation algorithm, will check whether the threshold for an initial location estimate has been met for the track and if so perform it. If the track selected for correlation already has a location, then a successful correlation will trigger an update in location, as described in detail in Chapter 11.

9.2 Overlapping Parameter Ranges

It might be thought that in an RF environment where there are no more than two or three radars with interleaved pulses (i.e., most RF environments) (see Chapter 3) the ESM can accurately correlate intercepts to tracks. However, there are so many radar types that are similar to each other that, even where there is a low radar pulse density, it is likely that there will be pulses received from radars with overlapping parameter ranges. For example, in the RF bands of 3.03 to 3.08 GHz and 9.30 to 9.45 GHz (the RF bands in which ship radars operate), there are hundreds of different types of radar. Even in the RF band of 5 to 6 GHz (the RF band in which many weather radars operate), there are several tens of different radar types, all with very similar parameter ranges.

As an illustration of the overlap in parameters, Figure 9.5 shows the extent of the RF ranges for just 10 ship radars in the I-band frequency range. The overlap in RF for these radar types is considerable.

The overlap in the PRI ranges of these 10 radar types is even more confusing, as shown in Figure 9.6 and the PW ranges do not offer any assistance in separating out the radar types, as shown in Figure 9.7.

Combining PRI and PW ranges in a single plot in Figure 9.8 shows the overlap in both these parameters and hence the difficulty in separating these radar types based on parameter matching.

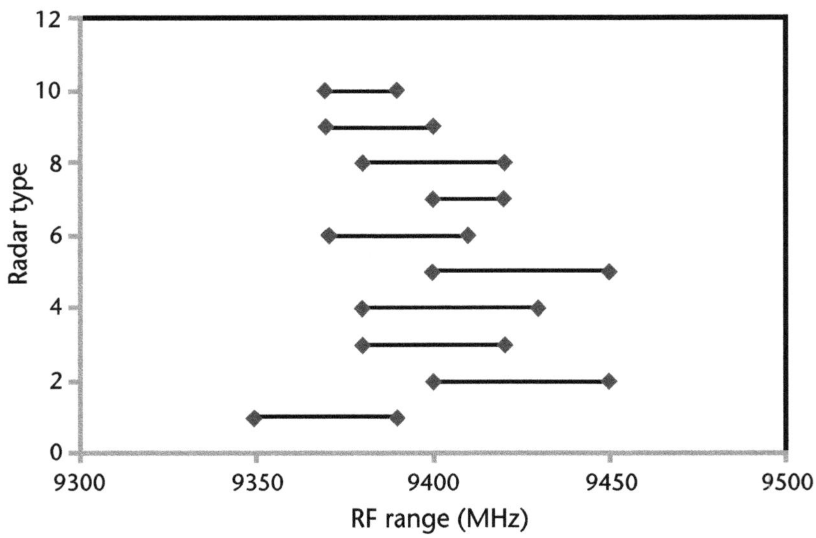

Figure 9.5 RF ranges of 10 ship radar types.

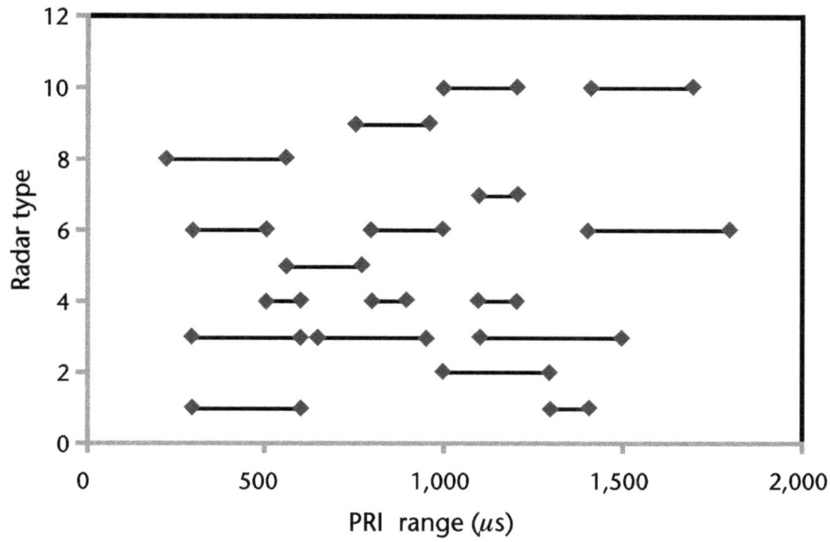

Figure 9.6 PRI ranges of 10 ship radar types.

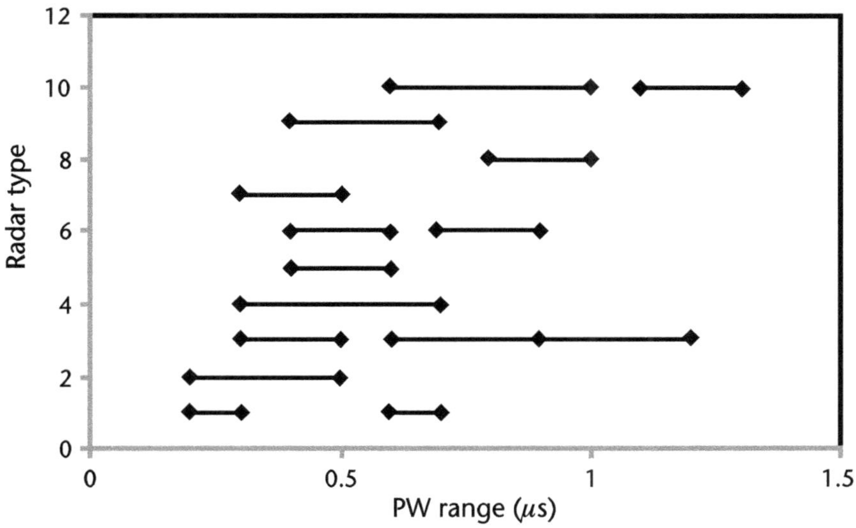

Figure 9.7 PW ranges of 10 ship radar types.

The final parameter that may help to distinguish the various radar types is the scan period, but as Figure 9.9 shows, there is still considerable overlap for these 10 ship radar types.

In summary, the overlap in all the parameters for these 10 radars means that the ESM will have considerable difficulty in correlating intercepts from different radars into the correct tracks. In reality, there are many more types of I-band radar, so it really is a very difficult task for the ESM system to accomplish.

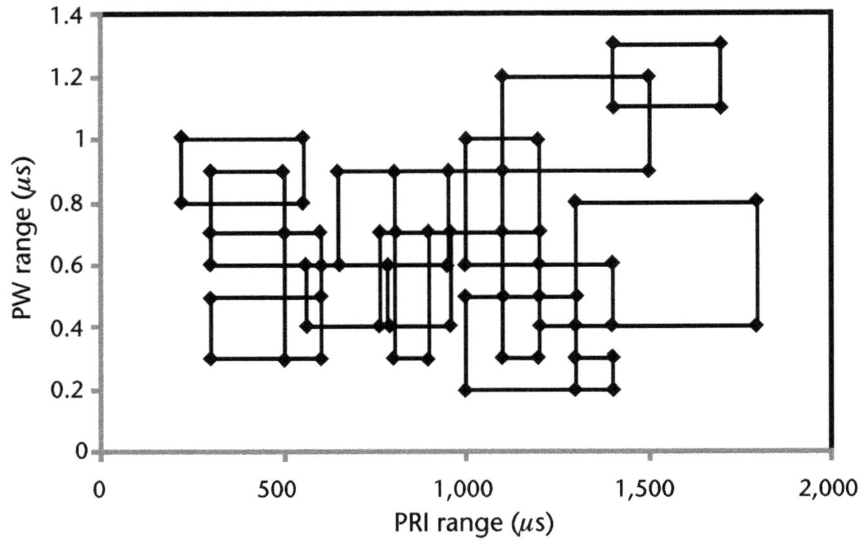

Figure 9.8 PRI versus PW for 10 ship radar types.

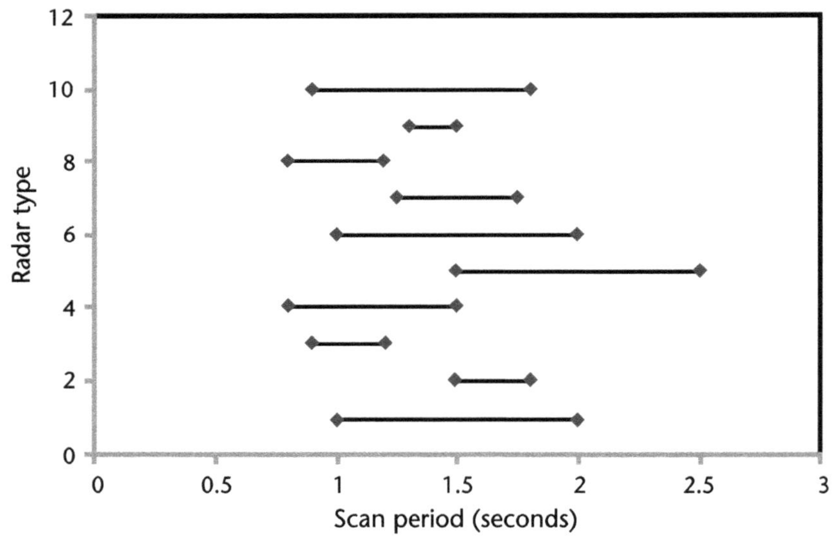

Figure 9.9 Scan periods of 10 ship radar types.

9.3 Overmerging of Tracks

The difficulty in separating intercepts from the different radars into individual ESM tracks is the opposite problem to multiple track creation for a single radar. Overmerging occurs most frequently for radars that operate in narrow RF bands such as those that ship radars use. The two radar bands, one at about 3 GHz and the other at about 9 GHz, are particularly susceptible to this phenomenon as there are lots of situations where there are likely to be many examples of radars with very similar parameters in a relatively small geographic area.

Figure 9.10 shows a simulation of pulse data to illustrate the problem of overmerging of radars. The simulation begins by considering radars at RF between 3.04 and 3.08 GHz in a limited DOA range of 100° to 135°. This type of pulse density is typical of a busy shipping environment such as the Straits of Malacca or the English Channel. In Figure 9.10, the amplitude profiles of all the radars in the specified RF and DOA ranges are shown with pulses from 14 different radars plotted during a time period of 2.5 seconds.

Figure 9.11 shows the DOA of all the pulses in the specified RF and DOA ranges against pulse number. It is possible to identify specific groups of pulses by eye for each individual radar in both the amplitude and DOA graphs. Two of the radars have scan peaks that occur almost simultaneously, with the pulses from one radar interleaved with those of the other radar. These radars are numbered 9 and 10 on Figures 9.10 and 9.11.

When the data is presented in a graph using pulse number as the x-axis and analyzing the data by eye, it is fairly obvious which pulses are from each radar. However, when the data is presented as the DOA against time, it is not as easy to determine which radar the pulses are from.

A simulation of a simple intercept-to-track correlation process carried out using a longer sequence of pulses leads to four tracks being created and updated for the 14 radars. Figure 9.12 shows the DOA against time of the pulses with the ESM tracks and their updates also shown.

The radars all had PWs of between 0.3 and 0.7 μs and either fixed PRI or simple stagger PRI sequences with elements between 650 and 850 μs.

An explanation of how two intercepts from different radars get correlated into the same track starts by considering the parameters of the intercepts

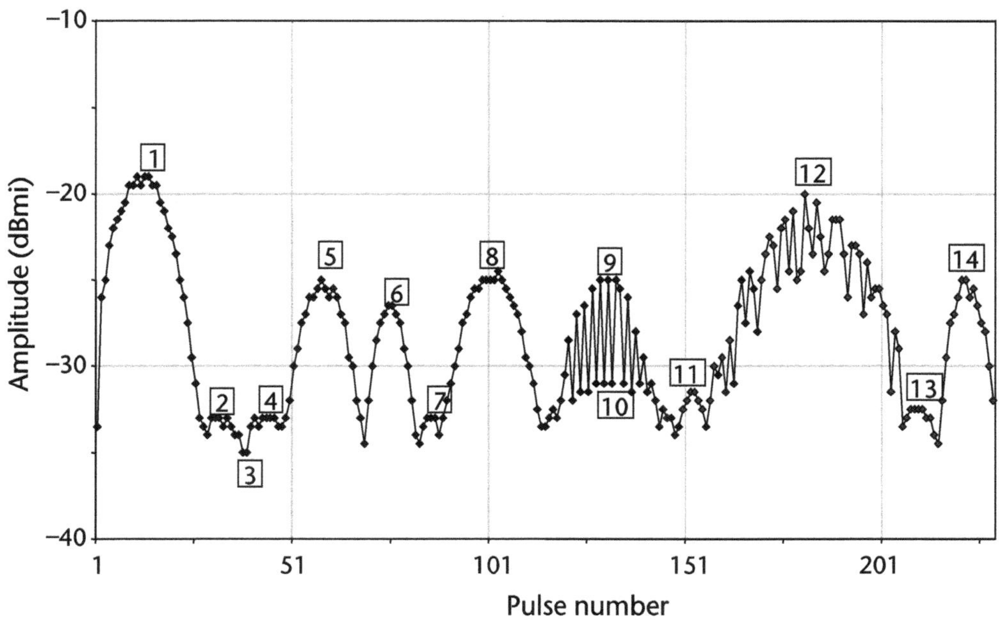

Figure 9.10 Amplitude of pulses from radars in the RF range of 3.04 to 3.08 GHz and in DOA range 100° to 135° for 2.5 seconds.

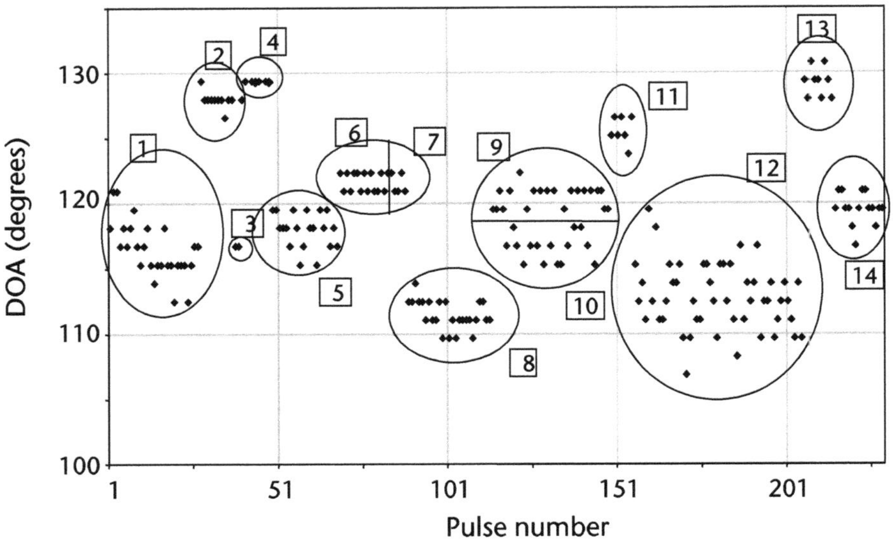

Figure 9.11 DOA of pulses from radars in the RF range 3.04 to 3.08 GHz and in the DOA range of 100° to 135° for 2.5 seconds.

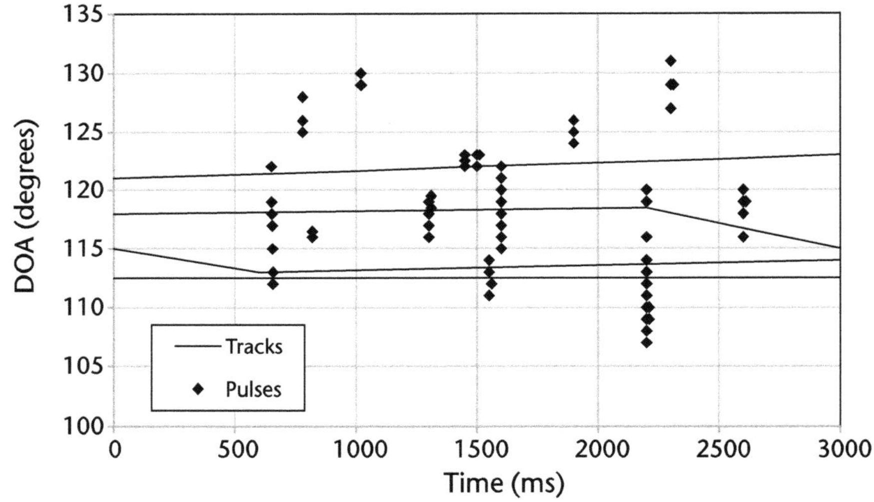

Figure 9.12 DOA of pulses and ESM tracks from a simulation of shipborne F-band radars in the DOA range of 100° to 135° for 2.5 seconds.

from the two radars. The first radar has RF of 3,048 MHz and a fixed PRI of 720 μs. The calculated DOA for the intercept from this radar is 117° with the spread in DOA of 5°. The second radar has an RF of 3,049 MHz, a staggered PRI with elements of 700, 720, and 740 μs, and pulses with a mean DOA of 117.5° and a DOA spread of 5°. It is almost impossible for conventional ESM systems to recognize that these two sets of pulses are from different radars and to create and maintain two different tracks for them.

In ESM systems where track reidentification takes place after the intercept to correlation process, overmerged tracks are likely to be assigned ambiguous or unknown identities. This is due to the incorrect determination of track parameters, which are updated from the intercept parameters from different radars.

9.4 Effect of Identity on ESM Track Formation

When an intercept has been constructed from a set of pulses, the ESM processor then either:

- Classifies the intercept using its library of radars and correlates the identified intercept to existing tracks of the same identity or creates a new track;
- Correlates the intercept using parameters other than identity to an existing track and possibly changes the identification of the track based on the parameters of the intercept.

In the first case, where radar identity is used as correlation parameter, a wrong identity will mean that there is no chance that a subsequent intercept from the same radar that is given a different identity will be able to correlate to the existing track, leading to multiple tracks in existence simultaneously for the same radar.

In the second case, the ESM keeps switching the identity of existing tracks and probably sets up alerts as various tracks change to being identified as threats.

If there are multiple entries in the ESM radar library with overlapping parameters and identity is used as a correlation parameter (for intercepts to tracks), then there will be multiple tracks in existence for a single radar, as shown in the example in Figure 9.13. This shows six tracks created and updated for one radar. Each track is assigned a different identity and is updated when an intercept from the radar is identified to match the track.

If the ESM tracks can change identity according to new intercept data there will be many "confused" tracks, as shown in the example in Figure 9.14, where only a handful of tracks are uniquely identified. The rest, shown in gray on the plot, are constantly having their identity changed.

In Figure 9.14, there are three radars that have unique identities and many other radars that have identities that change when new intercepts, matched on parameter ranges are allowed to correlate to the tracks. The situation in Figure 9.14

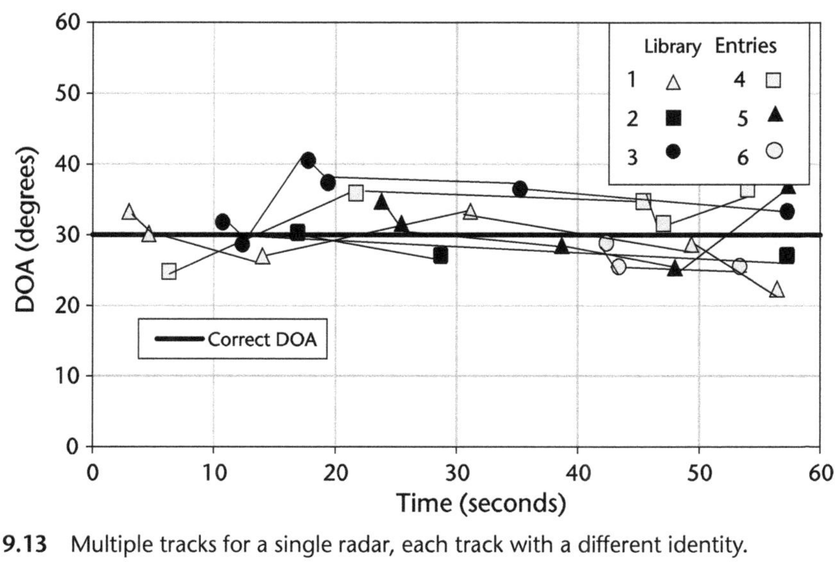

Figure 9.13 Multiple tracks for a single radar, each track with a different identity.

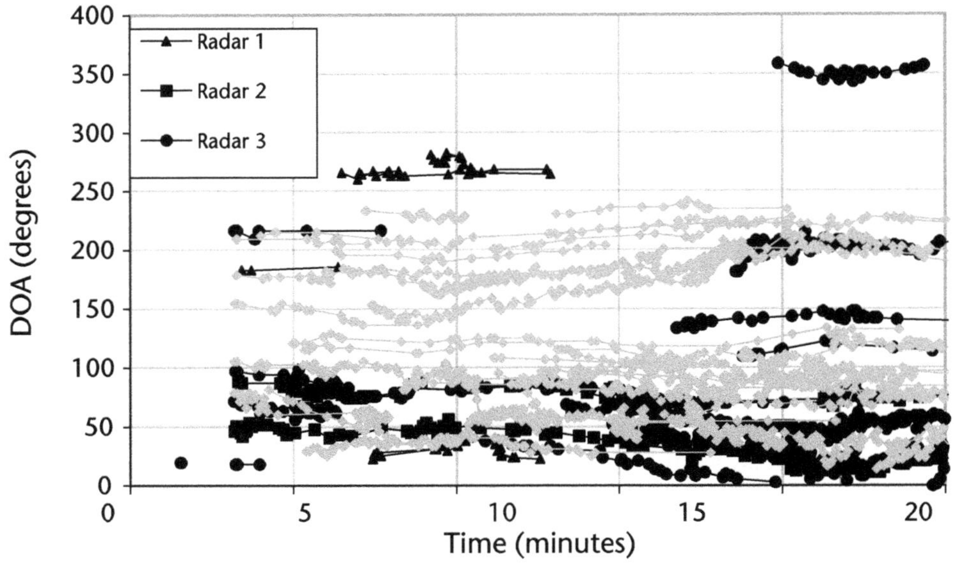

Figure 9.14 ESM tracks that change identity through the lifetime of the track.

will not cause as much multitracking as the case where identity is used as a correlation parameter, but the constant switching of track identity will cause confusion on the operator's screen.

9.5 Identification Ambiguity

The difficulty in determining the correct identity of a track means that the ESM identification process often arrives at the conclusion that there is more than one possible identity for a single track. In this case, the track identity is ambiguous. Because of this uncertainty in identification, the ESM system almost always has the capacity to keep several different identities for each track in existence simultaneously. The number of ambiguous identities allowed for each track is usually 4 or 5, but there are ESM systems that can keep 20 possible identities for each track. The identity chosen for display to the operator should be the radar type with the highest threat level. In cases where the threat levels of all the possible identities for a track are the same, there is usually a favored identity, perhaps based on the likelihood of a particular radar type operating in the specific RF environment of the mission.

Sometimes it is possible to have access to the all the ambiguous identities during a sortie so that the experienced ESM operator can make a judgment about how likely the ESM is to have chosen the correct identity. All the possible identities should be recorded for post-flight analysis and updating the ESM radar library.

9.6 Solutions for Overmerging of Tracks

Overmerging is a real challenge for ESM systems, but there are some measures that can be taken to improve this problem. The first is the use of parameter dispersions and tolerances as criteria for intercept to track correlation. Parameter dispersions are a measure of the spread of the pulse parameter values that have been used to generate the intercept, with the mean value reported as the intercept value. Parameter tolerances are the difference allowed from the intercept parameter value and that of the track into which it is correlated (see Section 9.1).

An example of using parameter dispersions to decide if two intercepts are correlated considers an intercept having a mean PRI of 700 μs and a PRI dispersion of 0.002 μs and another intercept that also has a mean PRI of 700 μs but a PRI dispersion of 0.2 μs. With the difference in PRI dispersion between these intercepts, it can be deduced that it is likely that these two intercepts are not from the same radar. Parameter dispersions can also be used in the radar identification process and can be set as part of the radar data set in the ESM library (see Chapter 10).

The use of small parameter tolerances can also help to prevent overmerging, although care must be taken not to end up with multiple tracks for each radar due to having too small a tolerance for parameters such as RF and PRI.

Another way of preventing overmerging is by using a specific emitter identification (SEI) technique such as radar clock period estimation to limit the merging of intercepts into existing tracks (see Chapter 10).

Although publications on this subject are limited, there is a statistical technique for intercept-to-track correlation that was shared in the public domain by Trunk and Wilson [1]. Their algorithm uses a combination of Bayes and Neyman-Pearson procedures with time alignment to make the maximum use of the data. The a posteriori probabilities of an intercept that would associate with each track are calculated and then the decision rules select the association that has the largest probability. Different thresholds are set to accommodate the real scenario in which tracks are constantly initiated, updated, and dropped. The algorithm has two disadvantages; first, it requires a prediction or interpolation for measurements that are not time aligned, and second, the algorithm requires definitions of different threshold settings using Monte Carlo simulations.

Neural networks feature in a number of publications on intercept to track correlation. For example, in [2], initial clustering algorithms are used to separate pulses from different radars according to position-specific parameters of the input pulse stream when implementing a What-and-Where fusion strategy. Neural networks are then used to classify streams of pulses according to radar type, using their functional parameters.

Another possible application of neural networks for the timely and reliable recognition of radar signals is investigated in [3]. In particular, a large data set of intercepted generic radar signal samples was used for investigating and evaluating several neural network topologies, training parameters, input and output coding, and machine learning data transformations. Three case studies are discussed. In the first two cases, the radar signals are classified in two broad classes, either with civil or military application, based on patterns in their pulse train characteristics. In the third case, the algorithm is trained to distinguish between several more specific radar functions. Following the training of the neural networks, results of about 82%, 84%, and 67% were achieved on the test data sets.

References

[1] Trunk, G. V., and J. D. Wilson, "Association of DF Bearing Measurements with Radar Tracks," *IEEE Transactions on Aerospace and Electronic Systems*, Vol. AES-18, No. 4, 1987.

[2] Granger, E., et al., "A What-and-Where Fusion Neural Network for Recognition and Tracking of Multiple Radar Emitters," *Neural Networks*, Vol. 14, No. 3, 2001.

[3] Petrov, N., I. Jordanov, and J. Roe, "Radar Emitter Signals Recognition and Classification with Feedforward Networks," *Proc. 17th Int. Conference in Knowledge Based and Intelligent Information and Engineering Systems*, 2013.

CHAPTER 10

Radar Identification and ESM Radar Libraries

The correct identification of radars is a key function of ESM systems but is one of the major causes of poor performance. The identification process relies on the correct measurement of the radar parameters, and, while RF is usually measured with sufficient accuracy, other parameters such as PW and PRI suffer from errors in their measurement or calculation for a variety of reasons, as described in Chapter 14.

Adding to this problem, there are thousands of different types of radars in existence, many of which have overlapping parameter ranges. Most radars operate in a small number of RF ranges, as discussed in Chapter 2, with peaks in the number of radar types operating at 3.04 to 3.08 GHz and 9.2 to 9.45 GHz due to ship navigation radars. With so many radar types in narrow RF ranges, the likelihood of generating a radar library for the unambiguous identification of all radars of interest in a particular mission is very small.

Typically, many tracks are created from radars operating in these narrow RF ranges, with few of the tracks corresponding to sets of intercepts from unique radars. The ESM either produces too many tracks from the pulse stream of a single radar (see Chapter 14), or it merges pulses from several radars (see Chapter 9). The most usual picture of the RF environment from the ESM is a mixture of fragmented and overmerged tracks, giving a poor interpretation of the real RF environment.

One of the most difficult processing functions for the ESM system is therefore the creation and effective operation of the radar library. Current ESM libraries, sometimes called the Emitter Program Library (EPL) or Pre-Flight-Message (PFM), have limited capability in terms of the number of radar parametric mode lines available. In the environment where the ESM operates, there may be several hundred potential radar types needing several thousand parametric mode lines for their collective specification. However, the capabilities of current ESM libraries are seriously deficient. The most extensive library is currently capable of holding about 10,000 mode lines, but this is an unusually large ESM library with most ESM libraries having a capacity of only about 1,000 mode lines.

10.1 Radar Parameters for an ESM Library

Almost all radars can operate in more than one mode. Even simple fixed RF radars such as small boat radars have different combinations of PRI and PW according to the range at which they are operating. Each combination of PRI and PW is a separate mode and should be specified separately in the ESM library, The minimum data fields for the specification of a mode are RFmax, RFmin, PRI max, PRI min, PW max, and PW min. Other parameters that may be specified are scan period, 3-dB beamwidth, and ERP.

PW may be used as a parameter in resolving identity ambiguities between radars, but care has to be taken as the effects of multipath may alter the received PW or cause modulation in the pulse profile leading to pulse width measurement problems, as described in Chapter 14.

The use of scan period as an identifier causes problems for an ESM system not wide open to all RFs within its range at all times as there is a possibility that some radar scan peaks will be missed altogether.

As described in Chapter 2, some radars are capable of operating with a wide variety of PRI stagger values and it is not possible to specify them all exactly in an ESM library. However, some radars have relatively few PRI options and can be more tightly specified. Other data such as the mount type of the radar (aircraft, ship, land-based) and countries using the radar type can be used to derive a confidence level for the identification of the radar. The clock period and PRI stability are very useful parameters for distinguishing between radars, but information on these parameter values is not readily available for all radar types.

10.2 Data Records for an ESM Library

The ESM library specifies each radar as a series of modes with each mode containing the maximum and minimum value expected for each parameter. Sometimes the parameter dispersions are included in the mode. The mode lines in the ESM library may not necessarily be the same as the specified modes for the radar; almost always the ESM library will require more mode lines to describe the radar than the number of radar manufacturer specified modes.

An example of the ESM library modes specified for a typical Air Traffic Control (ATC) radar is given in Table 10.1. This radar has fixed RF and pulse width, and fixed and staggered PRI modes. There are 6 mode lines are needed to specify this radar.

The above example of ESM library specification for a single radar is an extremely simple one. In practice it is likely that many more modes are needed for the specification of a single radar type, as will be seen later in Section 10.7.2.

10.3 Library Matching Using Parameter Weighting

This method of library matching has the radar modes set up with weights attached to each parameter, as follows:

Table 10.1 Example of Mode Lines for a Typical ATC Radar

Mode	*RF min (MHz)*	*RF max (MHz)*	*RF Type*	*PRI min (μs)*	*PRI max (μs)*	*PRI Type*	*PW (μs)*
1	2,750	2,900	F	800	900	F	2.0
2	2,750	2,900	F	1100	1200	F	2.0
3	2,750	2,900	F	1400	1500	F	2.0
4	2,750	2,900	F	800	900	S	2.0
5	2,750	2,900	F	1,100	1,200	S	2.0
6	2,750	2,900	F	1,400	1,500	S	2.0

- Minimum weight for wide parameter ranges not unique to the particular radar;
- A weight somewhere between minimum and maximum for those parameters having a narrower range, but that cannot alone be used to uniquely identify a radar;
- Maximum weight for parameters that can uniquely identify a radar.

There are some radars operating in relatively sparse RF ranges so the RF weighting is therefore greater for these radars than for those where there are many possible radar types in a particular RF range. PRI ranges are not usually exclusive, especially as most radars can operate using more than one PRI range. There are some radars that have unique PWs. These PWs are usually long (>10 μs) and the pulses are chirped. A higher weighing on PW could therefore be specified in the mode line for these radars. However, care must be taken when using PW as a matching parameter as long, chirped pulses are particularly susceptible to multipath interference (see Chapter 14).

Table 10.2 shows three examples of the specification of radar modes with parameter weighting included in the mode lines. Data for an intercept is also shown in the table so that the library matching process can be demonstrated.

The library matching process assigns the identity with the highest total weight score to the ESM intercept or track. For example, an intercept has an RF of 3,080 MHz, "matching" two library entries but assigned different scores for each of them. The intercept has a PRI of 1,795 μs and matches all 3 library entries. The pulse width is 13 μs, which only matches one of the library entries (CCCC) having

Table 10.2 Examples of Radar Modes with Parameter Weighting

Radar Name	*AAAA*	*BBBB*	*CCCC*	*Intercept*
RF range (MHz)	3,010 to 3,120	2,910 to 2,950	2,900 to 3,100	3,080
RF weight (1 to 100)	35	45	65	
PRI range (μs)	200 to 3,000	1,500 to 1,800	1,780 to 1,820	1,795
PRI weight (1 to 100)	10	25	75	
PW range (μs)	0.04 to 1.25	0.8 to 2.0	14.0 to 15.0	13
PW weight (1 to 50)	10	20	50	
Scan range (seconds)	1 to 3.5	3.6 to 4.6	3.0 to 5.8	3.5
Scan weight (1 to 50)	10	20	40	
Total score (for intercept)	55	25	230	

a high weighting of 50 as it is unique to the types of radar expected to be in the RF environment of the trial. The total weights for each library entry for this intercept show that the CCCC identity is assigned to the intercept. The range of weights for RF and PRI (1 to 100) are greater than for PW and scan period (1 to 50) as RF and PRI are deemed to be more important in the identification process than PW and scan period.

10.4 Library Matching Using Parameter Scoring

In this simple method of library matching, each intercept is compared with every mode line in the ESM library and a score is calculated. Matching radar types are ranked in order and the one with the highest score is selected as the identity of the intercept.

Figure 10.1 shows the parameter scoring process. In this example the radar's measured RF and PW ranges lie wholly within the ESM library ranges for the particular radar, but the calculated PRI lies only partially within the specified range.

A score lower than the maximum 5 would be set for the PRI match of this radar to the ESM library mode line with which it is being compared.

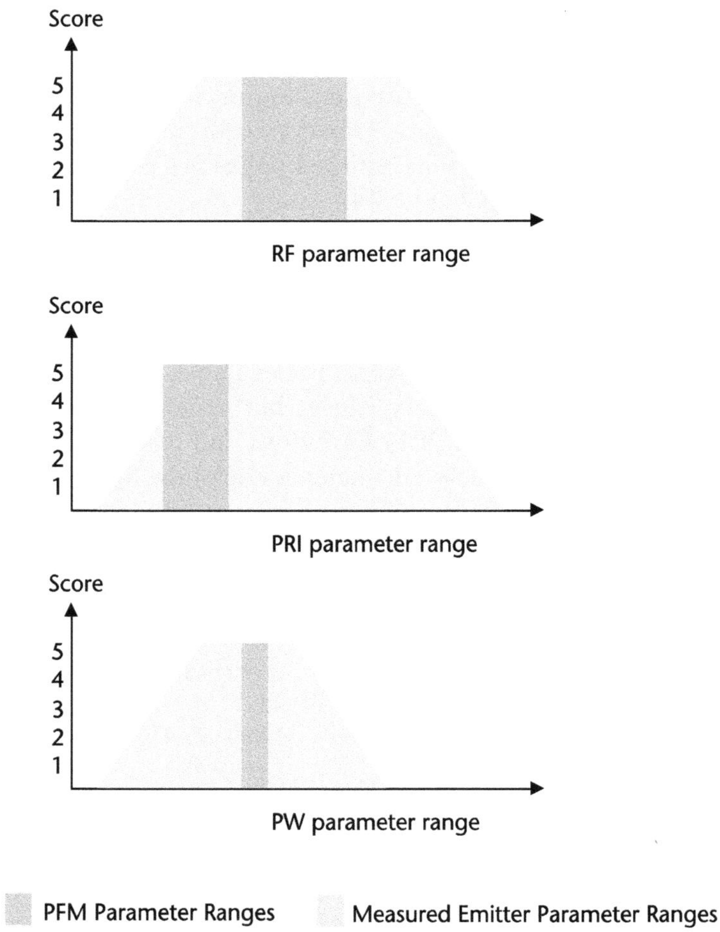

Figure 10.1 Scoring for library matching on 3 parameters.

10.5 Library Matching Using Parameter Tolerances

A more sophisticated library matching technique uses tolerances for the various parameters. In ESM systems where identity is used as an intercept to track correlation parameter the tolerances may be set individually for each radar type. Otherwise they may be given general values such as 5 MHz for RF and 50 μs and 1 μs for PRI and PW, respectively.

Starting with an intercept, the matching process generates a list of possible identities by taking only those that are within the parameter tolerances to decide that correlation to an existing track is possible. An example of the case where identity is used as a correlation parameter is shown in Table 10.3. The intercept has been identified as AAAA so attempts are made to correlate it into all tracks with the same AAAA identity; there are four tracks in this example.

The parameter tolerances for correlation are: 5 MHz for RF, 20 μs for PRI, 0.1 μs for PW, 5° for the DOA. The percentage difference of the tolerance for each intercept and track is shown next to the track parameters (e.g., track 2 has an RF difference of 2 MHz between the intercept and the track). This is 40% of the RF tolerance.

After calculating the percentage tolerance differences for the four tracks, track 2 is chosen as the one for intercept to track correlation as this track has the smallest percentage differences overall between the intercept and track parameters, according to the tolerances being used.

If the intercept had been given more than one possible identity (usually the case) then tracks with all the identities listed in the intercept are tested as potential correlation candidates. Similarly tracks often have several identities (perhaps up to five for this type of correlation) so all tracks that have the identity of the intercept as one of their possible identities take part in the correlation process.

A second example of the intercept to track correlation process using the same track data is given in Table 10.4. In this case identity is not used as a correlation parameter. Track 3 is not considered this time as its PRI falls outside the tolerance.

Track 2 is the preferred track for correlation. As identity is not used as a correlation parameter, there is more scope for tracks to have a larger list of alternative identities. As many as 20 possible identities for each track are sometimes available in the ESM systems. All this data can be presented to the experienced ESM operator to help make sense of what the ESM is seeing.

It should be noted that these examples are simplified versions of the intercept to track correlation process as other factors, particularly PRI type and RF type and possibly parameter dispersions, are also taken into account in most current ESM systems.

Table 10.3 Example of Intercept-to-Track Correlation Using Identity for Correlation

	Identity	*RF (MHz)*	*PRI (μs)*	*PW (μs)*	*DOA (°)*	*Track Order*
Intercept	AAAA	2,940	2,000	1.0 (0%)	30	
Track 1	AAAA	2,942 (40%)	2,000 (0%)	1.0 (0%)	31 (20%)	2
Track 2	AAAA	2,940 (0%)	2,005 (25%)	1.0 (0%)	30 (0%)	1
Track 3	AAAA	2,944 (80%)	2,050 (> 100%)	1.0 (0%)	30 (0%)	4
Track 4	AAAA	2,942 (40%)	1,995 (50%)	1.0 (0%)	29 (20%)	3

Table 10.4 Example of Intercept to Track Correlation without Using Identity for Correlation

	Identity	*RF (MHz)*	*PRI (μs)*	*PW (μs)*	*DOA (°)*	*Track Order*
Intercept		2,940	2,000	1.0 (0%)	30	
Track 1	AAAA	2,942 (40%)	2,000 (0%)	1.0 (0%)	31 (20%)	2
Track 2	AAAA	2,940 (0%)	2,005 (25%)	1.0 (0%)	30 (0%)	1
Track 3	BBBB	2,944 (80%)	2,050 (>100%)	1.0 (0%)	30 (0%)	Not considered
Track 4	CCCC	2,942 (40%)	1,995 (50%)	1.0 (0%)	29 (20%)	3

10.6 User Interfaces for Reprogramming ESM Libraries

Since library matching is such a difficult task it is necessary to update the ESM library as often as possible. Not only do the storage constraints on ESM libraries mean that there is a limit to the number of mode lines that they can contain, but different missions mean that different versions of the ESM library containing different radar types must be produced. The process of updating the ESM library is often called reprogramming. The need for reprogramming means that a suitable user interface should be available for updating the ESM library.

Figure 10.2 shows an example of a user interface for editing a mode line in a simple ESM library where parameter weighting is used in the correlation process.

The user interface shown in Figure 10.2 contains just maximum and minimum values for each mode and a weighting for each parameter. A more sophisticated user interface for editing an ESM library entry for a single radar type is shown in Figure 10.3.

The user interface shown in Figure 10.3 allows for parameter dispersions and extra parameters such as scan period and polarization to be specified for the radar. RF type and PRI type are also available and flags can be set to allow various parameters to be used in the correlation process (or not). For example, an intercept

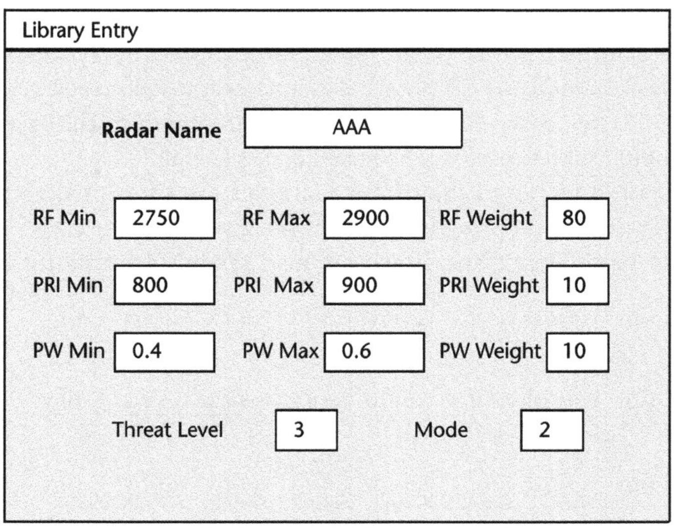

Figure 10.2 Simple user interface for reprogramming an ESM library.

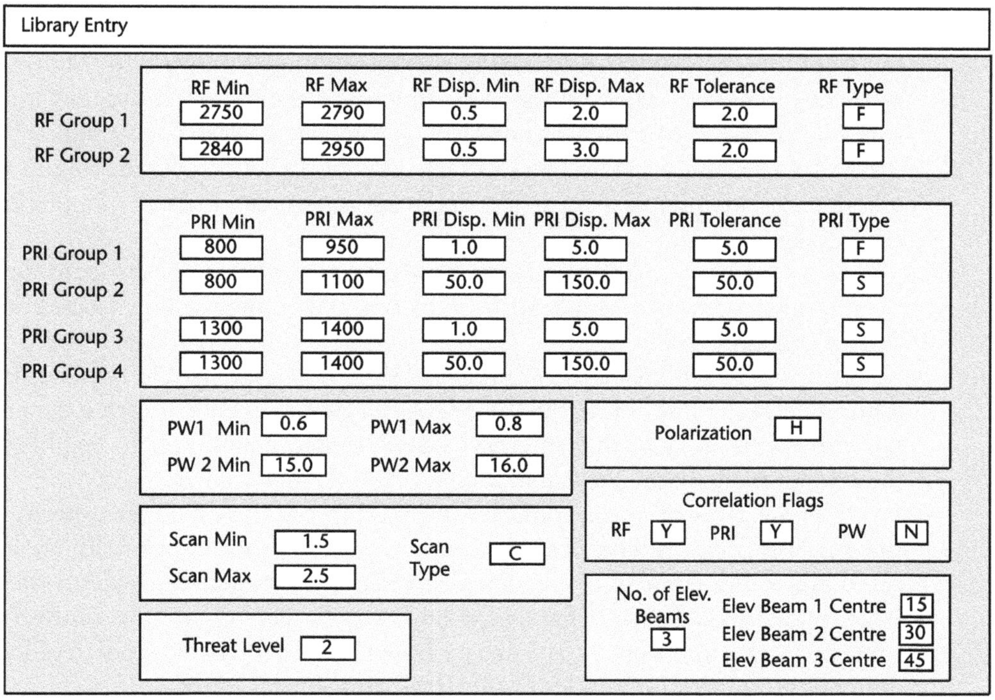

Figure 10.3 Complex user interface for reprogramming an ESM library.

with a complex PRI may be allowed to correlate into a track that has other parameters matching the intercept if PRI is not used as correlation parameter. However, the PRI correlation flag does not control the intercept to track correlation process for intercepts that have a PRI other than complex.

10.7 Methods for Optimizing Library Matching

Some techniques for improving radar identification are discussed here but it should be noted that future radars such as active electronically scanned array (AESA) and multiple-input multiple-output (MIMO) radars will require new techniques for radar fingerprinting.

Traditional methods for library matching are limited in their effectiveness so some ideas for optimising radar identification include:

- Using generic identifications for radar types that are not particularly of interest (e.g., ship navigation radars or weather radars);
- Analyzing the way in which individual ESM systems "see" different radar types and constructing library entries accordingly;
- Using a layered approach to dealing with radars of similar types and about which more detail is known (a priori environment knowledge);
- Using extra parameters such as parameter dispersions, tolerances, and clock period estimation.

10.7.1 Use Generic Library Entries for Some Types of Radar

In Chapter 9, the issue of overlapping parameter ranges is discussed. Many different types of radar that carry out the same function have parameter ranges that are so similar that the ESM system cannot reasonably be expected to distinguish between them. To save space in the ESM library, and to prevent multitracking in the case where radar identity is used as a correlation parameter, the use of generic library entries is proposed.

As an example of how this might work, a start is made by considering the overlapping parameter ranges for just 10 types of common ships I-band radars, as shown in Figure 10.4.

To specify each PRI and PW range for these 10 radars would require more than a hundred mode lines in the ESM library. In any case, there is such extensive overlap in these mode lines that the ESM system would be unlikely to be able to select the correct identity for the radar.

A single class of radar could be specified in the library to encompass most of the possible ships radars. This prevents the ESM from continually changing its mind about the identity of a particular radar and also saves space in the library. A generic ShipNav radar type could be set up to encompass the overlapping parameter ranges and Figure 10.5 shows how this new PRI/PW mode would include many of the individual modes of the 10 I-band ships' radars.

In the example shown in Figure10.4, there are some unique modes for the radars and these could still be specified separately in the ESM library. In reality, there are not just 10 types of shipborne I-band radars (there are hundreds), so combining the modes of these similar radars removes the many ambiguous track identities that are so common when each radar type is specified separately in the ESM library.

A similar generic radar type could be used for other nonthreat radar types, such as ATC radars, typically operating at 2 to 3 GHz, and weather radars, at about 5 GHz.

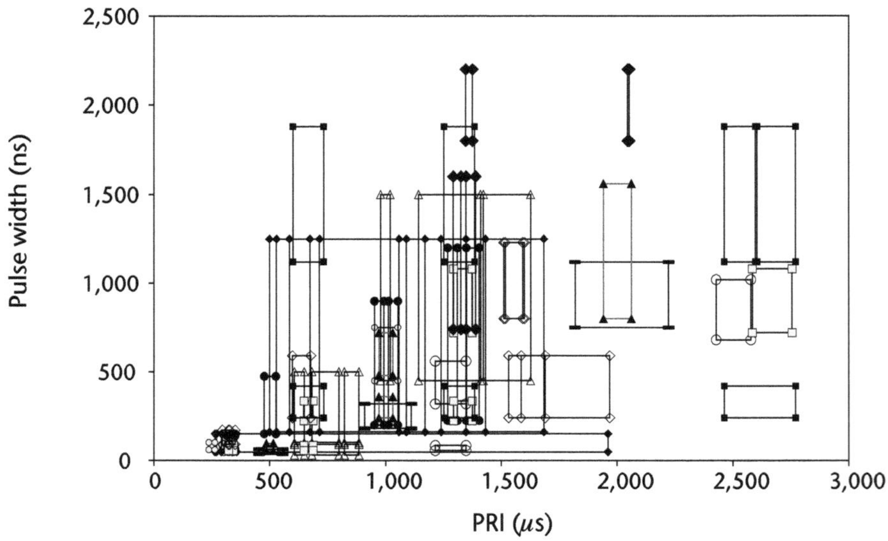

Figure 10.4 PRI and PW ranges for 10 I-band radars.

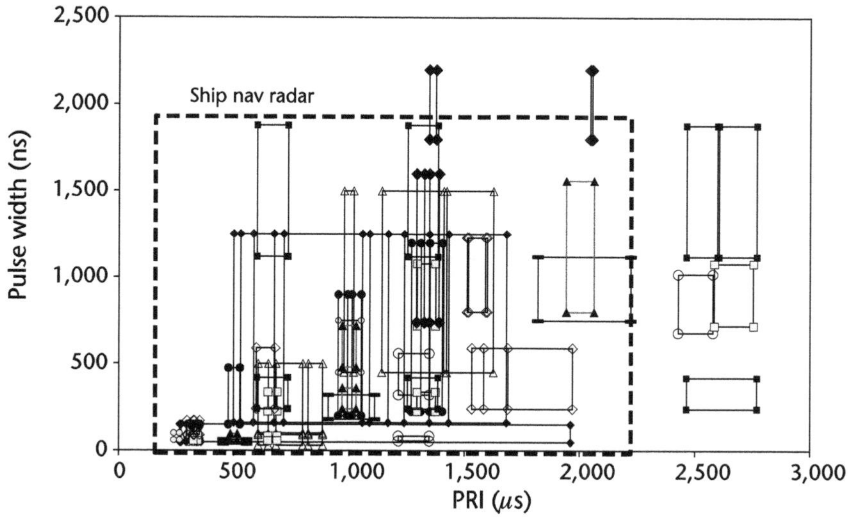

Figure 10.5 PRI and PW ranges for a generic ShipNav radar.

10.7.2 Library Matching Using the Way That ESM Systems See Radars

Traditionally, ESM libraries have been programmed with information provided in radar manufacturer specifications. Every mode line, even those rarely used, has been put into the ESM library. A better way of specifying a radar is to analyze the intercepts recorded from operating the ESM equipment against known radars to determine how the ESM sees radars. For example, a radar with a complex PRI sequence will rarely have enough pulses detected by the ESM to allow a proper determination of the PRI. In addition, it may be that the pulse group repetition interval (PGRI) is easier to find than the individual elements. Because pulses are often missed from the pulse sequence of most radars, even those with simple PRI sequences, PRI harmonics should be incorporated into the ESM library.

Consider an example of a radar Type X having the following parameter ranges: PW = 0.3 to 0.6 μs, and PRI = 700 to 1,100 μs.

Figure 10.6 shows how the ESM sees radar Type X in 400 ESM tracks created for the radar. This example is a simulation of what might be recorded during a mission.

Although the manufacturer specification of the PW range extends from 300 to 600 ns, when the radar is specified in the ESM library, the PW ranges (the left entries on the plot) should be extended to include the PW truncation seen in the actual data. An extra PRI range should be included in the library to take account of the tracks seen at range 2,200 to 2,800 μs (PRI harmonics) from this type of radar.

Another example of a radar Type Y has the following parameter ranges: PW = 0.1 to 0.2 μs, PRI = 500 to 700 μs. Figure 10.7 shows how the ESM might see this radar.

In this case, the PW range in the library is probably appropriate, but the PRI range should be increased significantly because it appears that the ESM has difficulty in determining the PRI for radar Type Y.

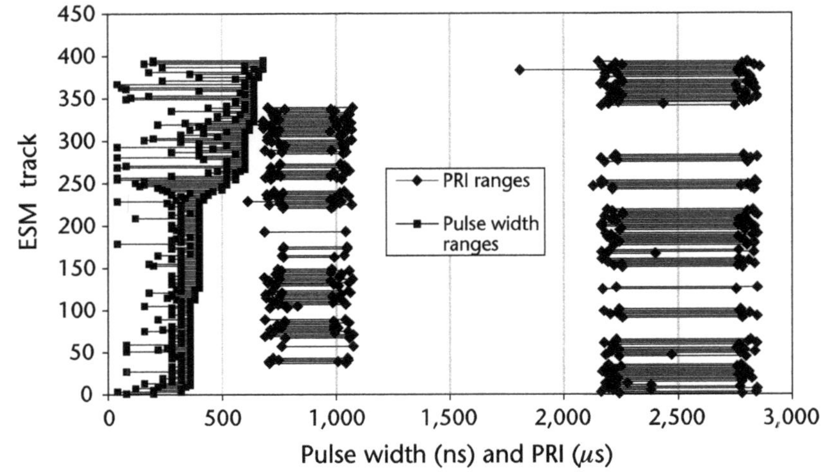

Figure 10.6 Simulation of parameters of ESM tracks: example 1.

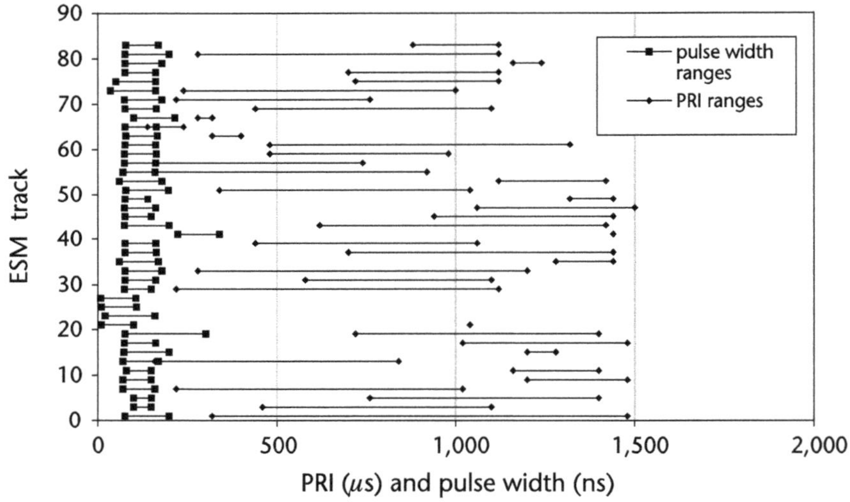

Figure 10.7 Simulation of parameters of ESM tracks: example 2.

Intercepts with PRIs that have been calculated as a harmonic of the true PRI are a major source of extra tracks. PRI harmonics occur when there are pulses missing from the pulse stream received from a radar (see Section 14.5). Mode lines should therefore be put into the ESM library to take account of harmonics. It is sometimes difficult for the ESM to determine individual PRI elements for radars that have multi-element PRI sequences, so a mode line should also be put in the library containing the PGRI. An example of a simple description of a radar type appeared in Table 10.1. The way in which this radar type is specified should be augmented to take account of harmonics and the PGRI, so that the ESM library entry for this radar type appears as in Table 10.5. There were 6 mode lines in the simple description of the radar but when the harmonic and PGRI modes are added 15 mode lines are needed to specify the radar type.

Table 10.5 Library Mode Lines Including Harmonic and PGRI Modes for a Typical ATC Radar

Mode	*Min RF (MHz)*	*Max RF (MHz)*	*RF Type*	*PRI Min (μs)*	*PRI Max (μs)*	*PRI Type*	*PW (μs)*	*Notes*
1	2,750	2,900	Fixed	800	900	Fixed	2.0	
2	2,750	2,900	Fixed	1,100	1,200	Fixed	2.0	
3	2,750	2,900	Fixed	1,400	1,500	Fixed	2.0	
4	2,750	2,900	Fixed	800	900	Stagger	2.0	
5	2,750	2,900	Fixed	1,100	1,200	Stagger	2.0	
6	2,750	2,900	Fixed	1,400	1,500	Stagger	2.0	
7	2,750	2,900	Fixed	1,600	1,800	Fixed	2.0	Harmonic
8	2,750	2,900	Fixed	2,200	2,400	Fixed	2.0	Harmonic
9	2,750	2,900	Fixed	2,800	3,000	Fixed	2.0	Harmonic
10	2,750	2,900	Fixed	1,600	1,800	Stagger	2.0	Harmonic
11	2,750	2,900	Fixed	2,200	2,400	Stagger	2.0	Harmonic
12	2,750	2,900	Fixed	2,800	3,000	Stagger	2.0	Harmonic
13	2,750	2,900	Fixed	8,000	9,000	Stagger	2.0	PGRI
14	2,750	2,900	Fixed	11,000	12,000	Stagger	2.0	PGRI
15	2,750	2,900	Fixed	14,000	15,000	Stagger	2.0	PGRI

10.7.3 A Layered Approach to Identification

Assigning priorities to radars in the ESM library so that the most common type of radar has the lowest priority means that other less common radars with overlapping parameter ranges but whose parameters are more specific are given higher priority. Their identities are chosen in preference to the more common radar type.

For example, consider that there are 30 radars of type X in an RF environment and 2 radars of Type Y. Both radar types have the same RF range, PW and fixed PRI in the range 1.5 to 2.3 ms. The two Type Y radars have specific known PRIs of 1.85 ms and 2.15 ms, respectively. When intercepts are created at these two PRIs, they can be identified ambiguously as Type X and as Type Y; however, the Type Y identification takes preference in this case. All other intercepts with the same RF and PW, but with PRIs other than 1.85 ms and 2.15 ms are able to be uniquely identified as Type X.

10.7.4 Use of Measured Parameter Dispersions

The dispersions of parameters such as RF and PRI calculated for pulses that contribute to an intercept can be very different for different radar types or can appear to be so because of ESM processing limitations. Some examples of typical RF and PRI dispersions for intercepts are shown in Table 10.6. Some intercepts have very small dispersions; for example, radar 3 has an RF dispersion of 0.2 MHz and radar 2 has a PRI dispersion of only 0.004 μs. Use can be made of these precise dispersions in the intercept to track correlation and identification processes.

PRI dispersions of tens or hundreds of microseconds are usually due to PRI calculation errors or PRI classification errors. Jittered PRIs may not have been correctly classified or staggered PRIs may not have had their elements correctly extracted in the deinterleaving process. However, even these larger dispersions can

Table 10.6 Examples of Typical RF and PRI Dispersions

Intercept	*RF (MHz)*	*RF Dispersion (MHz)*	*PRI (μs)*	*PRI Dispersion (μs)*
1	2,893.6	2.3	1,870	120.1
2	2,897.3	2.9	1,880	0.004
3	2,894.5	0.2	1,890	40.5
4	2,893.1	2.4	1,920	119.5
5	2,899.2	2.9	1,880	0.004
6	2,893.2	0.2	1,875	39.2

be made use of for correlation and identification as the ESM is likely to always see intercepts from these radars in the same way.

To program the ESM library with realistic parameter dispersions some data must be recorded for the radar types of interest and the dispersions must be calculated in the first instance using this data. Subsequently, the values of parameter dispersions that are put into the library can be improved as more data becomes available for analysis following trials.

Two examples of the calculation of PRI dispersion using intercept data where the intercepts are known to be from the same radar are given in Table 10.7. In the first example, there are wide tolerances as this radar has a staggered PRI. The second example shows very small dispersions as this radar has a fixed, very stable PRI.

Radar 1 has a dispersion of 28.231 μs, whereas radar 2 has a much smaller dispersion of only 0.001 μs. The parameter ranges of these two radar types are similar so using dispersion as a matching parameter is very useful to discriminate between them.

Dispersion can be estimated by taking the standard deviation (σ) of as many intercepts as possible that are known to come from a specific radar using the usual formula:

$$\sigma = \sqrt{\left[\sum\left(x_{(1-n)} - m\right)^2 / (n-1)\right]} \tag{10.1}$$

where $x_{(1-n)}$ are the intercept PRIs, m is the mean PRI and n is the number of intercepts.

Table 10.7 Examples of PRI Dispersions for Intercepts from Two Separate Radars

Radar 1		*Radar 2*	
Intercept	*PRI (μs)*	*Intercept*	*PRI (μs)*
1	1,400.123	1	1,400.001
2	1,426.243	2	1,400.000
3	1,433.345	3	1,400.003
4	1,475.03	4	1,400.000
5	1,410.567	5	1,400.001
6	1,457.334	6	1,400.001

Intercept to track correlation can then be carried out using the parameter range ± dispersion ± tolerance as the parameter limits (e.g., an intercept has a PRI of 1,400 μs with a dispersion of 20 μs and a PRI tolerance of 50 μs). This intercept would match tracks with PRI from 1,330 to 1,470 μs, as long as other parameters are also in the correct ranges.

10.7.5 Use Sensible Tolerances for ESM Correlation and Library Matching

Using generic values for tolerances (especially for PRI matching) leads to tolerances that are too wide for some radar types. While not strictly an identification issue, the only place that tolerances for individual radar types can be set is within the radar library, hence the description of this issue here. For ESM systems using identification as a correlation parameter, changing the tolerance according to the radar type can help with the intercept to track correlation process.

An example of the use of PRI tolerance in intercept-to-track correlation begins by considering intercepts from three separate radars shown on Figure 10.8.

The three radars on Figure 10.8 are easily separated by eye, but because their PRI values, of about 732 μs, are within 1 μs of each other, it is not so easy for the ESM to separate them.

If the PRI tolerance set in the ESM library is ±1 μs, the ESM tracks can jump between the radars, as shown in Figure 10.9, assuming that their RFs are also very similar.

When the PRI tolerance is set to be ±0.2 μs, the intercepts from the individual radars contribute to separate ESM tracks, as shown in Figure 10.10.

10.7.6 Use A Priori Information

The limit on the number of mode lines available in the ESM library means that the ESM must be reprogrammed before a mission is undertaken in a new geographical area. Knowledge of local radar types and their specific parameters (if known)

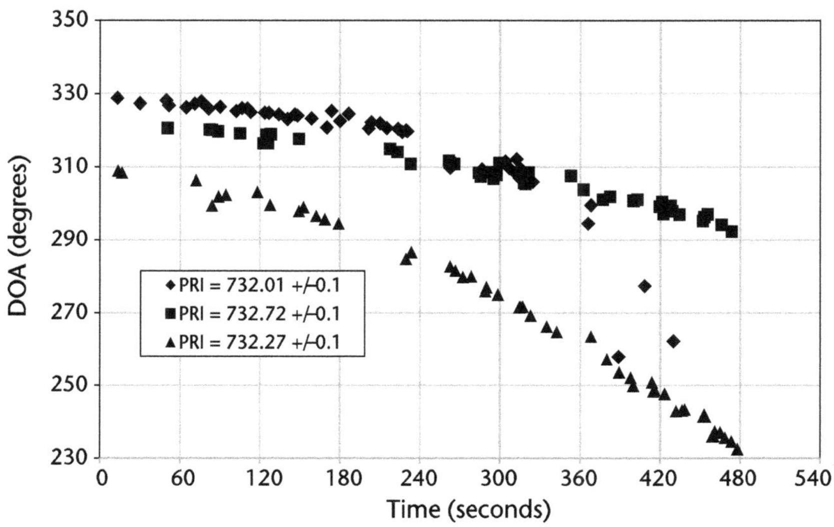

Figure 10.8 Intercepts for three radars with similar PRIs.

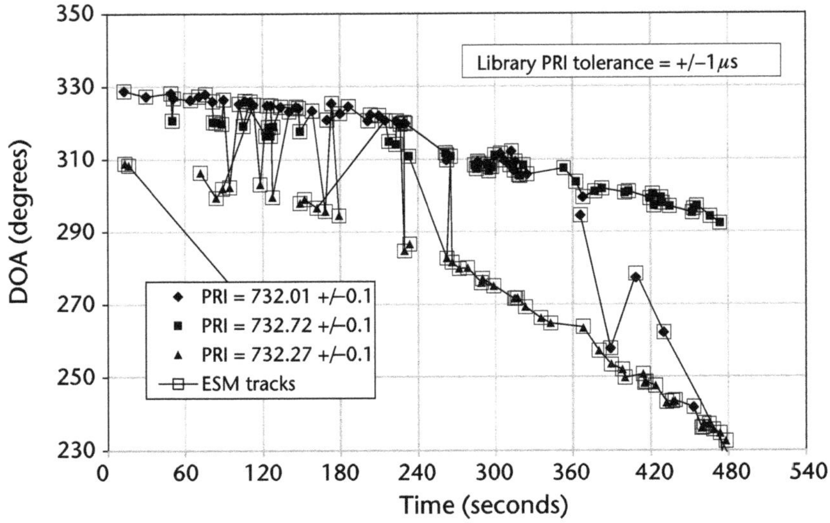

Figure 10.9 Intercept-to-track correlation for PRI tolerance of 1 μs.

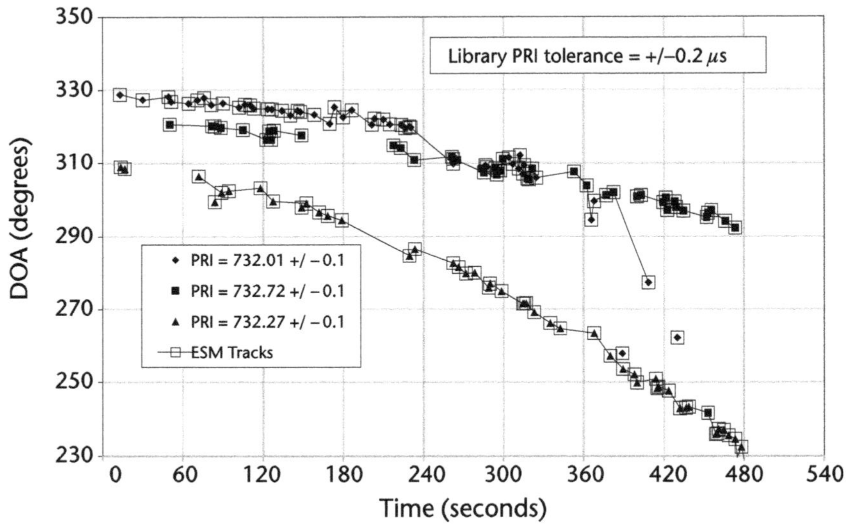

Figure 10.10 Intercept to track correlation for PRI tolerance of 0.2 μs.

should be included in the library. The ESM library could also contain data on radar locations. Some ESM systems already use location data to confirm radar identities. They also use this information to reduce multitracking by suppressing tracks that appear at the wrong DOA or fixes at the wrong location for radars that have well-known parameters.

10.7.7 Use of Specific Emitter Identification Data

One method of specific emitter identification (SEI) is to determine the clock period of the radar. There are several methods available to determine the radar clock period; but, these techniques are classified so they cannot be described in detail in

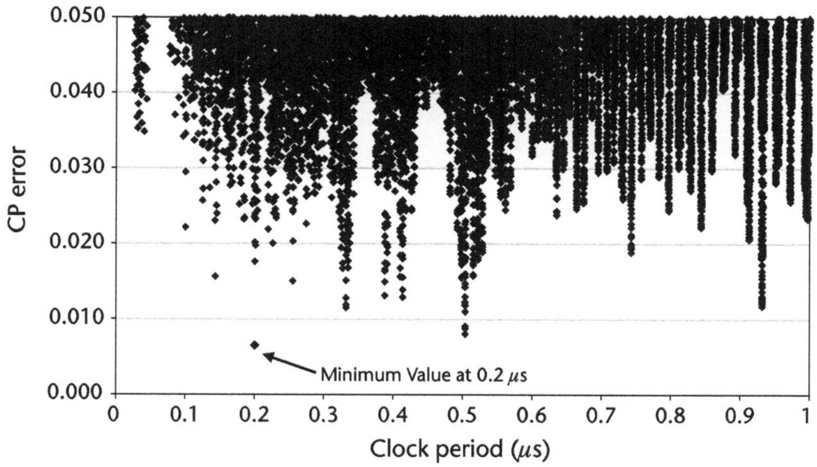

Figure 10.11 Estimation of radar clock period.

this book. However, an indication of the use of estimating the radar clock period to allow SEI is given in Figure 10.11. This method arrives at the minimum value of clock period error, which is repeatable for the specific radar for all sets of pulses recorded from this radar.

CHAPTER 11

Radar Location Estimation

Many types of ESM systems attempt to locate the radars from which they are receiving signals: pulses or CW. For a single ESM platform, the only source of information on the pulsed radar location comes from the DOA measurements of the pulses that combine to form an intercept. For CW signals, the measured DOA at the time of LOB creation is available. The intercept or CW DOA measurements supply LOBs from the ESM platform to the radar and it is these LOBs, as shown in the example in Figure 11.1, that are used in the estimation of radar location.

11.1 Location Calculation

The most usual method for radar location calculation uses several LOBs recorded as the ESM platform moves with respect to the radar. Figure 11.2 shows the geometry of a simple example where there are three LOBs forming two triangles with the radar location being at the point of intersection of the triangles.

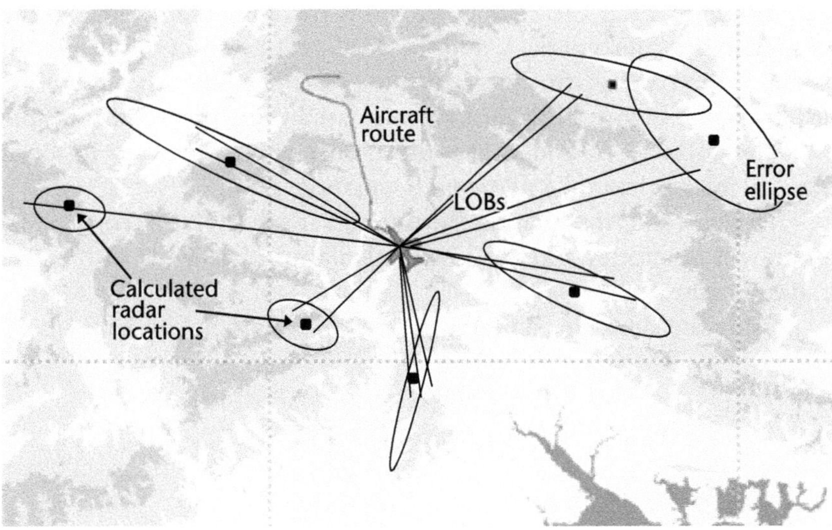

Figure 11.1 Example of LOBs from the ESM platform to radars.

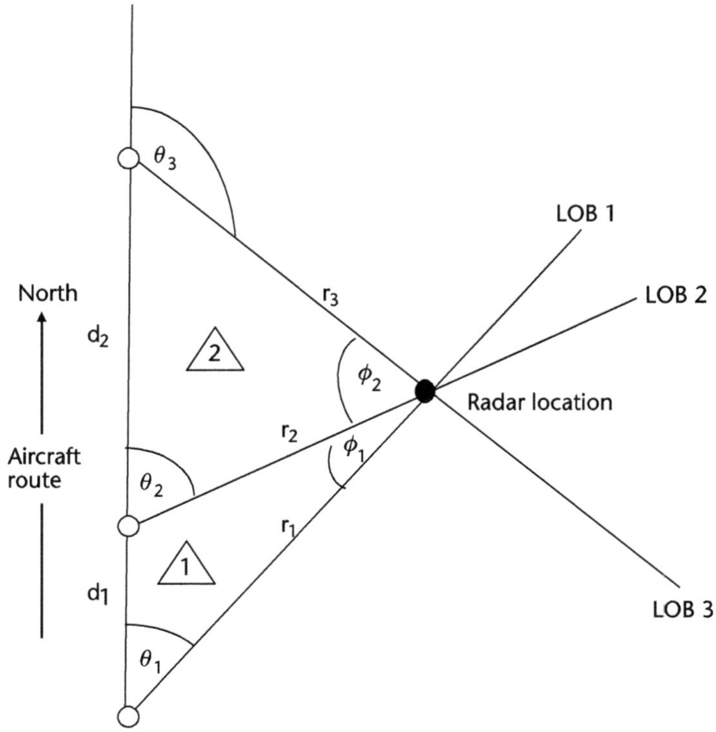

Figure 11.2 Geometry of radar location calculations.

In Figure 11.2, the ESM platform is traveling due North so the DOA, as recorded in the intercept data, and the AOA (with respect to the aircraft nose) are the same. If the ESM platform is travelling on a heading other than North the AOAs (θ_1, θ_2, θ_3) in Figure 11.2 must be calculated using the aircraft heading.

The distances d_1 and d_2 that the aircraft travels between LOBs are known values, calculated from the platform navigation data.

The ranges to the radar at the time that each LOB is recorded (r_1, r_2, r_3) are the unknown quantities and it is their calculation that leads to the estimation of the radar location.

Using the sine rule ($\sin(A)/a = \sin(B)/b = \sin(C)/c$) on triangle 1:

$$\sin(\theta_1)/r_2 = \sin(180-\theta_2)/r_1 = \sin(\phi_1)/d_1 \tag{11.1}$$

where

$$\phi_1 = 180-\theta_1-(180-\theta_2) = \theta_2-\theta_1 \tag{11.2}$$

so

$$\sin(\theta_1)/r_2 = \sin(180-\theta_2)/r_1 = \sin(\theta_2-\theta_1)/d_1 \tag{11.3}$$

$$r_1 = d_1\sin(180-\theta_2)/\sin(\theta_2-\theta_1) \tag{11.4}$$

and

$$r_2 = d_1 \sin(\theta_1) / \sin(\theta_2 - \theta_1) \tag{11.5}$$

Similarly using triangle 2, r_3 can be found:

$$\sin(\theta_2) / r_3 = \sin(180 - \theta_3) / r_2 = \sin(\phi_2) / d_2 \tag{11.6}$$

$$r_3 = d_2 \sin(\theta_2) / \sin(\phi_2) \tag{11.7}$$

Once the range to the radar has been established, its location can be found using the DOA and the ESM platform location at the time that the DOA was measured.

The coordinates x_{radar_rel} and y_{radar_rel} are calculated relative to the ESM platform position at the time of calculation and may be expressed in meters or kilometers

$$x_{radar\_rel} = r_1 \cos(90 - DOA_1) = r_1 \sin(DOA_1) \tag{11.8}$$

$$y_{radar\_rel} = r_1 \,\mathrm{c}\sin(90 - DOA_1) = r_1 \cos(DOA_1) \tag{11.9}$$

To express the radar location in latitude and longitude the following equations are used:

$$Lat_{radar} = Lat_{ESM} + y_{radar\_rel} / degs_{\_lat} \tag{11.10}$$

$$Lon_{radar} = Lon_{ESM} + x_{radar\_rel} / degs_{\_lon} \tag{11.11}$$

where $degs_{\_lat}$ and $degs_{\_lon}$ are the number of degrees in latitude and longitude that are represented by x_{radar_rel} and y_{radar_rel} in kilometers.

$$degs_{\_lat} = 111120\mathrm{m} \; or \; 111.2\,\mathrm{km} \tag{11.12}$$

$$degs_{\_lon} = 111120 / \cos(lat)\mathrm{m} \; or \; 111.2 / \cos(lat)\,\mathrm{km} \tag{11.13}$$

All lines of longitude are great circles but lines of latitude (except for the equator) are not. Equation (11.13) shows the correction that must be made to the number of degrees of longitude represented by a kilometer at the particular latitude. (Note: 360° = 3,600 nm, 1° = 60 nm at the equator.)

In the simple example shown above, the ESM platform is traveling due North so the values θ_1, θ_2, θ_3 for each LOB are the DOA values. When the platform maneuvers during the gathering of the LOB data, a geometry such as that in Figure 11.3 is seen.

The sine rule can still be used in the calculation of the ranges but the initial calculation of the angles is more complicated.

From triangle 1,

$$\sin(\theta_1)/r_2 = \sin(a)/r_1 = \sin(\phi_1)/d_1 \tag{11.14}$$

and from triangle 2,

$$\sin(\theta_2)/r_3 = \sin(b)/r_2 = \sin(\phi_2)/d_2 \tag{11.15}$$

θ_1 and θ_2, the AOAs, at LOB1 and LOB2 are found from the aircraft heading and the DOAs at the LOB times. The following algorithm is used to calculate AOA:

If ($DOA > heading$) Then
    $AOA = DOA - heading$
Else
    $AOA = 360 - heading + DOA$
End If

The geometry needed to find angles a and b is shown in Figure 11.4. h_1, h_2, and h_3 are the aircraft headings at the time of recording LOB1, LOB2, and LOB3.

Angles a and b (in degrees) are found using angle c as follows

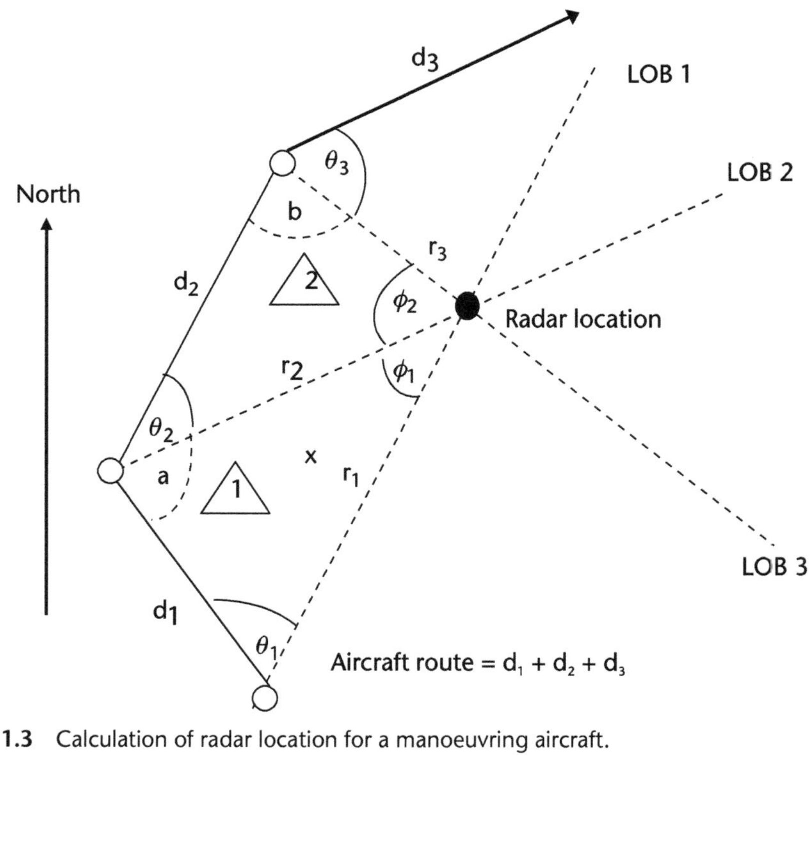

Figure 11.3 Calculation of radar location for a manoeuvring aircraft.

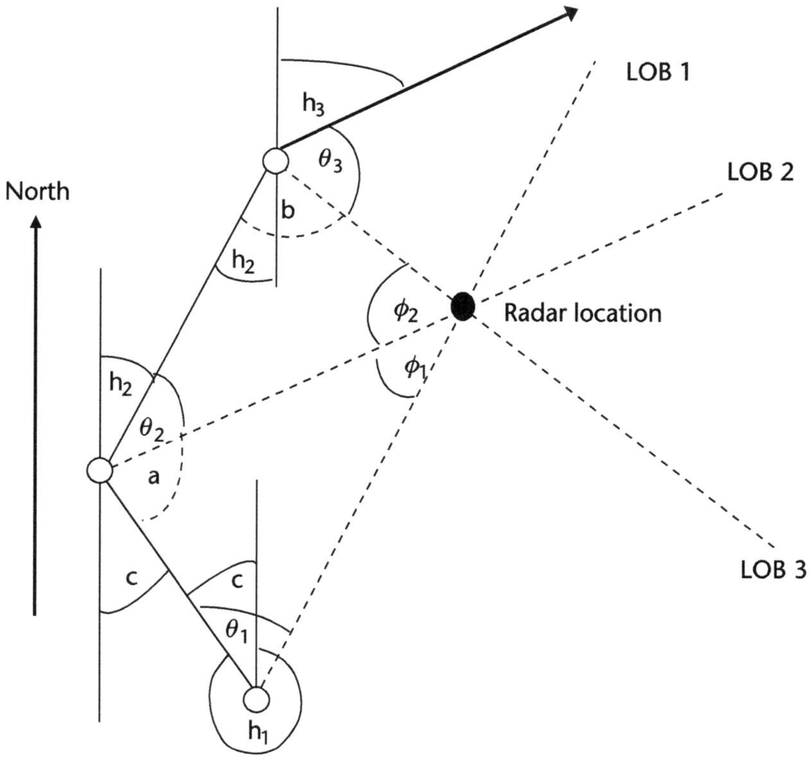

Figure 11.4 Angles needed in the calculation of radar location.

$$c = 360 - h_1 \tag{11.16}$$

$$a = 180 - h_2 - c - \theta_2 \tag{11.17}$$

$$b = 180 - h_3 - \theta_3 + h_2 \tag{11.18}$$

From (11.14),

$$r_1 = d_1 \sin(a) / \sin(\phi_1) \tag{11.19}$$

where

$$\phi_1 = 180 - a - \theta_1 \tag{11.20}$$

x_{radar_rel}, y_{radar_rel}, lat_{radar}, and lon_{radar} can be calculated using (11.8) to (11.11).

In a perfect world, the location of the radar can be found using just two LOBs, but in reality the expected errors in DOA mean that several LOBs are needed for the location calculation and there must be a reasonable inter-LOB-angle (ILA) and radial distance (RD) traveled by the ESM platform for the calculation to be started. Figure 11.5 defines the ILA and RD.

In existing ESM systems, there are typically at least six LOBs and ILA of 12° needed before a location estimation can be carried out. Sometimes a RD that must be flown by the ESM aircraft past the radar is also specified. Typically, this distance is a minimum of 10 km. If a location for a particular threat is needed, the aircraft should fly on a course to optimize the intercepts from that radar (i.e., flying past the radar with the radar intercepts arriving on the beam of the aircraft). If the aircraft flies directly towards or directly away from the radar, it is not possible for a location estimate to be made as all the LOBs have the same DOA.

The geometry of a perfect radar location estimation is shown in Figure 11.6. There are six LOBs shown here and the radar is located at the intersection of all six LOBs.

In reality, there are always errors in DOA, so the situation looks like the geometry shown in Figure 11.7. There are nine points at which the LOBs intersect, so nine estimates of radar location are made, all of which are in error.

The estimate of radar location in Figure 11.7 is taken as the average of the individual locations and an error ellipse is calculated to show the confidence in the location.

11.2 Location Error Ellipse

The location estimation algorithm generates several possible values of radar location. An error ellipse must therefore be constructed to encompass all the possible

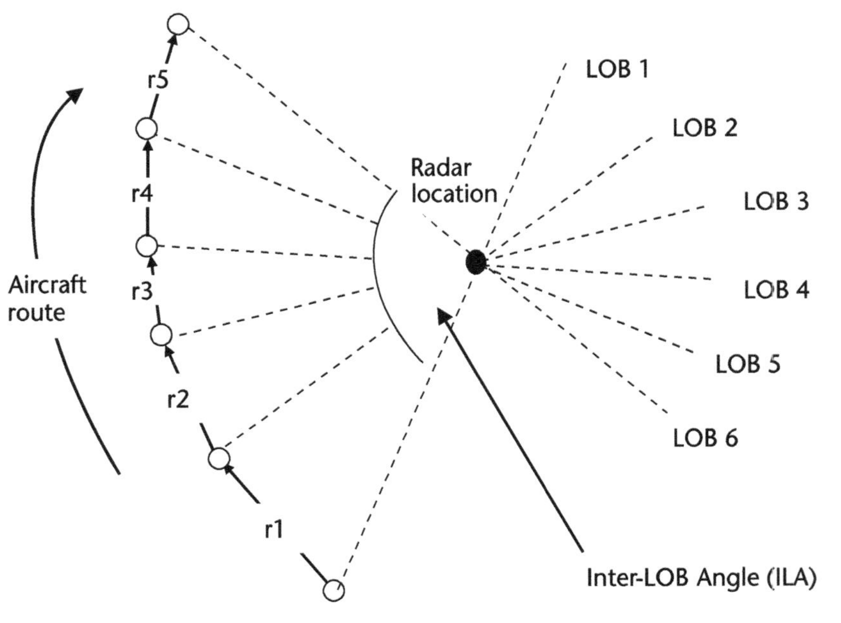

Figure 11.5 ILA and RD definition.

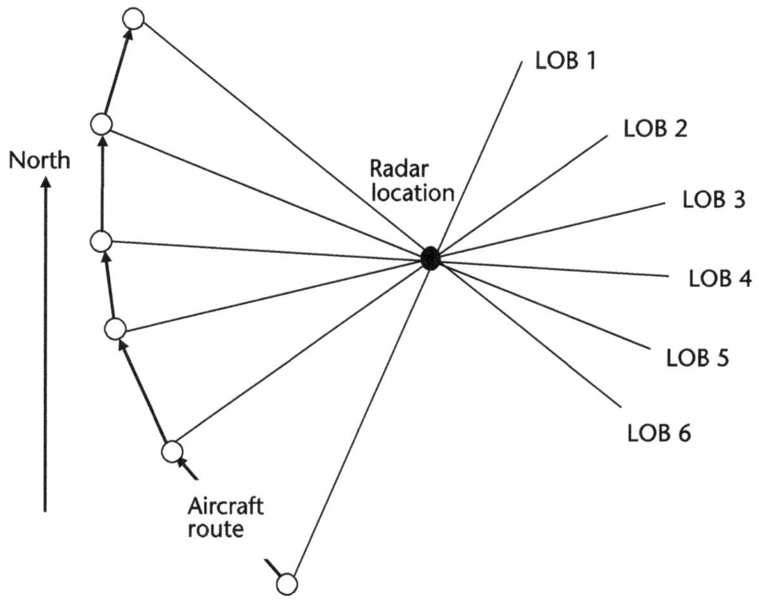

Figure 11.6 Example of a perfect radar location estimation.

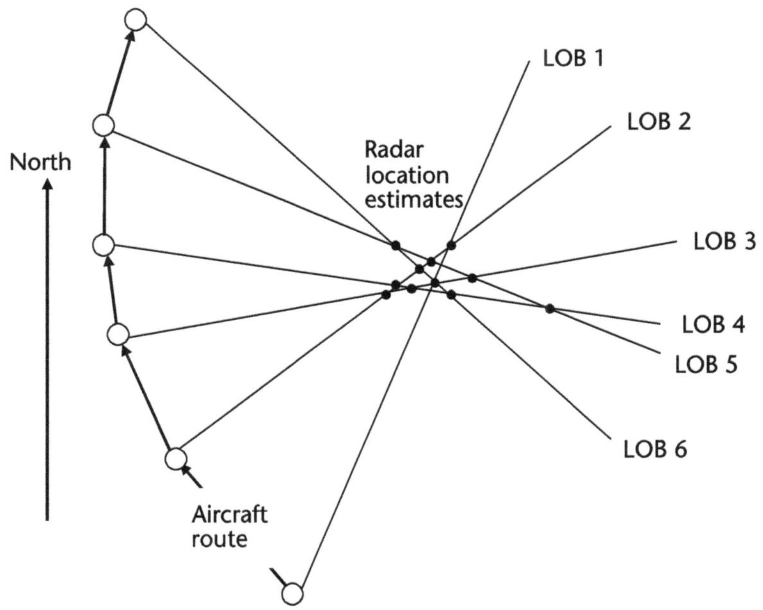

Figure 11.7 Realistic radar location estimates.

solutions. Figure 11.8 shows how the radar locations in the example on Figure 11.7 would be encompassed by such an error ellipse.

The parameters needed to draw the ellipse are the major and minor axes, the ellipse orientation and the confidence level required for the location data. The confidence level is used in the calculation of the magnitude of the major and minor axes. Figure 11.9 shows the geometry of the ellipse.

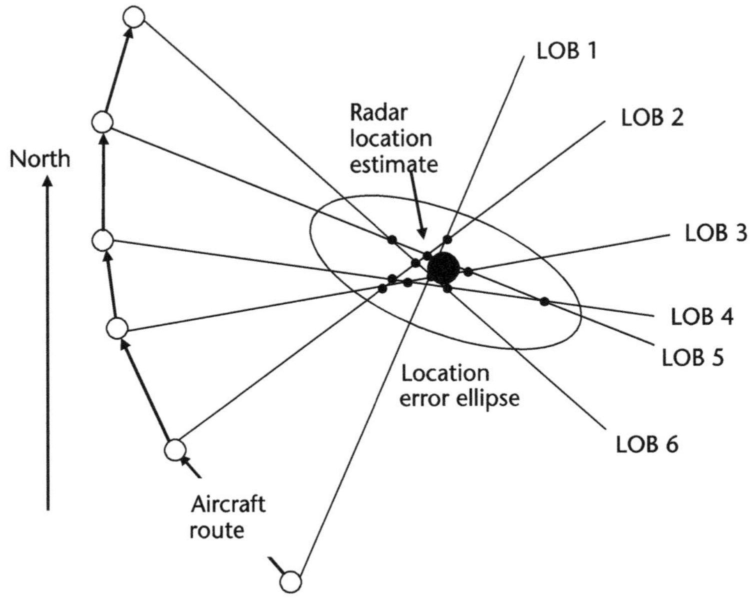

Figure 11.8 Radar location estimations with error ellipse.

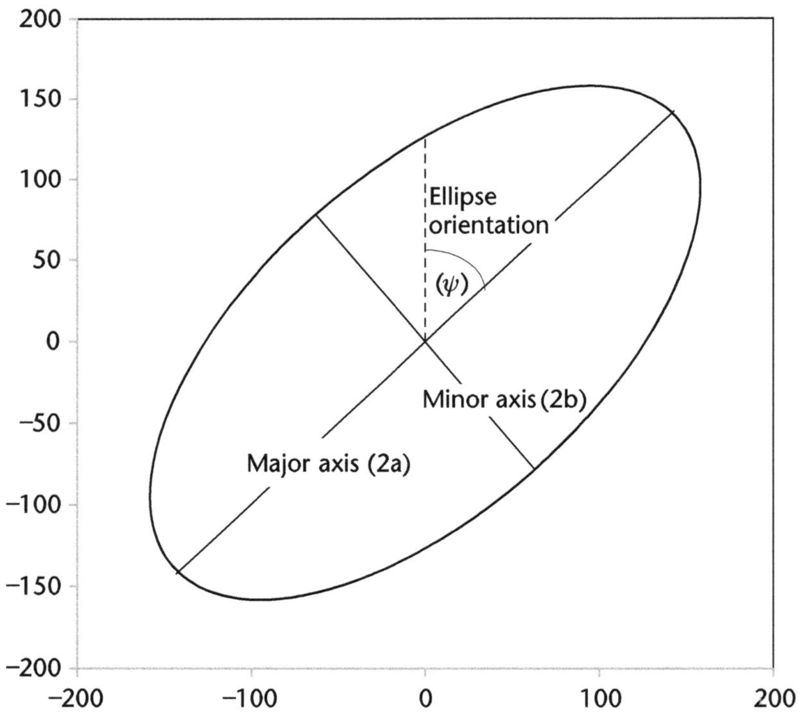

Figure 11.9 Geometry of error ellipse.

As an example of an error ellipse, Figure 11.10 shows a 99% confidence ellipse for a set of two-dimensional (2-D) normally distributed data samples. This confidence ellipse defines the region that contains 99% of all samples drawn from the underlying Gaussian distribution of the data points. The latitudes and longitudes of

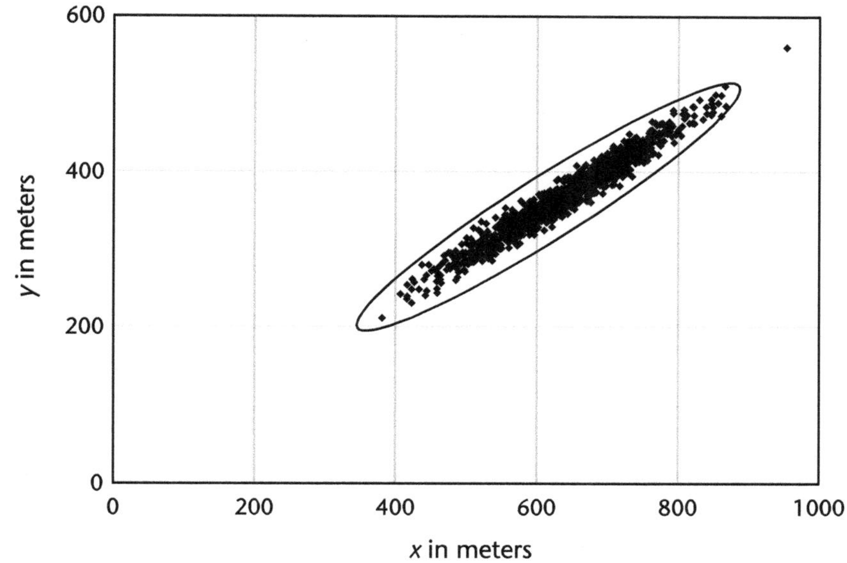

Figure 11.10 Error ellipse for normally distributed data.

the radar location estimations may not be normally distributed as biases may have been introduced into their calculation due to DOA errors and DOA resolution. However, the method described here is a good approximation of that used in ESM systems to calculate a location error ellipse.

The equation of an ellipse with a major axis of length $2a$ and a minor axis of length $2b$, centered at the origin, is defined as:

$$(x/a)^2 + (y/b)^2 = 1 \quad (11.21)$$

The axes a and b are defined by the standard deviations of the data, σ_x for the x data points and σ_y for the y data points.

The standard deviation of the set of n data points $(x_1, x_2, x_3, \ldots, x_n)$ is given by:

$$\sigma_x = \sqrt{\left[\sum \left(x_i - x_{mean}\right)^2 / (n-1)\right]} \quad (11.22)$$

Equation (11.21) now becomes

$$\left(x/\sigma_x\right)^2 + \left(y/\sigma_y\right)^2 = A \quad (11.23)$$

The value of A is chosen so that the scale of the ellipse represents the required confidence level. For example, an ellipse that has a 95% confidence level of encompassing all the location estimates has $A = 5.991$. The values of A can be found from tables [1] if the number of degrees of freedom is known. Table 11.1 shows an extract of the confidence level table needed for use in the radar location error ellipse. In this case there are 2 degrees of freedom and Table 11.1 shows the values of A needed for various confidence levels.

Table 11.1 Multipliers for Ellipse Confidence Levels

Confidence Level	*A*
90%	4.605
95%	5.991
99%	11.210

A 99% confidence interval corresponds to $A = 11.210$ and 99% of the data points fall inside the ellipse defined as:

$$(x / \sigma_x)^2 + (y / \sigma_y)^2 = 11.21 \tag{11.24}$$

In cases where the data is not uncorrelated (x representing latitude and y representing longitude values in this case) and a covariance exists, the error ellipse is orientated by an angle ψ, shown on Figure 11.9 and estimated by the following equation:

$$\tan 2\psi = 2\sigma_{xy} / \left(\sigma_x^{\ 2} - \sigma_y^{\ 2}\right) \tag{11.25}$$

where σ_x is the standard deviation of the x values, σ_x is the standard deviation of the y values and σ_{xy} is the covariance of the two data sets, calculated as follows:

$$\sigma_{xy} = \left[\sum_{I=1}^{n} (x_i - x_{mean})(y_i - y_{mean})\right] / [n-1] \tag{11.26}$$

11.3 How to Draw an Error Ellipse

An ESM track that has a radar location associated with it contains the latitude and longitude of the location estimation together with the major (a) and minor (b) ellipse axes and the orientation (ψ) of the error ellipse.

For the purposes of ESM performance analysis, it may be necessary to draw the error ellipse using the data recorded in the ESM track data.

The algorithm for calculating x and y to plot the ellipse is as follows:

For $n = 1$ to 360
  $x' = a \sin(n)$
  $y' = b \cos(n)$
  $x = (x' \cos(\psi) - y' \sin(\psi))\ degs_{\_lon} + lon_{\_radar}$
  $y = (x' \sin(\psi) + y' \cos(\psi))\ degs_{\_lat} + lat_{\_radar}$
Next n

where $degs_{\_lat}$ and $degs_{\_lon}$ are as specified in (11.12) and (11.13) and $Lat_{\_radar}$ and $Lon_{\_radar}$ are the estimates of latitude and longitude of the radar.

11.4 Radar Location in Practice

The initial estimate of the radar location may be in error by many kilometers. However, this error should reduce as more LOB data becomes available for use in the calculations. Figure 11.11 shows the progression of a typical location fix that starts out with an error of 20 km reducing to 200m after 12 updates to the location fix.

The ESM system typically reevaluates the location fix and error ellipse when a new intercept (LOB) is available. When a suitable number of LOBs are available (typically between 20 and 50), the LOBs that were previously included in the calculation but are subsequently deemed the most inaccurate are excluded from the location calculation.

Once the fix has been made, some ESM systems change their intercept-to-track correlation process to allow LOBs to correlate into the ESM track if they pass through the error ellipse. This correlation condition has the potential to cause over-merging of tracks in cases where the error ellipse is large, as discussed in Section 11.5.4.

11.5 Issues with Radar Location

There are several sources of inaccuracy in the radar location process. The magnitude of the DOA error, the initial orientation of the ellipse relative to the aircraft, the intercept to track correlation process once a location fix has been made and the existence of radars with similar parameters all cause problems in the calculation of radar location.

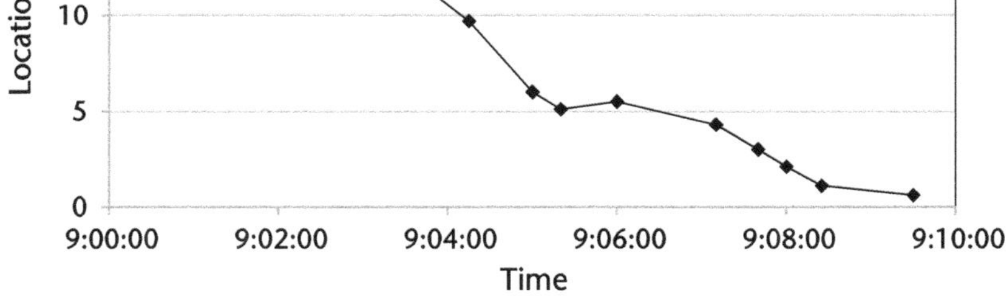

Figure 11.11 Example of a reducing location error.

11.5.1 Magnitude of DOA Error

The accuracy with which DOA is calculated is a major factor in the accuracy of the radar location. An example with two perfect LOBs, the first at DOA 30° and the second at DOA 45° is shown in Figure 11.12. The distance traveled between the two LOBs is 10 km and the exact radar location is calculated and plotted on Figure 11.12. Varying the DOA of one of the LOBs by just 1° leads to an error in location of 1.9 km. The incorrect location of the radar is also plotted.

The error in the range to the radar for LOB 1 errors up to 5° is shown in Figure 11.13. As can be seen, the error in range is significant and is up to 14% for a 5° error in LOB 1. The percentage error in range is the same for any distance traveled between the recording of LOB 1 and LOB 2.

11.5.2 Intercept-to-Track Correlation

Some ESM systems change the rules for intercept-to-track correlation once a track has a location fix associated with it. This often results in the DOA tolerance for correlation being changed, sometimes to such an extent that correlation of intercepts with a wide range of DOAs into the track is now possible. An example appears in Figure 11.14.

This overcorrelation has the potential to cause major errors in the picture of the RF environment, so it is highly recommended that the DOA correlation criterion is not changed once a location fix has been made. Overcorrelation is particularly problematic when very large error ellipses are created because of sparse data, as described in Section 11.5.4.

The opposite problem of overcorrelation of LOBs into large error ellipses occurs when long, narrow ellipses are created. New genuine LOBs from the radar may not correlate into the ellipse (track), leading to multitracking and the possibility that extra tracks and even extra locations and error ellipses are created for a single radar, as shown in the example in Figure 11.15.

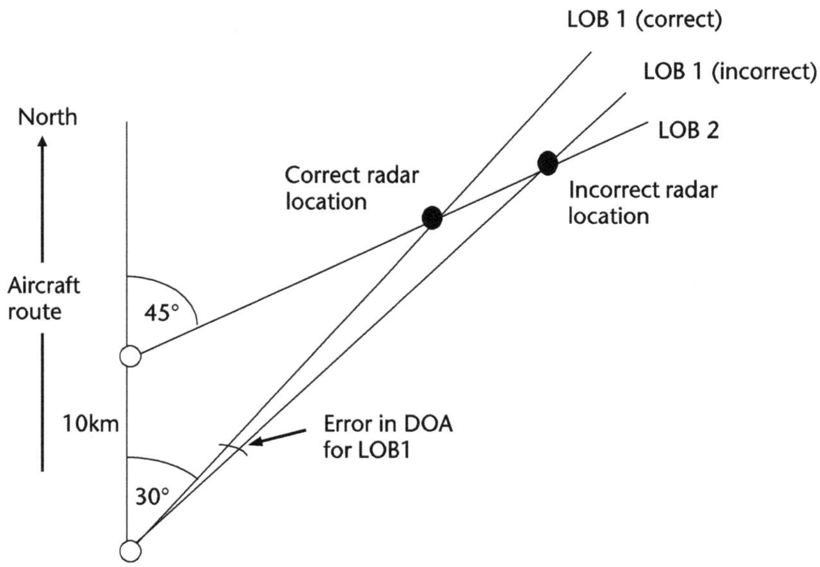

Figure 11.12 Effect of DOA error on location estimation.

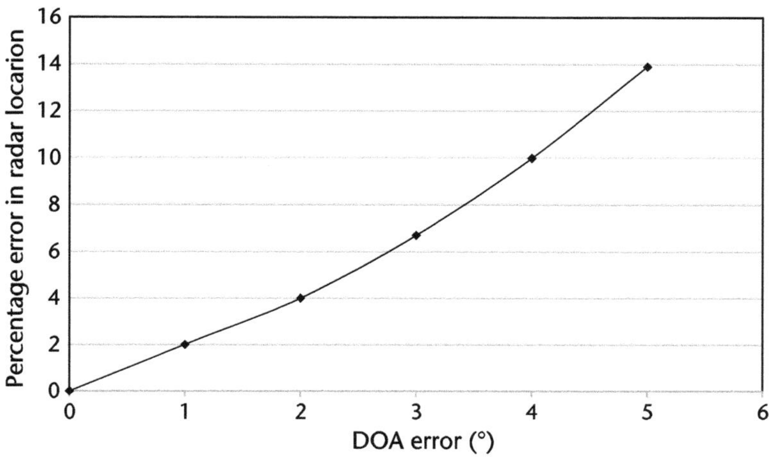

Figure 11.13 Percentage error in radar location versus DOA error in one of two LOBs.

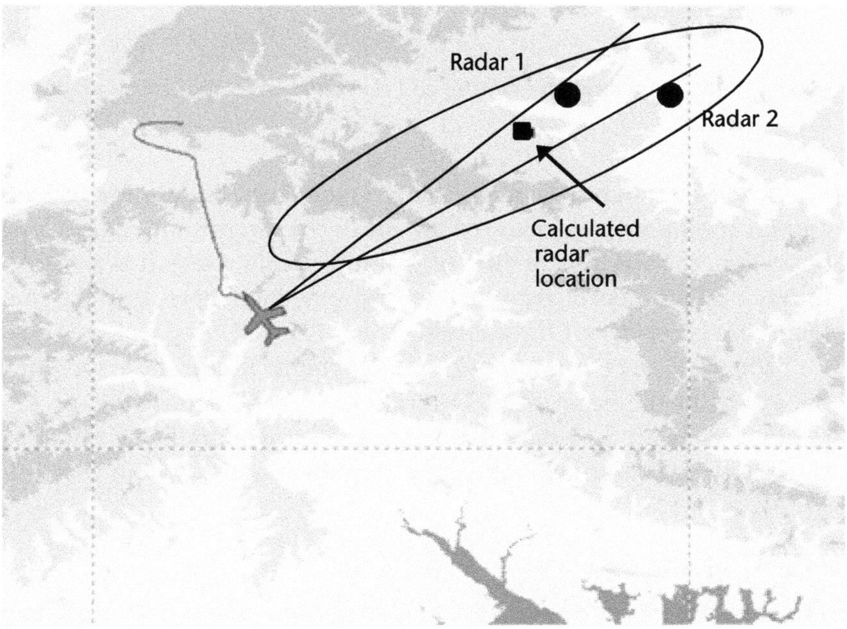

Figure 11.14 Errors in correlation to an error ellipse.

11.5.3 Orientation of the Error Ellipse

In practice the geometry of the error ellipse depends on the orientation of the aircraft with respect to the radar when the initial estimate of location is made. The algorithms governing the size of the ellipse and the lack of suitable track updates due to the LOB to ellipse correlation criteria usually do not allow major changes to the ellipse orientation in the short term. Therefore these initial ellipses are likely to persist after the orientation of the aircraft relative to the radar has changed.

Figure 11.16 shows the aircraft at three different points on its route and the error ellipses that are generated at each of the route points.

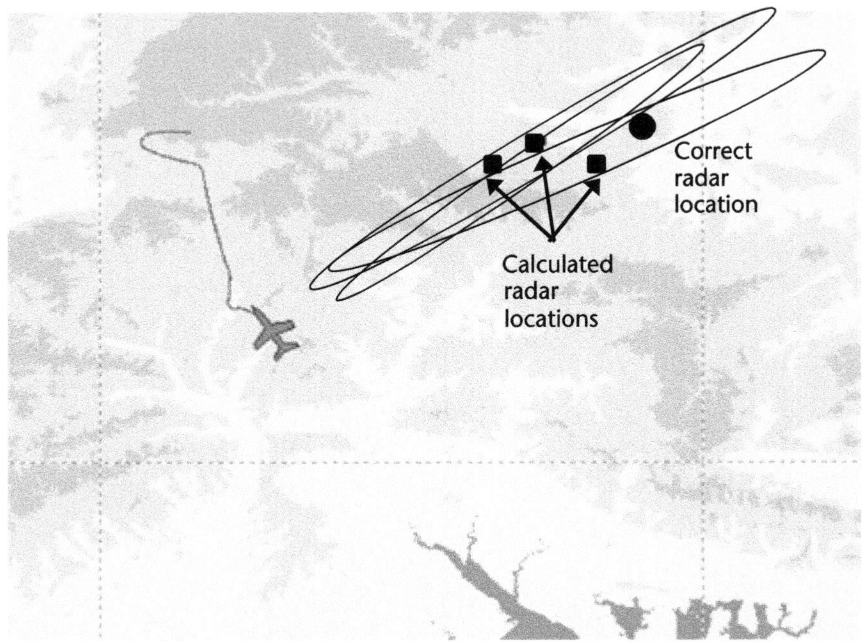

Figure 11.15 Extra location estimates and elongated error ellipses for a single radar.

When the aircraft is at positions 1 and 3 the radars for which location fixes and error ellipses are generated are aft of the aircraft and have elongated error ellipses (i.e., there is a large ratio of major to minor axes). It is possible that intercept to track (ellipse) correlation will fail for new intercepts from the radars due to the long, thin geometry of the ellipses.

When the aircraft is at position 2 on Figure 11.16, the error ellipses created for the radars at the beam of the aircraft (i.e., at AOA near 90° or 270°) have lager minor axes relative to their major axes and it is more likely that correct intercept-to-ellipse correlation will take place for new intercepts from these radars.

11.5.4 Sparse Data Sets

The size of the error ellipse is affected by the amount of data available for its estimation. When there are many data points (more than 1,000), the error ellipse usually accurately represents the area in which 95% (or 99% depending on the choice of confidence level) lie. However, when there are far fewer data points, the error ellipse is likely to be calculated to be much larger to allow the confidence that it can encompass all future location estimates. With very few data points (fewer than 10), a huge error ellipse may be calculated, like the one shown in Figure 11.17.

In practice, measures should be taken to artificially limit the size of the ellipse, especially if the intercept-to-track correlation process is different for tracks that have location fixes than for those tracks that are still LOBs. Otherwise, the problem of overmerging intercepts (LOBs) to the track (ellipse) will almost certainly occur.

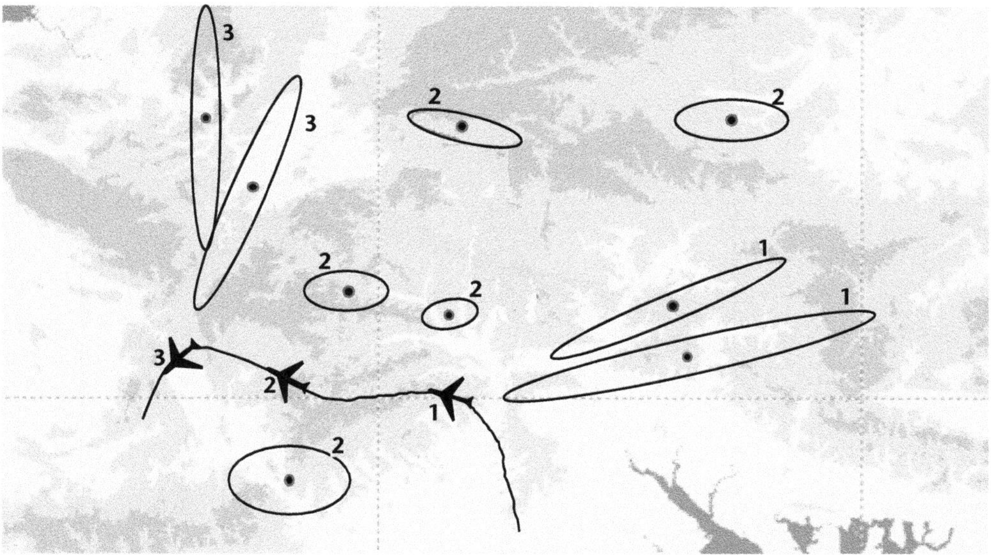

Figure 11.16 Error ellipse creation depending on aircraft orientation with respect to the radar.

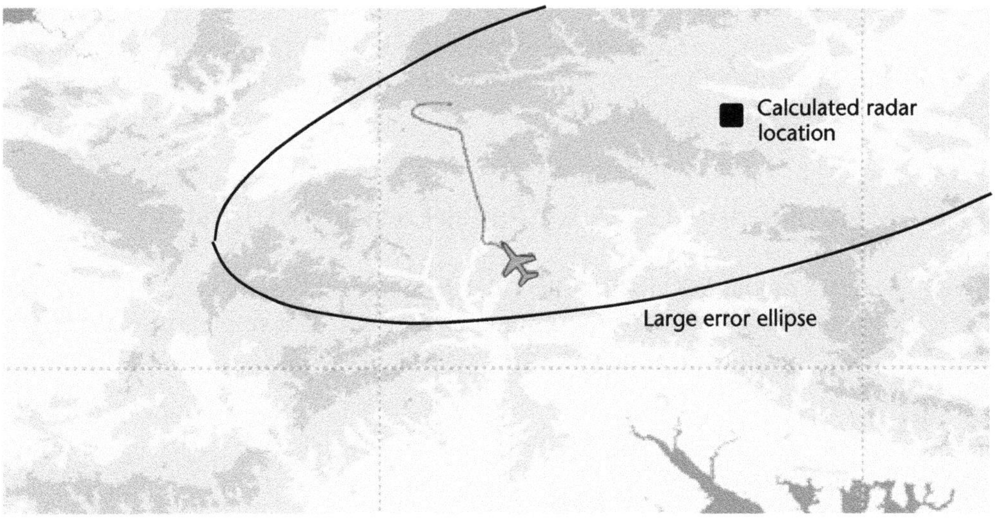

Figure 11.17 Examples of error ellipses for different numbers of data points.

11.5.5 Multiple Radars of the Same Type

If there are several radars of the same type with identical or very similar parameters, then intercepts from more than one radar may be correlated into a single location fix if the intercept-to-track correlation algorithm has relaxed the criterion for DOA correlation. An example of the radar location and error ellipse that may be calculated for radars of the same type and parameters at very different DOAs is given in Figure 11.18.

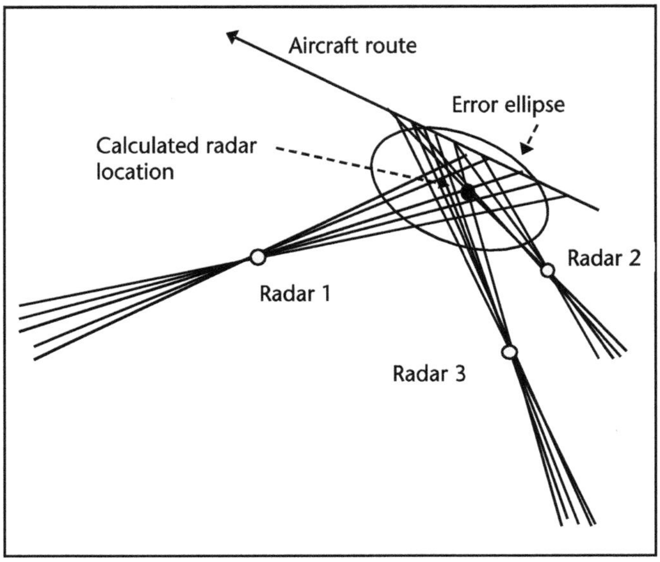

Figure 11.18 Effect of multiple radars of the same type contributing to radar location.

In Figure 11.18, there are LOBs received from three nearly identical radars. All the LOBs pass through the single error ellipse that has been created for the location fix that is not near any of the radars. The LOB to ellipse correlation is allowed to happen as the DOA tolerance for intercept-to-track correlation is ignored once the location fix has been made.

11.6 Multiplatform TDOA for Radar Location Estimation

Another method of finding the radar location makes use of the TDOA method [2] where measurements of the difference in TOA of a single pulse at three spatially remote locations are made, as shown in Figure 11.19.

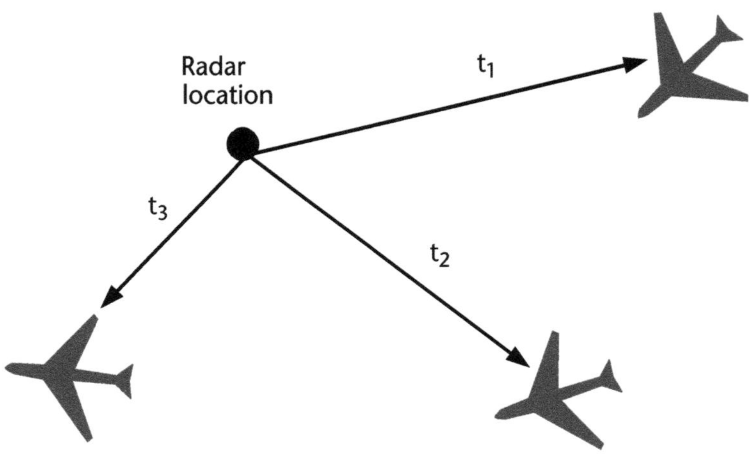

Figure 11.19 TDOA estimation of radar location using three separate platforms.

To achieve good radar location accuracy, large sensor separations and highly accurate timing sources are required. The three timing measurements each give rise to ranges to the radar and as the location of each platform is known precisely, the point at which the ranges intersect can be calculated.

As simultaneous measurements are needed at each sensor, the radar main beam is unlikely to be received at all sensor platforms, with sidelobe pulses being detected instead. Pulses that are received near the threshold of the ESM receiver experience TOA errors relative to those nearer the peak of the main radar beam (see Chapter 6). Multipath interference also causes timing errors, especially for pulses of relatively low amplitude. A 50-ns delay in time measurement, typical of that induced by multipath (see Chapter 15) equates to a range error of 15m.

A single TDOA measurement from two airborne platforms provides a hyperbolic curve containing the radar location that is formed by the intersection of a cone with the surface of the Earth. A second measurement between a third airborne platform and either of the first two platforms provides similar hyperbolic lines, the intersection of which gives the radar location. This method requires a precise knowledge of the distance between the airborne platforms and a precision system clock to synchronize TDOA measurements. GPS can be used to accurately locate the platforms, while a caesium clock is generally used to provide high time precision. The use of GPS-enhanced emitter location techniques using multiple platforms can, in principle, give substantial improvements in position accuracy over methods involving a single platform.

The use of inmanned air vehicles (UAVs) as ESM platforms is becoming more prevalent [3]. For UAV separations of a few kilometers, using high-quality GPS receivers, with highly synchronized clocks, a position accuracy of <10m should be possible. However, multipath effects are likely to degrade accuracy due to pulse timing inaccuracy.

11.7 FDOA Measurement of Radar Location

Relative movement of the ESM platform to the radar shifts the signal received by the ESM in the frequency domain (the Doppler shift). Another method to determine radar location uses the difference in Doppler shift resulting when two ESM platforms are traveling with different radial velocities relative to the radar. This is known as the frequency direction of arrival (FDOA) method [4]. The frequencies at the two platforms are:

$$f_1 = f_0 - \left[(r_{11} - r_{12}) / \lambda T \right] \tag{11.27}$$

and

$$f_2 = f_0 - \left[(r_{21} - r_{22}) / \lambda T \right] \tag{11.28}$$

where rij is the distance of ESM receiver i from the radar at time tj, $T = t_2 - t_1$, and λ is the wavelength of the radar.

The FDOA is defined by:

$$\Delta f = f_1 - f_2(r_{22} - r_{21} - r_{12} + r_{11}) / \lambda T \tag{11.29}$$

Equation (11.29) defines a surface in three-dimensional space. If the radar is on the surface of the earth, the intersection of two surfaces defines a curve on the Earth on which the radar lies. If a second measurement of FDOA is made, two curves can be obtained and the intersection of the two curves provides an estimate of the radar location. FDOA is appropriate for stationary radars, while TDOA can be used for either fixed or mobile radars.

11.8 Amplitude in Radar Location Estimation

The relationship between the amplitude of pulses received from a radar (P_r) and the range to the radar (R) is:

$$R^2 = (P_t\lambda^2) / (16\pi^2 P_r) \tag{11.30}$$

In theory, it is possible to estimate the range to the radar by measuring the amplitude of the pulses (specifically the pulse amplitude at the peak of the main beam) if the radar ERP is known. However, amplitude is not a reliable parameter, due to the effects of multipath (see Chapter 15), so it cannot be used to provide anything other than a rough estimate of the range to the radar.

An example of the effect of multipath on the amplitude profiles of consecutive scans of a radar of ERP of 76 dBW operating at 3 GHz is shown in Figure 11.20. The radar is at a range of 25 km, so the ESM should receive a peak amplitude of –24 dBmi if there are no multipath effects. This is the case for scan 1 on Figure 11.20. However, if the amplitude of the peak pulse from scan 2 (peak amplitude –30 dBmi) is used to calculate the radar range, a value of 50 km results. Similarly, the ranges calculated for scan peaks 3 and 4 are 16 km and 40 km, respectively.

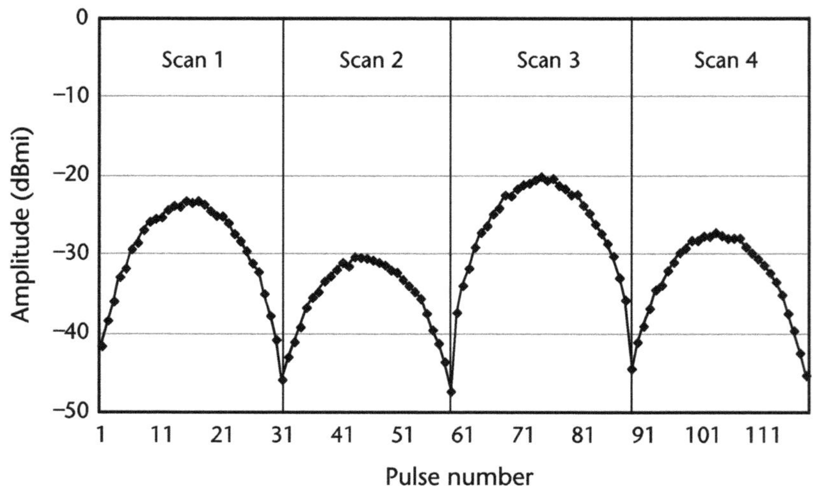

Figure 11.20 Amplitude differences for consecutive scans.

The change in calculated range as the received amplitude changes is shown in Figure 11.21 for radars of ERP of 76 dBW and 80 dBW operating at 3 GHz.

For both radars shown in Figure 11.21, the decrease in calculated range per dBmi increases as the received amplitude decreases. For example, for the radar with ERP of 80 dBW, the change in range when the amplitude reduces from −25 to 26 dBmi is 2.5 km. However, when the amplitude reduces from −35 to −36 dBmi, the change in range is 11 km.

Both the radars in Figure 11.21 have the same RF, 3 GHz. Figure 11.22 shows how the calculated range to the radar changes for two radars of ERP of 80 dBW. One of the radars is operating at 3 GHz and the other operates at 9 GHz.

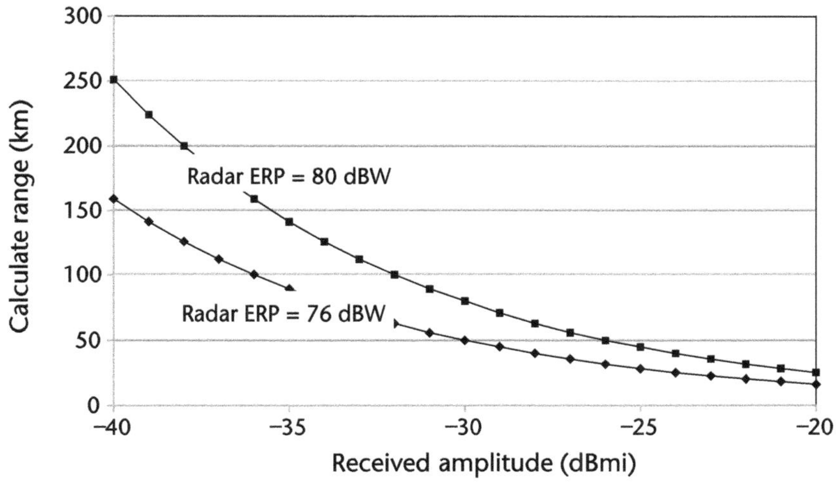

Figure 11.21 Calculated range for peak amplitude for radars of ERP of 76 dBW and 80 dBW.

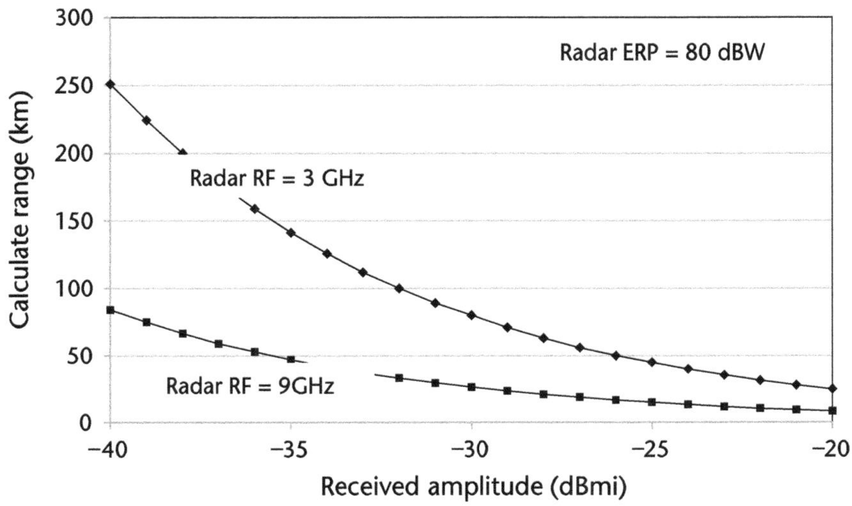

Figure 11.22 Calculated range for peak amplitude for radars at 3 GHz and 9 GHz with ERP of 80 dBW.

Although the range to the radar is less for all received amplitudes in dBmi for the 9-GHz radar, the difference in range per dBmi increases in both cases for lower amplitudes.

11.9 The Use of Extended Kalman Filters for Range-Finding

The reality of DOA measurements means that while a series of measurements may be centered on the correct value, individual measurements differ from this by random amounts. A good system for passive ranging makes use of its own observations but also uses a priori target information to produce a better estimate of range to a radar. The Kalman filter allows prior information to be used along with measurements to update range calculations and estimate the accuracy of such information.

In ESM systems, the extended Kalman filter [5] for the 2-D angle-only emitter location problem may be used based on an initialization procedure combining a priori probability density function (pdf) information with single measurement a posteriori pdf information in a manner that is more efficient than the traditional Kalman filter.

References

[1] https://people.richland.edu/james/lecture/m170/tbl-chi.html.

[2] O'Neill, S., *Electronic Warfare and Radar Systems Engineering Handbook,* U.S. Naval Air Warfare Center Weapons Division, 2013.

[3] Du, H. -J., and J. Lee, "Radar Emitter Localization Using TDOA Measurements from UAVs and Shipborne/Land-Based Platforms," *Proc. of the RTO SCI Symposium on Multi-Platform Integration of Sensors and Weapons Systems for Maritime Applications*, 2002.

[4] Okello, N., *Emitter Geolocation with Multiple UAVs*, Melbourne Systems Laboratory, Department of Electrical and Electronic Engineering, http://fusion.isif.org/proceedings/fusion06CD/Papers/137.pdf.

[5] Guerci, J. R., R. A. Goetz, and J. Dimodica, "A Method for Improving Extended Kalman Filter Performance for Angle-Only Passive Ranging," *Proc. IEEE Transactions on Aerospace and Electronic Systems*, Vol. 30, No. 4, 1994, pp. 1090–1093.

CHAPTER 12

ESM Performance Analysis

To establish that an ESM system is creating a good picture of the RF environment, there are several aspects of system performance that should be tested. The accuracy in calculating a radar DOA and its location and the timely creation of ESM tracks are among the issues where confidence in the ESM system can be established through ESM performance analysis.

12.1 Data Recording and Required Data Capacity

There are several sets of data that must be recorded to carry out post-mission analysis of ESM system performance. Some of the data sets, such as track and pulse data, are specific to the ESM system and some data comes from other parts of the mission system. Platform navigation data is needed and location data for fixed and moving radar targets (Automatic Identification System (AIS) data) should be recorded if it is available. The essential datasets needed for ESM performance analysis are:

- ESM pulse data;
- ESM intercept data;
- ESM track data;
- ESM status, built-in-test (BIT) and alarm data;
- Platform navigation data (GPS and compass);
- Ground truth in the form of fixed radar locations;
- AIS data to provide ground truth for ship radars;
- ESM radar library.

Other data sets such as radar tracks, sonar contacts and magnetic anomaly detector data are useful when considering how the ESM data is fused with data from other sensors.

12.1.1 ESM Pulse Data

Each PDW consists of at least the following data fields: TOA, RF, PW, amplitude, and DOA. Table 12.1 shows the memory requirements for each data field in a typical PDW.

Table 12.1 Data Fields and Memory Requirements for a PDW

Parameter	*Resolution and Range*	*Memory Requirement*
Pulse number/header	1 (to 1 billion)	32 bits
Frequency (RF)	1 MHz (1 MHz to 32 GHz)	16 bits
TOA	1 μs (0 to 24 hours)	38 bits
PW	50 ns (0 to 30 ms)	17 bits
Amplitude	0.5 dB (–60 to 10 dB)	9 bits
DOA	0.25° (0° to 360°)	12 bits
DTOA*	For each of 4 antennas	
	1 ns (0 to 512 ns)	10 bits
Status Bits	Measurement qualities, pulse modulation, and so forth	12 bits

*For TDOA systems only.

For a typical ESM system, the total memory requirement for a PDW is 146 bits, rounded up to 20 bytes. In Chapter 3, the radar pulse density was discussed with the conclusion that, in a high (nonthreat) pulse density environment, there may be as many as 2,000 pulses per second detected by the ESM. Using this value as the basis for calculation leads to memory storage requirement for pulse data of about 150 megabytes per hour.

12.1.2 ESM Intercept Data

Table 12.2 shows the data fields needed for intercept data. The storage requirements for this type of data are not nearly as extensive as for the pulse data as typically there are only a few intercepts created per second on average.

12.1.3 ESM Track Data

Table 12.3 shows the data fields typically found in ESM track data. At the time of track creation the ESM does not have enough information to be able to generate a location for the radar. The data fields for latitude, longitude and error ellipse are not populated until there is sufficient line-of-bearing (LOB) data to allow for the estimation of radar location (see Chapter 11). In the meantime the DOA is available. Sometimes once a fix has been made, the track DOA is no longer updated as the intercept-to-track correlation process makes use of intercept LOB to track the location fix and does not correlate the DOAs of the intercept and the track.

12.1.4 ESM Status, BIT, and Alarm Data

This data contains records of ESM start-up and reset times. It also contains data from built-in-test (BIT) procedures. System alarms such as those that occur when the ESM track table is full are recorded in the ESM status dataset.

Table 12.2 Intercept Data Fields

Parameter	*Resolution and Range*	*Notes*
Intercept Number	1 (1 to 10,000,000)	In a mission of 6 hours 5 intercepts/second on average can be created
TOA	1 μs (0 to 24 hours)	
Frequency (RF)	1 MHz (1 MHz to 32 GHz)	
RF dispersion	0.001 MHz (0.001 to 100 MHz)	This field may be replaced by maximum and minimum RF levels
RF type		Fixed, agile, hopper
RF agility extent	0.001 MHz (0.001 to 500 MHz)	
RF hopper levels		Up to 20 levels
PRI	1 μs (1μs to 300 ms)	
PRI dispersion	0.01 μs to 100 μs	
PRI levels		Up to 30 levels
PRI type		Fixed, stagger, jitter, switch and dwell
PW	50 ns (0 to 30 ms)	
Amplitude	0.5 dB (–60 to 10 dB)	The peak amplitude of pulses making up the intercept
DOA	0.25° (0 to 360°)	
DOA dispersion	0.25° (0° to 7.5°)	
Number of pulses	1 (1 to 4,000)	Most ESMs collect buffers of pulses with a buffer size limit
Status flags		Measurement qualities

Table 12.3 Track Data Fields

Parameter	*Resolution and Range*	*Notes*
Track number	1 (1 to 20,000)	In a mission of about 6 hours, this allows 1 track to be created per second
Time of creation	1 μs (0 to 24 hours)	Usually reported in hh:mm:ss.000 format
Identity		There may be several (as many as 20) alternative identities for the track; the first identity in the list is usually the most likely identity according to the ESM
Frequency (RF)	1 MHz (1 MHz to 32 GHz)	
PRI	1 μs (1 μs to 300 ms)	
PW	50 ns (0 to 30 ms)	
DOA	0.35 (0° to 360°)	
Latitude	Degrees, minutes, seconds	May be expressed in alternative formats
Longitude	0 to 90°N or S, 0° to 180°E or W	See Section 12.1.5
Error ellipse major axis	1m (1m to 100 km)	
Error ellipse minor axis	1m (1m to 100 km)	
Error ellipse orientation	0.35° (0° to 360°)	
Location fix quality	No fix, low, medium high quality	
Number of intercepts	No limit in theory	

12.1.5 Platform Navigation Data

The platform navigation system should have the means to record the platform location, altitude and heading on a timely basis. About 1-second time intervals are good enough for ESM analysis. The format of each parameter and the memory requirements for recording are shown in Table 12.4.

Latitude and longitude may appear in any of the following formats: decimal degrees, dd.xxxx N, ddd.xxxx E, degrees and decimal minutes, dd:mm.xx N, ddd:mm.xx E, or degrees, minutes and seconds dd:mm:ss N,ddd:mm:ss E. When using decimal degrees, latitude (S) and longitude (W) are recorded as negative decimal numbers. The resolution varies between formats, but those specified here all give resolutions to tens of meters. Lines of longitude are all great circles, but latitude lines are not. This means that latitude and longitude have different resolutions as the distance in longitude per degree decreases with increasing distance from the equator. Multiplying the distance by cosine (latitude) is a good enough approximation of the factor needed to reduce the distance per degree of longitude at a particular latitude.

In addition to the recording of navigation data, information on the platform attitude must be made available to the ESM during the mission. Platform heading data is needed for the calculation of DOA. Ideally, pitch and roll should also be available to the ESM system so that DOA can be more accurately calculated. In practice, there are so many sources of DOA error that this extra level of processing using pitch and roll does not lead to a significant improvement in DOA. However, heading, essential for DOA calculation, should be made available to the ESM system at least once a second.

12.1.6 Ground Truth

Ground truth in the form of radar site names, latitude, longitude, and altitude and radar types together with known parameters: RF, PRI elements, PW, ERP, beam and scan types, and polarization are needed for analysis. Sometimes this radar-specific data is stored in the ESM system so that it can be used for radar identification.

Table 12.4 Memory Requirements for Platform Navigation Data

Parameter	*Resolution and Range*	*Memory Requirement*
Time	1 second (0 to 24 hours)	18 bits
Latitude*	10m (0° to 90°N or S)	22 bits + 1 sign bit
Longitude*	0° to 180°E or W)	22 bits + a sign bit
Altitude	1 ft (1 to 40,000 ft)	17 bits
Heading	0.25° (0° to 360°)	12 bits
Pitch	0.5° (–90° to +90°)	1 sign bit + 9 bits
Roll	0.5° (–90° to + 90°)	1 sign bit + 9 bits

* Calculated assuming latitude and longitude have a resolution of 0.0001°. Latitude and longitude may be expressed in alternative formats as described below.

12.1.7 AIS Data

The introduction of AIS data worldwide has meant that ground truth can easily be established for ships. Identification, position, course, speed, and navigation status are transmitted. There are two classes of AIS transmitter: class A for commercial and fishing vessels and class B for pleasure craft.

Class A regulated vessels transmit AIS data at 2- to 10-second intervals while under way and at 3-minute intervals while at anchor. Class B, nonregulated vessels, such as private yachts transmit at approximately 30-second intervals while under way at speeds greater than 2 knots and at 3-minute intervals while at anchor.

12.1.8 ESM Radar Library

The ESM radar library data is loaded into the ESM system before each mission and should be deleted at the completion of the mission so that classified data (for it is almost always classified) is not kept on the aircraft. The ESM library is usually in the form of a number of mode lines for each radar type containing parameter ranges. Each mode line represents a specific combination of RF range, PRI range, and PW range for a radar. Typical current ESM libraries contain only 1,000 mode lines. In practice, this is not nearly enough data to adequately cover all the modes of the radars likely to be detected in a mission, so this is one of the main failure mechanisms for ESM systems. More information on this aspect of ESM performance and typical formats used in ESM libraries is given in Chapter 10.

12.2 ESM Performance Visualization

The first stage in the visualization of the ESM performance is the replay of the recorded track data on a suitable mapping tool. An example of such a map is shown in Figure 12.1.

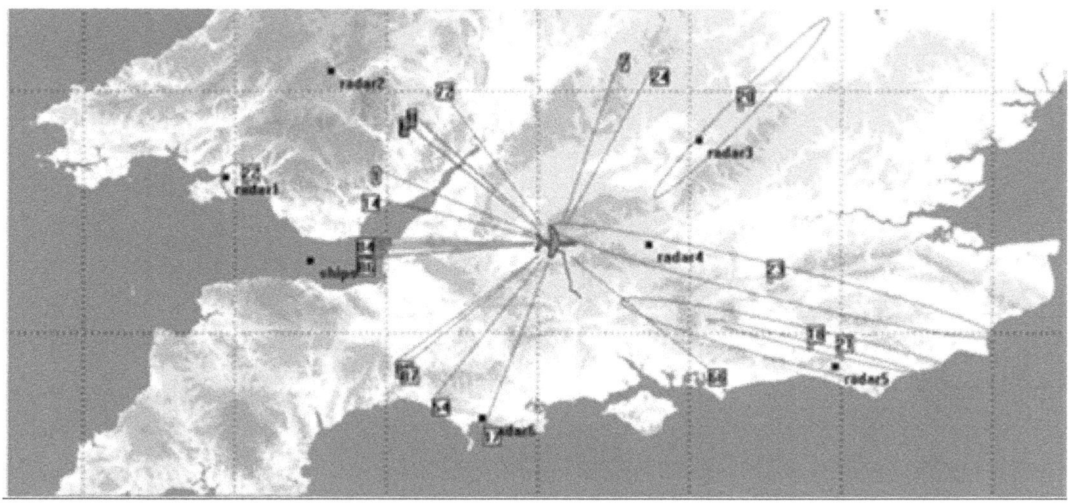

Figure 12.1 Example of an ESM replay map.

The basic visualization shown in Figure 12.1 consists of a map background, radar targets, and the aircraft route. The aircraft is moved along its route at a speeded-up rate so that the visualization can be carried out more quickly than in real time. The final components of the replay are the ESM tracks and the ESM location fixes and error ellipses, if these are calculated by the ESM.

The track format in Figure 12.1 shows the origin of all tracks at the center of the aircraft (i.e., the track origins move with the aircraft). An alternative format, in Figure 12.2, shows the track origins at the time of creation or track update. This format makes it easy to see which tracks are being updated frequently. Those that are not updated are left behind by the aircraft symbol as it makes its way along the aircraft route.

This second map/track format is a better indication of the actual RF picture that the ESM system has derived and is more closely aligned to what is available to the ESM operator onboard the aircraft. However, it does make for a more confusing display. Ideally, the mapping tool would support either way of displaying the tracks so that the analyst can choose which they prefer.

The mapping tool should have some filtering available so that the analyst can choose tracks for display based on their parameters or on track identification. A table of tracks being displayed to show the track parameters is a useful addition to the map itself.

12.3 DOA Performance Assessment

The DOA performance of an ESM system is determined by how close the ESM track DOA is to the correct DOA of the radar. Underlying the track DOA updates are the individual pulse DOAs. The analysis of this data is very important in determining the DOA performance of the ESM and in diagnosing issues that arise from the analysis.

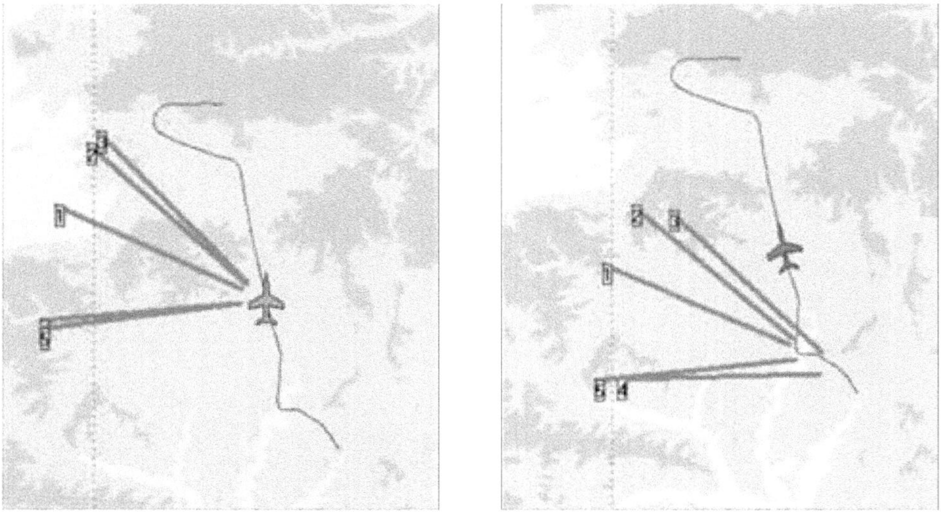

(a) (b)

Figure 12.2 Examples of a replay map with different track origins.

The first stage in the DOA analysis is to determine the correct DOA profile of the radar. The radar location is known and this information, together with the recording of the GPS data from the platform is used to calculate the correct DOA profile.

The two-step formula for calculating great circle DOA is as follows:

$$1.\ \cos(D) = \left(\sin\left(lat_1\right)\sin\left(lat_2\right)\right) + \left(\cos\left(lat_1\right)\cos\left(lat_2\right)\cos\left(\Delta_L\right)\right) \tag{12.1}$$

$$2.\ \cos(C) = \left(\sin\left(lat_2\right) - \left(\sin\left(lat_1\right)\right)\cos\left(D\right)\right) / \left(\cos\left(lat_1\right)\sin\left(D\right)\right) \tag{12.2}$$

where lat_1 = ESM platform latitude in degrees, lat_2 = radar latitude in degrees, Δ_L = radar longitude – platform longitude, D = distance along path in degrees of arc, and C = radar bearing, to be corrected as follows:

$$If\ \sin\left(\Delta_L\right) is\ positive, DOA = C \tag{12.3}$$

$$If\ \sin\left(\Delta_L\right) is\ negative, DOA = 360 - C \tag{12.4}$$

The DOA must be calculated for each latitude and longitude recorded in the platform navigation data. An example of a plot of the correct DOA profile of a radar is given in Figure 12.3. The DOA is plotted from the start time of interest, 14:00:00 in this case, to the end time at 14:10:00. The units of the x-axis are hh:mm:ss on Figure 12.3 as a 10-minute time span is plotted, but units of seconds or milliseconds may be used if shorter time periods are plotted.

12.4 ESM Track Analysis

Once the correct DOA of the radar has been established, the DOA of each intercept contributing to a track is plotted alongside the correct DOA profile. There are three ways in which the track will appear:

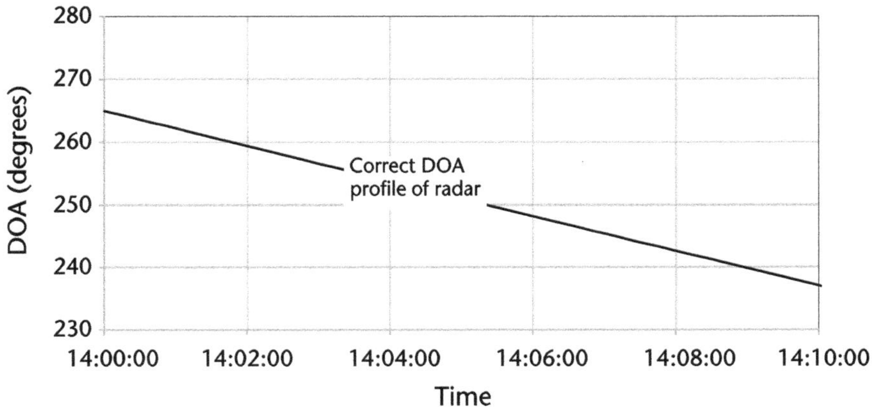

Figure 12.3 Example of a DOA profile for a radar.

- Those that are created and never updated will appear as a straight line on the DOA against time graph with a length indicating the lifetime of the track.
- Those that have few updates will appear as a set of discrete jumps in DOA.
- Those with frequent updates will have a profile that follows the correct DOA.

Examples of the three ways in which track data appears are shown in Figure 12.4.

The tracks that are updated on Figure 12.4 show discrete jumps in the DOA (i.e., the DOA of the latest intercept of the track is kept until a new intercept updates the track). However, while this is a true representation of the ESM picture, it does require some extra processing of the track data to visualize the track time plot. It is therefore easier just to plot the DOA against time directly from the recorded data. Figure 12.5 illustrates this point. The first plot has had extra data points inserted so that the true DOA profile is available, the second is the same track, but with the data taken directly from the recorded ESM track data.

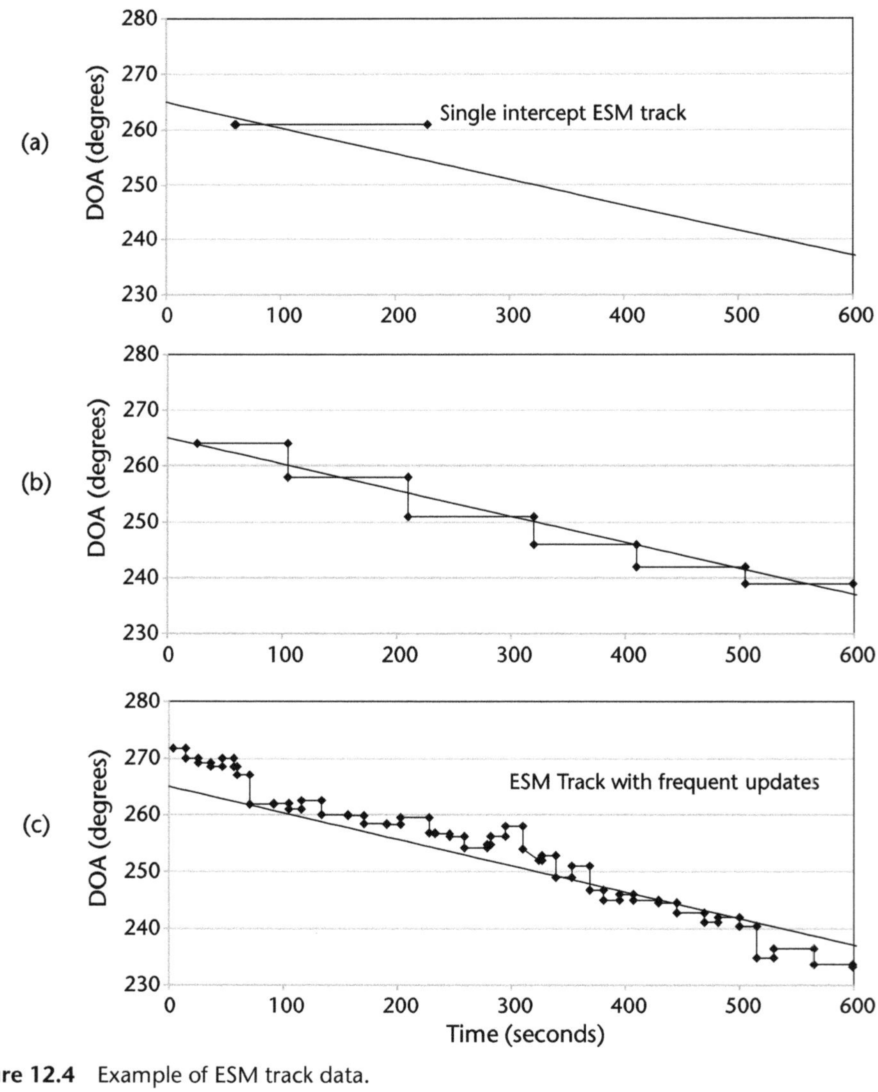

Figure 12.4 Example of ESM track data.

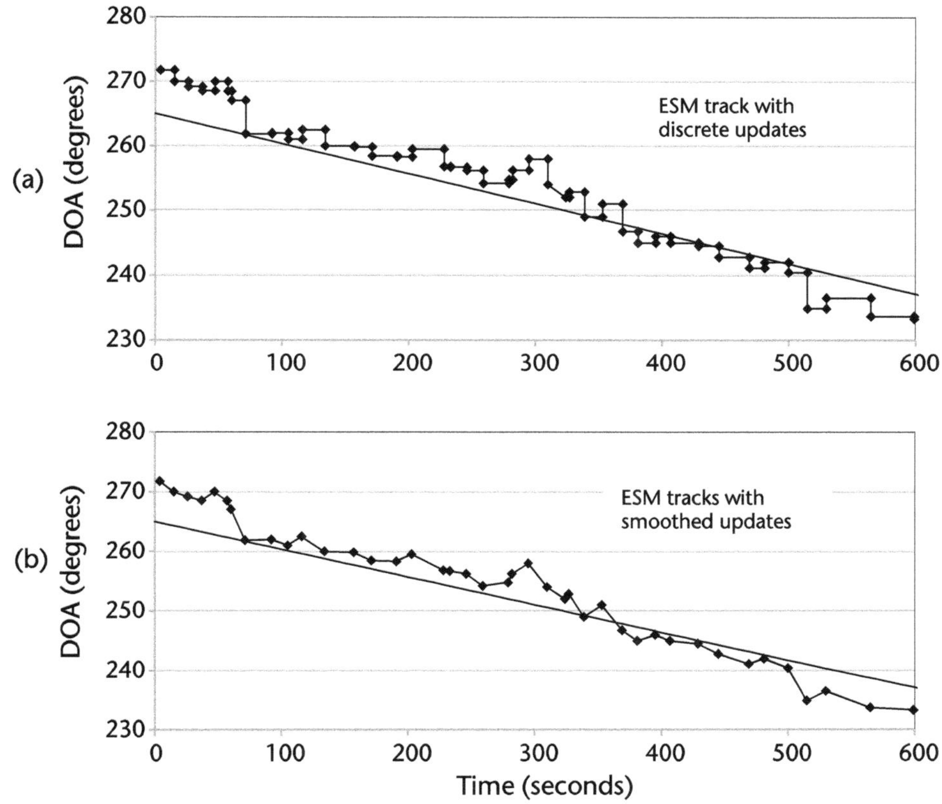

Figure 12.5 ESM track updates with discrete and smoothed update profiles.

The track DOA is the ESM output shown to the operator and may be used in the fusion of data from other sensors to provide an overall tactical picture. It is therefore useful to be able to determine an overall accuracy for this data. This is usually expressed as the root mean square error defined by:

$$rms_{\_intercept\_DOA\_error} = \sqrt{\left(\left(x_1^2 + x_2^2 + x_3^2 \cdots x_n^2\right) / n\right)} \tag{12.5}$$

where x_1, x_2, x_3 ... are the DOA errors of the individual intercepts and n is the number of intercepts.

It is also useful to be able to visually inspect the DOA performance by plotting not only the correct DOA of the radar and the DOA of the track, but also by plotting lines indicating the DOA requirement. For example, in Figure 12.6 the DOA accuracy is required to 5°. Plotting the correct DOA of the radar and lines at the correct DOA ±5° shows in this case that the track DOA is at the limit of the required DOA performance for much of its lifetime.

Another example of a track that deviates from the DOA requirement is shown in Figure 12.7. In this case, the rate of change of the correct DOA profile indicates that the radar is close to the ESM platform and the ESM is having difficulty in coping when the ESM platform changes course significantly during the time period of the plot.

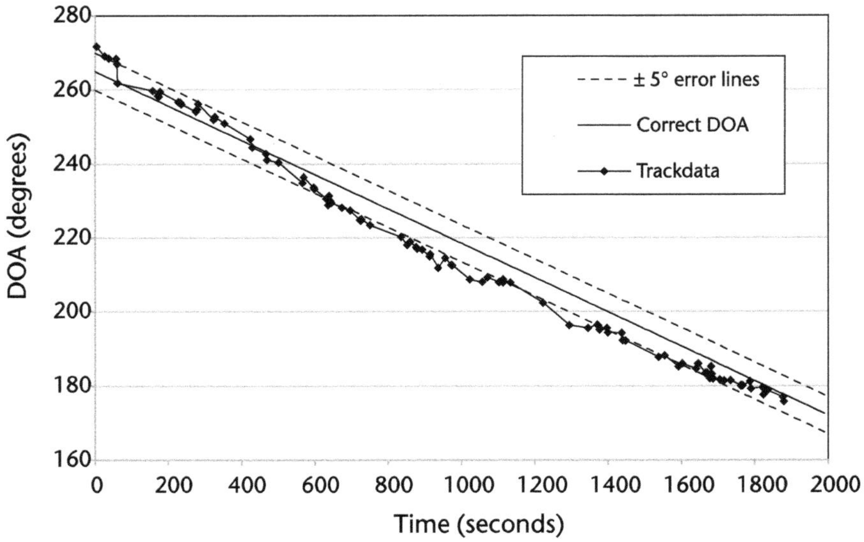

Figure 12.6 Example of an ESM track that lies within the required DOA performance limits.

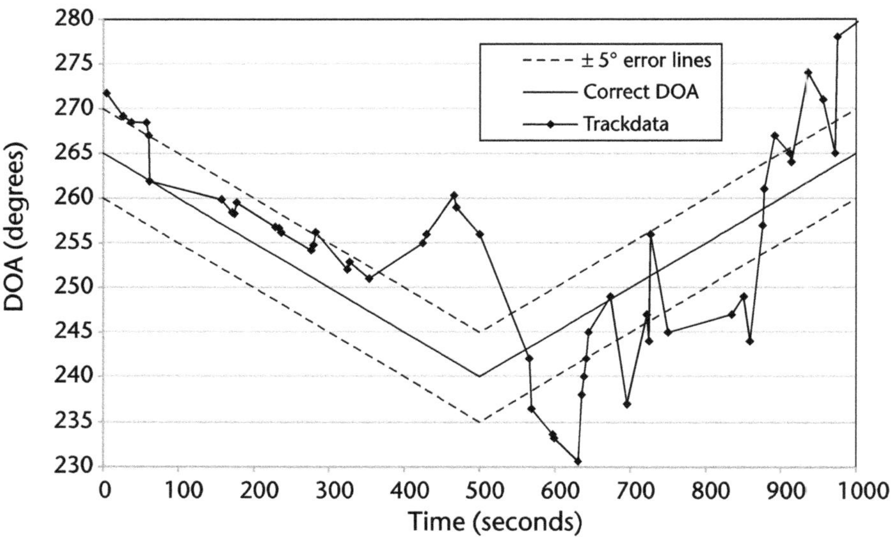

Figure 12.7 Example of an ESM track that does not meet the DOA requirement.

Plotting the DOA of the tracks in this way is useful for two reasons, apart from confirming the DOA performance of the ESM system. The first reason is that, if there are many tracks seen with poor DOA performance, then steps can be taken to improve DOA. Chapter 14 gives some ideas on how this can be accomplished.

The second reason for knowing the DOA performance and quantifying it by using these plots is that operator expectations can be managed. This is because ESM systems are rarely as accurate in the DOA as stated in their manufacturers' specifications.

12.5 Pulse Data Analysis

The most effective way to determine ESM performance is to analyze the radar pulse data. To do this, it is necessary to extract the data from the medium in which the pulse data has been recorded on the ESM platform and to transfer it into a database or spreadsheet. An example of an Excel spreadsheet containing consecutive PDWs is shown in Table 12.5.

When carrying out radar pulse data analysis it is desirable to extract pulses from a single radar so that its features such as the DOA and amplitude profiles and the PRI sequences can be examined in detail.

To choose the pulses from a specific radar, the pulse data set must first be filtered by RF. The RF of the radar of interest should be known and a range of a few megahertz either side of the known RF is chosen so that all the pulses that are likely to be from the particular radar are selected. Table 12.6 shows an example of RF filtered pulse data.

Once the RF filtered pulse data set is available, the pulses from the radar of interest are selected for further analysis using the timing data available in the PDWs. A column is inserted (the rightmost column shown in bold in Table 12.6) so that the time since last pulse (*tslp*) can be calculated for each pulse (*n*) using the following equation:

Table 12.5 Example of an Excel Spreadsheet Containing Radar Pulse Data

RF (MHz)	*TOA*	*PW (μs)*	*Amp (dB)*	*DOA (°)*
2,939	9:43:24.00000	0.64	–33.5	352.75
9,370	9:43:24.00103	0.32	–38.5	60.25
2,750	9:43:24.00107	0.85	–24.5	183.50
2,939	9:43:24.00128	0.64	–33.5	352.75
9,370	9:43:24.00206	0.32	–37.5	60.50
2,750	9:43:24.00207	0.85	–24.0	183.50
2,938	9:43:24.00264	0.64	–33.0	352.50
2,750	9:43:24.00307	0.86	–235	183.50
9,371	9:43:24.00309	0.32	–35.5	60.50
2,939	9:43:24.00387	0.60	–32.5	352.75
2,750	9:43:24.00407	0.85	–23.0	183.75
9,370	9:43:24.00412	0.32	–32.5	61.00
2,750	9:43:24.00507	0.85	–22.5	183.25
9,370	9:43:24.00515	0.32	–35.5	60.75
2,939	9:43:24.00518	0.60	–32.5	351.50
2,750	9:43:24.00607	0.85	–21.5	183.50
9,370	9:43:24.00621	0.32	–37.5	61.25
2,939	9:43:24.00643	0.64	–32.0	354.25
2,750	9:43:24.00707	0.86	–20.0	183.50
9,371	9:43:24.00724	0.32	–38.0	61.50
2,938	9:43:24.00770	0.64	–33.0	352.00

Table 12.6 Example of a RF Filtered Pulse Data Set

RF (MHz)	*TOA*	*PW (μs)*	*Amp (dB)*	*DOA (°)*	*tslp (μs)*
2,939	9:43:24.000000	0.64	−33.5	352.25	
2,939	9:43:24.001275	0.64	−33.5	352.25	1,275
2,938	9:43:24.002640	0.64	−33.0	352.25	1,365
2,939	9:43:24.003870	0.64	−32.5	352.75	1,230
2,939	9:43:24.005175	0.64	−32.5	351.50	1,305
2,939	9:43:24.006425	0.64	−32.0	354.50	1,250
2,938	9:43:24.007700	0.64	−33.0	352.75	1,275
2,938	9:43:24.009065	0.64	−34.0	352.75	1,365
2,938	9:43:24.010295	0.64	−33.5	357.25	1,230
2,938	9:43:24.911600	0.64	−34.0	357.25	1,305
2,939	9:43:24.012850	0.64	−35.5	352.75	1,250
2,938	9:43:24.014125	0.64	−34.5	352.75	1,275
2,938	9:43:24.014490	0.60	−34.0	352.00	1,365

$$tslp = \left(TOA_n - TOA_{n-1}\right)M \tag{12.6}$$

In (12.6), M is used to scale the tslp to microseconds. This is the most useful unit in which to express the tslp as it gives enough time definition to allow the discrimination of PRI between radars without showing numbers that are too large for easy pattern recognition by eye.

If there are pulses from only one radar in this filtered data set, the tslp sequence is the same as the PRI sequence for the radar. However, if there are pulses from more than one radar, the tslp sequence contains extra values that do not fit the PRI sequence for the radar of interest.

A method of plotting DOA similar to that used for the track data is used for the pulse data (i.e., the correct DOA of the radar is plotted using the known radar location and the aircraft navigation data). Each pulse has a DOA associated with it or for TDOA systems a set of TOA differences for each antenna. The formula linking TDOA with DOA is given in Section 6.1. Each calculation of DOA from TDOA gives an ambiguous value for DOA as the correct DOA may lie either side of the baseline linking the two antennas. The use of a second baseline resolves the ambiguity.

Figure 12.8 shows pulse data plotted against the correct DOA of three radars. There is often more than one radar in a particular RF range in the pulse data so if the DOA performance is being tested, it is not really necessary to separate all the pulses into individual data sets as long as the radars are separated in the DOA.

In Figure 12.8, apart from a few outliers, it is easy to see the pulses from each of the three radars. The DOA outliers on this plot can be examined individually to see which they radar they belong to, if it is deemed necessary.

Individual scans of the radar provide further insight into the ESM performance. Figure 12.9 shows amplitude and DOA profiles of pulses from a single radar scan. In this case, the pulse DOA has been correctly calculated and the amplitude profile is as expected for the range to the radar. However, amplitude cannot be relied upon as a measure of the range to a radar where the radar location is not known as

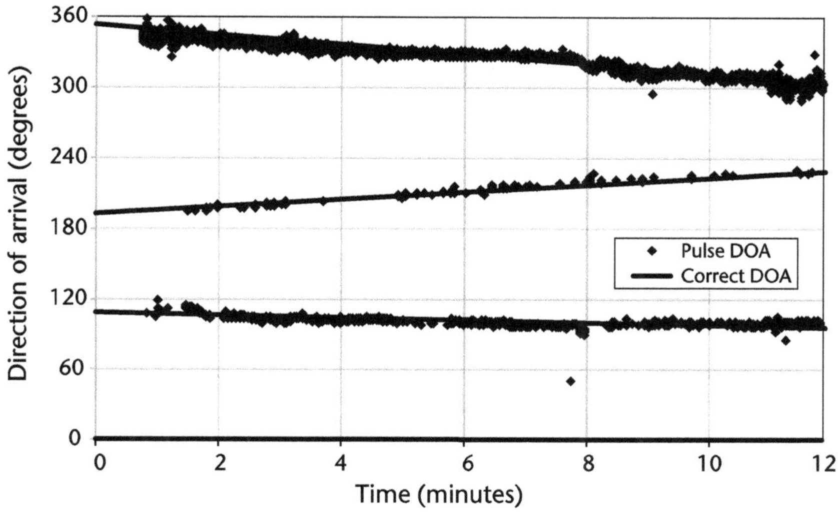

Figure 12.8 Accurate pulse DOA.

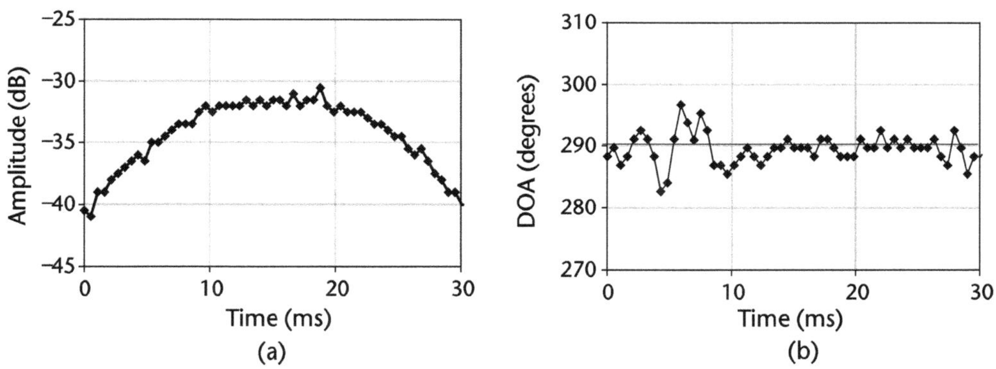

Figure 12.9 Amplitude and DOA profiles of a single radar scan.

multipath effects will almost always alter the received amplitude of the pulses (see Chapters 11 and 15).

The DOA profile of the pulses shows whether the performance in DOA for the particular radar is good enough or whether there are issues that must be addressed in the way the system operates.

The graphical method of viewing DOA performance described here is often sufficient to decide whether the ESM has adequate DOA accuracy. However, to arrive at a quantitative assessment of the DOA performance, the error in DOA for each pulse should be found and used in the calculation of root mean square (rms) error.

The correct DOA of the radar is available as a set of data calculated for each time interval using the platform navigation data and the radar location. The pulse data set must be aligned with this correct DOA data so that the DOA error can be calculated for each pulse. The difficulty in carrying this out is that the correct DOA appears at regular intervals (typically once per second), whereas the pulse data has a variable time difference between each pulse, depending on the PRI sequence and the radar scan period.

The following algorithm can be used to align these two data sets:

```
nav_dataset = 0
For pulse = 1 To n_pulses (do this for each pulse)
   read pulse_time and pulse_DOA
   100 nav_dataset = nav_dataset + 1 (do this for each nav data set)
   read nav_time
   If (pulse_time ≤ nav_time) Then
      Read correct_DOA at nav_time
      Calculate DOA_Error = Abs(pulse_DOA - correct_DOA)
      Goto 200 (Next pulse)
   Else
      Goto 100 (nav_dataset = nav_dataset + 1)
   End If
200 Next pulse
```

The DOA error for each pulse is then found using the following equation:

$$DOA_{\_error} = correct_{\_DOA} - pulse_{\_DOA} \tag{12.7}$$

The rms DOA error is calculated using this equation:

$$rms_{\_DOA\_error} = \sqrt{\left(\left(x_1^2 + x_2^2 + x_3^2 \cdots x_n^2\right)/n\right)} \tag{12.8}$$

where n is the number of pulses and x_n is the nth pulse DOA error.

12.6 AOA Analysis

The DOA analysis shows how good the ESM system is at determining where the radar is with respect to North; however, the performance in DOA may vary according to the orientation of the platform with respect to the radar. This is defined as the angle of arrival (AOA). Figure 12.10 clarifies the difference between DOA and AOA.

The objective in carrying out AOA analysis is to arrive at a plot like the one in Figure 12.11 showing the rms angle error around the aircraft. In this plot, data has been used from only one radar so the rms error for only a 120° segment is available for the calculation.

To produce a plot like the one in Figure 12.11, it is necessary to start with the DOA of the pulses as the DOA is recorded in the PDWs. The DOA error is calculated using the method described in Section 12.5. The next step is to use an algorithm similar to that used in the calculation of DOA error to align the PDW

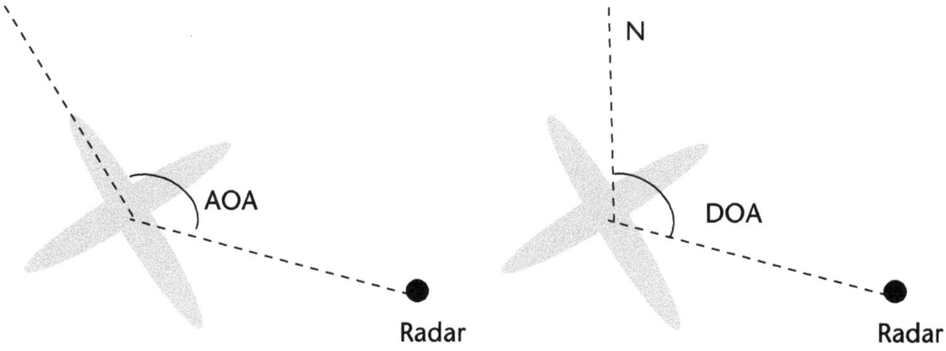

Figure 12.10 Definition of the DOA and AOA.

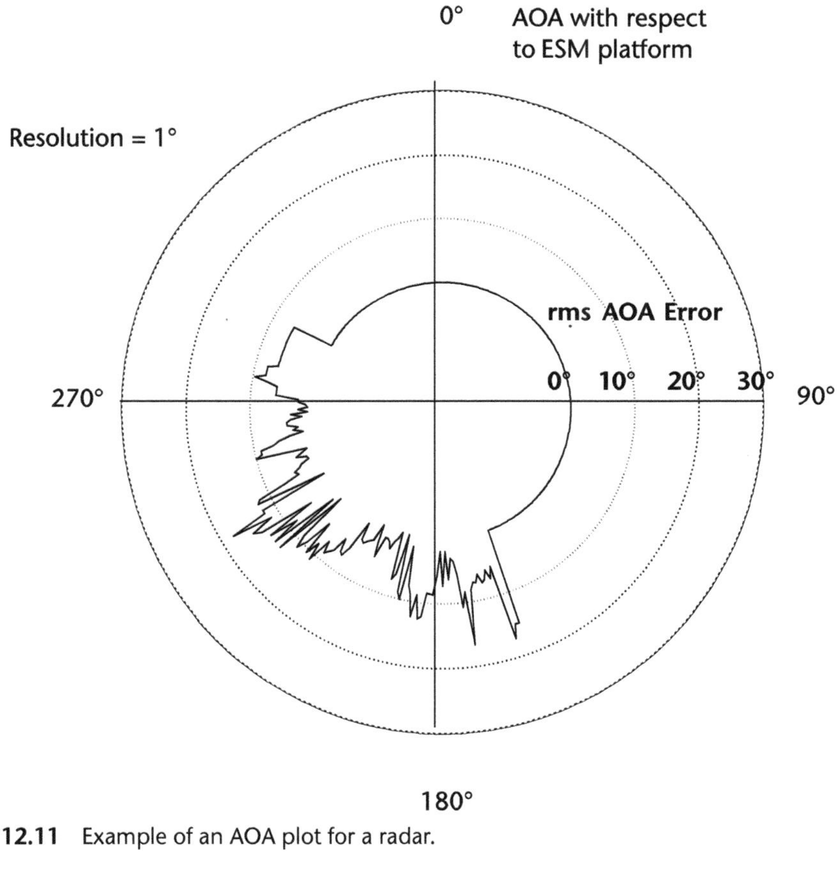

Figure 12.11 Example of an AOA plot for a radar.

and heading data (recorded in the navigation data sets). The AOA for each pulse is calculated using the following algorithm:

```
If (DOA > heading) Then
    AOA = DOA – heading
Else
    AOA = 360 – heading + DOA
Endif
```

Table 12.7 Examples of Pulse Data AOA and DOA Error

Pulse	*DOA (°)*	*DOA_error (°)*	*AOA (°)*
1	120	5	30
2	120	6	30
3	121	2	31
4	121	1	31
5	122	4	32
6	122	2	32
7	122	3	32
8	123	1	33
9	123	6	33
10	123	3	33

The data for each pulse should now look like the examples shown in Table 12.7. The aircraft heading is 90° for all the entries in this table.

A table of AOA error data can be calculated to the required angle resolution. A resolution of 1° will require a table with 360 entries; a 5° resolution will need a table of 72 entries.

Using the data in Table 12.7 to construct a 1° resolution table would produce the subset of the 360° entry table shown in Table 12.8.

The rms error is calculated for each AOA angular range using this equation:

$$rms_{\_AOA\_error} = \sqrt{\left(\left(x_1^2 + x_2^2 + x_3^2 \cdots x_n^2\right) / n\right)} \tag{12.9}$$

where n is the number of pulses in the AOA range (1° in this case) and x_n is the nth DOA error.

The calculation of the rms error around the platform is the only use for AOA data. If the raw AOA data is plotted, it is hard to make any sense of how well the ESM is performing in terms of bearing accuracy, as illustrated in Figure 12.12 showing time plots of both AOA and DOA.

It is time-consuming to produce an AOA error plot, so it is not something that needs to be done in every post-mission data analysis. However, these plots do give an indication of platform-based causes of DOA errors. They can also be constructed for intercepts where the volume of data is small relative to the volume of pulse data, so intercepts from several or even all radars in a mission can be included.

Table 12.8 Subset of 360° Data for AOA Versus DOA rms Error Plot

AOA (°)	*Contributing Pulses*	*Pulse DOA Errors (°)*	*rms Error (°)*
30	1, 2	5, 6	5.5
31	3, 4	2, 1	1.6
32	5, 6, 7	4, 2, 3	3.1
33	8, 9, 10	1, 6, 3	3.9

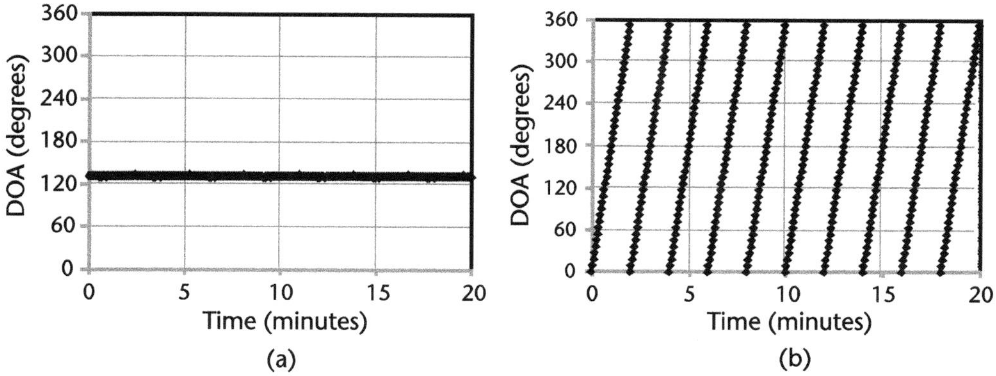

Figure 12.12 Time plots of AOA and DOA.

12.7 TDOA Histograms

TDOA ESM systems record TOA at each antenna. The DOA is represented as the TDOA on each baseline of the system (see Chapter 6). To get an indication of how accurate the DOA might be, it is necessary to produce a histogram of TDOA on the most populous baseline for the pulses from an individual radar recorded in a buffer (dwell) of pulses. If there is no DOA error in the pulse data, the TDOA histogram appears as in Figure 12.13.

However, it is very rare to see a TDOA histogram with no spread in TDOA at all. More usually, the TDOA histograms look like the ones in Figure 12.14.

In the first histogram in Figure 12.14, the spread in TDOA is 10 ns. This equates to a spread in DOA of 3°. The second histogram has an even greater spread in TDOA, of 25 ns, leading to a spread in the DOA more than 8° for this set of pulses. If the DOAs of the majority of the pulses on a baseline have been affected by

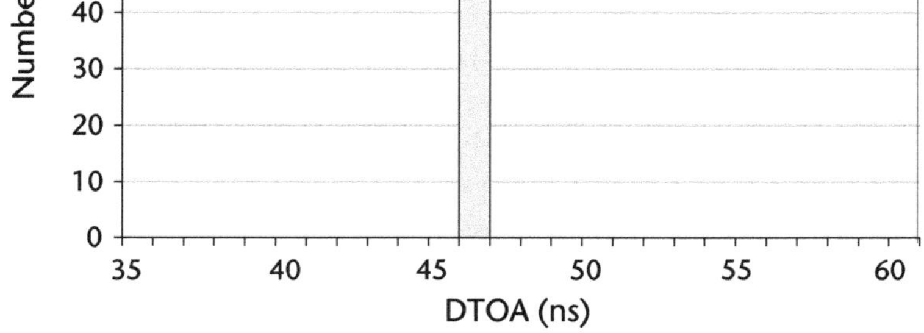

Figure 12.13 Example of a TDOA histogram with perfect pulse data.

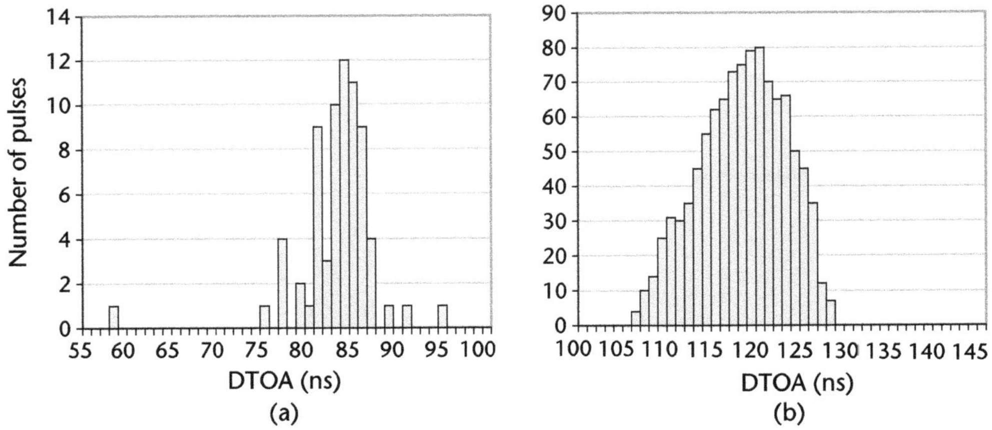

Figure 12.14 Histograms with TDOA spreads.

multipath, the peak in the TDOA histogram may be offset from the correct DOA. It is not immediately possible to tell by looking at the TDOA histogram that this is the case, but if there is a large spread in TDOA, that is an indication that there might be DOA errors for the pulses contributing to the histogram.

12.8 Parameter Histograms

Parameter histograms can be created for pulse data or for ESM track updates for a particular radar. For the pulse data, the PDWs must be filtered, as described in Section 12.5 to make sure that pulses from a single radar are used to construct the histogram. The most usual parameters for the histogram plots are RF and PW, so that an indication of the parameter measurement accuracy can be gained.

If the histogram is constructed using track data, then the ESM has decided which intercepts are correlated into the track so there is no filtering to be done by the analyst.

Figure 12.15 shows a couple of typical RF histograms, one for a set of pulses from a single radar, the other from the intercepts that make up an ESM track.

12.9 Location Accuracy

The accuracy of location estimation by the ESM can be determined if the location of the target radar is known. The latitude and longitude of the radar as calculated in the ESM geolocation algorithm and the correct radar location are compared to find the location accuracy. An example of the progression of the ESM location calculation for a radar is shown in Figure 12.16.

A more useful plot appears in Figure 12.17. This shows the difference between the correct location and the location fixes calculated by the ESM. This difference, R, is calculated using the following equation:

$$R = r\cos^{-1}\left[\cos(lat_1)\cos(lat_2)\cos(\Delta L) + \sin(lat_1)\sin(lat_2)\right](km) \tag{12.10}$$

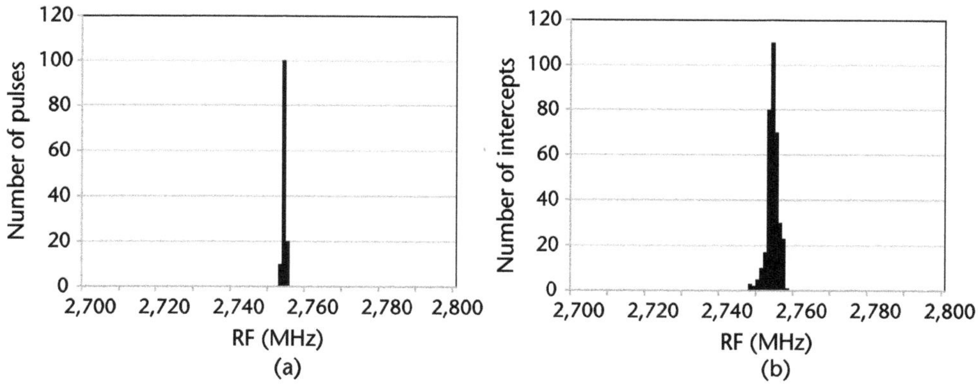

Figure 12.15 Examples of RF histograms.

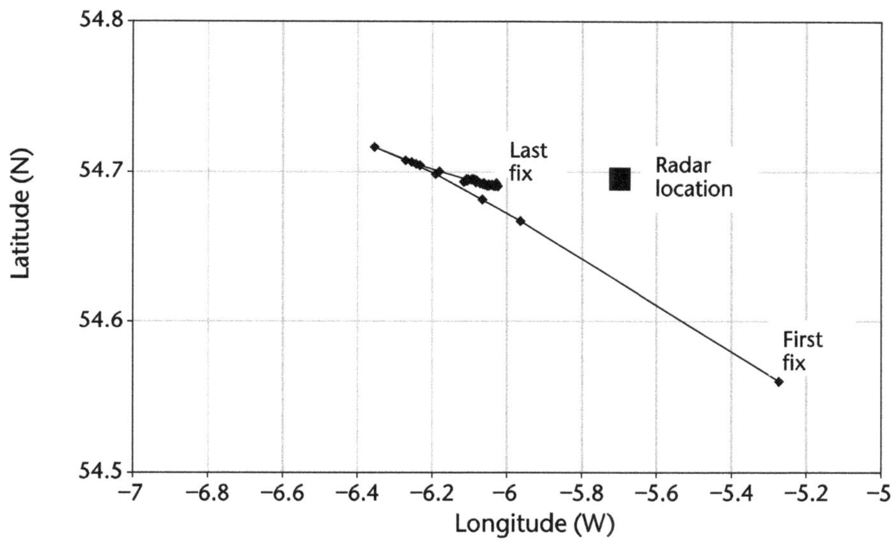

Figure 12.16 Correct radar location with ESM estimation.

where r is the average radius of the Earth in kilometers (6,371 km).

lat_1 and lon_1 are the latitude and longitude of radar location as calculated by the ESM. lat_2 and lon_2 are the correct latitude and longitude of the radar location.

$$\Delta L = lon_1 - lon_2$$

In the example shown in Figure 12.17, the error in the location of the radar quickly drops from over 30 km to a steady error of just greater than 2 km, before reducing further to under 1 km.

Part of the geolocation process is the calculation of an error ellipse (see Chapter 11). The error ellipse is specified with values for major axis (x), minor axis (y), and orientation (ψ), as defined in Figure 12.18.

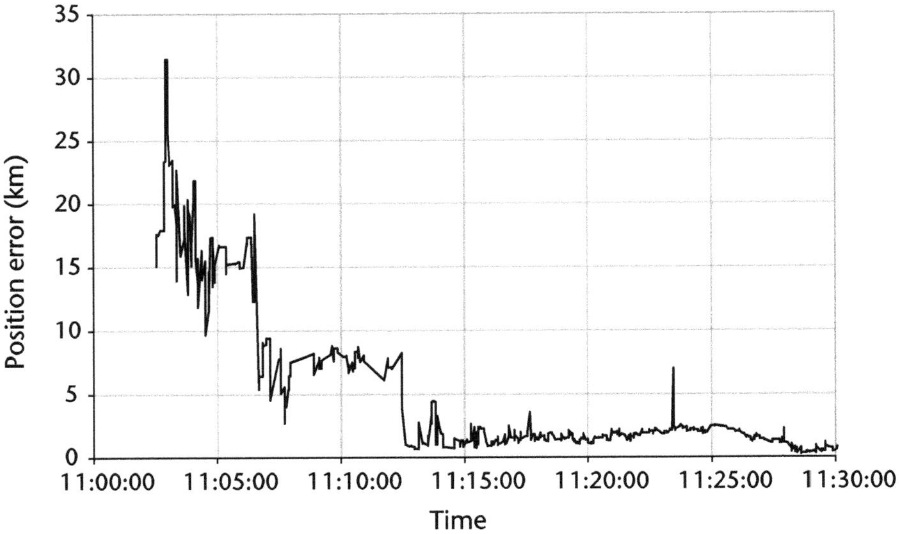

Figure 12.17 ESM performance for a radar location.

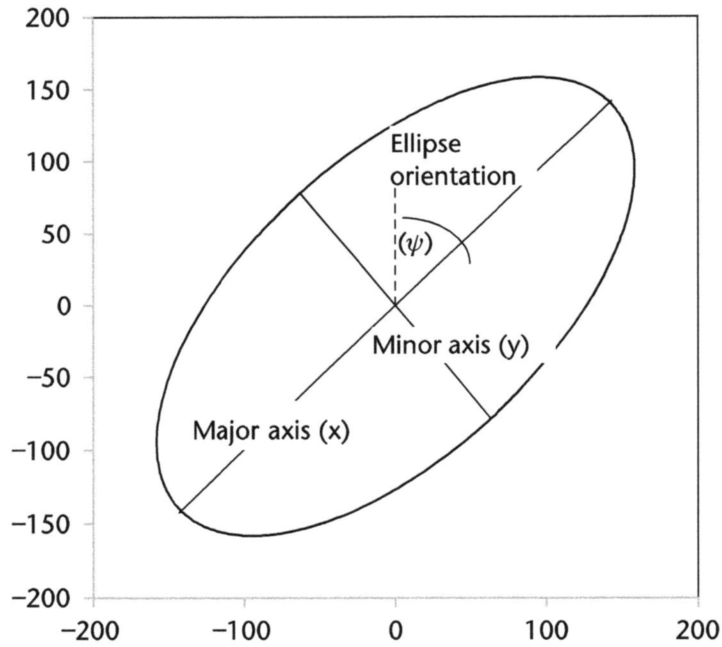

Figure 12.18 Error ellipse.

Combining the radar location and the error ellipse gives a plot like the one in Figure 12.19 where the positon error profile is plotted with the major axis of the ellipse drawn on either side of the location line. The correct radar position is also shown so that it can be seen whether the correct radar location lies within the error ellipse calculated by the ESM.

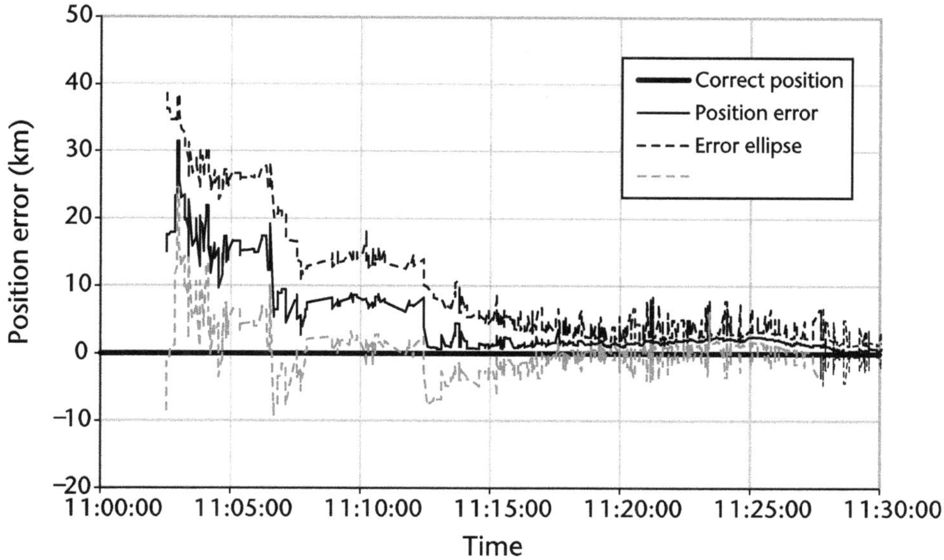

Figure 12.19 Radar location error with associated error ellipse.

12.10 Probability of Intercept

For a wide-open ESM system (i.e., one where all RFs in a wide range can be detected simultaneously), the probability of intercept should be 100% for all radars within the radar horizon (see Chapter 3). To determine the range at which the ESM can first see the radar, the maximum range at which the radar should be detected is calculated using the ERP of the radar and the sensitivity of the ESM system. A simple formula for this calculation is:

$$R = \sqrt{\left(P_t \lambda^2 / 16\pi P_r\right)} \tag{12.11}$$

where R is the range to the radar.

P_t is the radar power in watts, calculated from the ERP using the formula:

$$P_t = 10^{(ERP/10)} \tag{12.12}$$

P_r is the sensitivity of the ESM in watts converted from the specified sensitivity in dBmi

$$\mathrm{P}_r = 10^{(dBmi-30)/10} \tag{12.13}$$

λ is the wavelength of the radar in meters.

For example, a radar with ERP 80 dBW could be detected by an ESM system of sensitivity −60 dBmi at over 800 km. This range is far greater than the radar horizon for any reasonable platform altitude, so for many radars the calculation of detection range can stop at this point and the radar horizon can be taken to be the detection range.

However, the range calculated by (12.11) is the one at which the peak pulse from the radar beam is detected by the ESM system. As discussed in Chapter 3, the range at which the radar can be detected should be adjusted to account for the number of pulses that are needed to form an intercept. For radars with narrow azimuth beamwidths, this factor may significantly affect the ESM detection range. Appendix A gives details on how to calculate the detection range taking account of the radar beamshape.

The radar location is used in conjunction with the platform navigation data and the radar horizon to find the range at which each radar should come into view.

The equation for radar range from the platform is:

$$R = r\cos^{-1}\left[\cos(lat_1)\cos(lat_2)\cos(\Delta L) + \sin(lat_1)\sin(lat_2)\right](km) \tag{12.14}$$

where r is the average radius of the Earth in kilometers (6,371 km), lat_1 and lon_1 are the latitude and longitude of the ESM platform, lat_2 and lon_2 are the latitude and longitude of the radar location, and $\Delta L = lon_1 - lon_2$.

A graph like the one shown in Figure 12.20 is derived for the radars of interest. The range to each radar is plotted as a separate line on the graph, but only for the time spans that the radar would be detected by the ESM. In this example, radars 3, 4, and 5 should be seen for the whole of the mission. Radars 1 and 2 are only within detection range for part of the mission.

If the ESM system is mounted on an aircraft, then the aircraft altitude profile is needed to determine the radar horizon at the various stages of flight. If the ESM is mounted on a ship or is land-based, then the radar horizon will not change as the mission progresses.

The probability of intercept also depends on the RF scan strategy (see Chapter 4) as if the ESM is not active in the RF band of the radar when it comes into range of the ESM, then it will not detect the radar until the ESM is looking in the appropriate RF band.

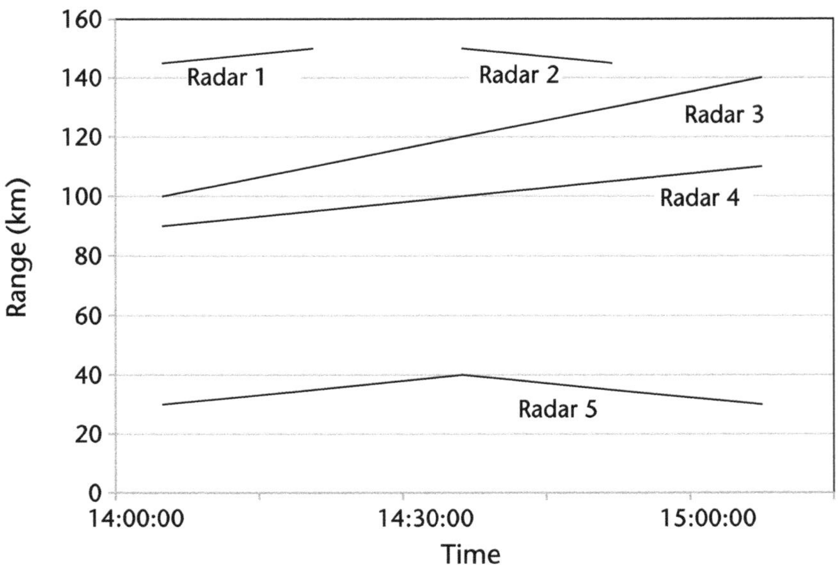

Figure 12.20 Radar ranges within the radar horizon.

12.11 Track Fragmentation

There are almost always multiple tracks created for a single radar. In the best case, there is one track in existence for a particular radar. However, it is extremely unlikely that one track will persist for one radar for the whole duration of the mission. At some point, the track will cease to be updated and be deleted from the track table. New intercepts for the radar will cause the creation of a new track. It is often the case that there are several tracks in existence simultaneously for a radar as shown in Figure 12.21, an example of multitracking for a single radar.

To see whether there are multiple tracks for a single radar, the tracks must be extracted from the ESM track data set. A filter on RF ± a few megahertz shows the tracks needed for the analysis. Examining the PRI of each filtered track shows that the track has been created from pulses of the target radar.

For example, the tracks in Table 12.9 have resulted from an RF filter of 2,740 ± 5 MHz. There are eight tracks here; some of them have several updates, and two of them are single intercept tracks.

The PRI values of tracks show that five of the tracks in Table 12.9 are likely to be for a single radar (track numbers 1, 3, 4, 5, 8) and the other three tracks (2, 6, and 7) are probably from a second radar. However, the parameters of both radars are similar making it difficult for the ESM to distinguish between them. The reason for multitracking for both radars in this example is the DOA difference between the tracks for each radar and also for each radar there is a track with a PRI harmonic of the correct PRI (tracks 3 and 6).

To carry out multiple track analysis, a good knowledge of the parameters of the specific radars seen in the mission data is needed. It takes considerable time to build up a knowledge of how a particular ESM system will see a radar, so usually the analysis of multitracking is limited to radars where the analyst knows the radar location and parameters extremely well. It is not necessarily a bad thing that only well-known radars are considered in an evaluation of ESM multitracking. It means

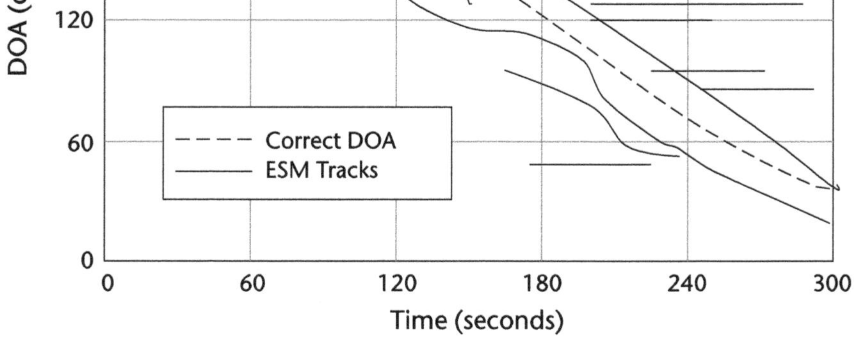

Figure 12.21 Examples of multitracking for a single radar.

Table 12.9 Example of the Extraction of ESM Tracks for a Single Radar

Track No.	*RF (MHz)*	*PRI (μs)*	*PW (μs)*	*DOA (°)*	*Updates*
1	2,942	1,020	0.9	135	1
2	2,941	1,000	0.7	142	20
3	2,940	2,040	0.9	143	5
4	2,940	1,020	0.7	133	10
5	2,940	1,020	0.8	135	1
6	2,940	2,000	0.9	139	5
7	2,939	1,000	0.7	141	10
8	2,941	1,021	0.8	140	4

that it is relatively easy to determine that multitracking is occurring and to take steps to mitigate this effect.

12.12 Accuracy of Identification/Ambiguity

This aspect of ESM performance is difficult to assess. The only way to test this is to use radars whose identities and parameters are well known to see if the ESM can correctly and unambiguously identify them.

The effects of using identity as a parameter for the correlation of intercepts into tracks or allowing track identity to be updated once correlation has taken place based on parameters alone are discussed in Chapter 9.

The problems of radar identification where there are so many radars with overlapping parameter ranges means that much of the time the ESM is making the correct choice by labelling tracks as ambiguous. It simply does not have enough information to make the correct identification decision. However, some ideas for improving the way that ESM radar libraries are specified are given in Chapter 10.

12.13 Automation of ESM Analysis

All of the techniques for ESM analysis described in this chapter are labor-intensive involving many hours of inspection of the data recorded from trials. However, software tools can be developed to automate some of the tasks. The calculation of the correct DOA and range to the radars can be carried out automatically and the filtering of track and pulse data can also be automated. These processes should be checked manually often as it is always possible that the wrong track or pulse data has been selected for analysis. Detailed analysis of pulses from individual scans must always be carried out by looking directly at the pulse data. There is no shortcut for this type of analysis.

Software tools can also be used to simulate the RF environment using known radar locations and parameters. The predicted route of the ESM platform can be used to produce an accurate simulation of the radar pulse data that the ESM should be expected to detect during an actual trial (see Section 13.6.2 for more details).

CHAPTER 13

ESM Testing and Trials

To ensure that an ESM system is producing a good picture of the RF environment, it is essential that system testing is undertaken. Data must be recorded so that a detailed analysis of the ESM system can be performed and bugs in the system can be sorted out. Identifying and fixing problems is usually an iterative process in ESM testing as sources of problems could be hardware, software, system setup, unintentional interference, background effects, or all of the above. Some aspects of performance can be assessed in a laboratory, but testing in a real-world environment, on dedicated trial ranges or with well-known radar scenarios, should be carried out prior to accepting ESM equipment into service. Operational Test and Evaluation (OT&E) should be continued throughout the lifetime of the ESM system to ensure that a consistent level of performance is maintained.

13.1 Laboratory Testing

The first stage of ESM equipment testing usually takes place in a laboratory environment. To test the full functionality of the ESM system, a systems integration laboratory (SIL) is needed. The SIL is a hardware in the loop facility with components and connectivity, including an advanced dynamic RF simulator (ADRS) as shown in Figure 13.1.

The RF generator produces RF signals for each pulse at each ESM antenna. These signals pass through an intermediate frequency (IF) module to the ESM receiver. The output from the ESM receiver is a set of PDWs that are passed to the ESM processor where the deinterleaving and intercept-to-track correlation processes are carried out.

Reasonably realistic threat scenarios can be generated including the simulation of real-life effects such as multipath. An advantage of this type of testing is that the scenarios are repeatable and can be configured for as many or as few threat radars as are needed to address specific issues with the ESM system.

Testing of the ESM system in the laboratory while in the development stage means that problems can be identified and addressed at an early stage, possibly before too much effort has gone into platform integration of the system. Although such facilities are expensive to implement, there are usually significant cost savings over doing real-life testing and trials.

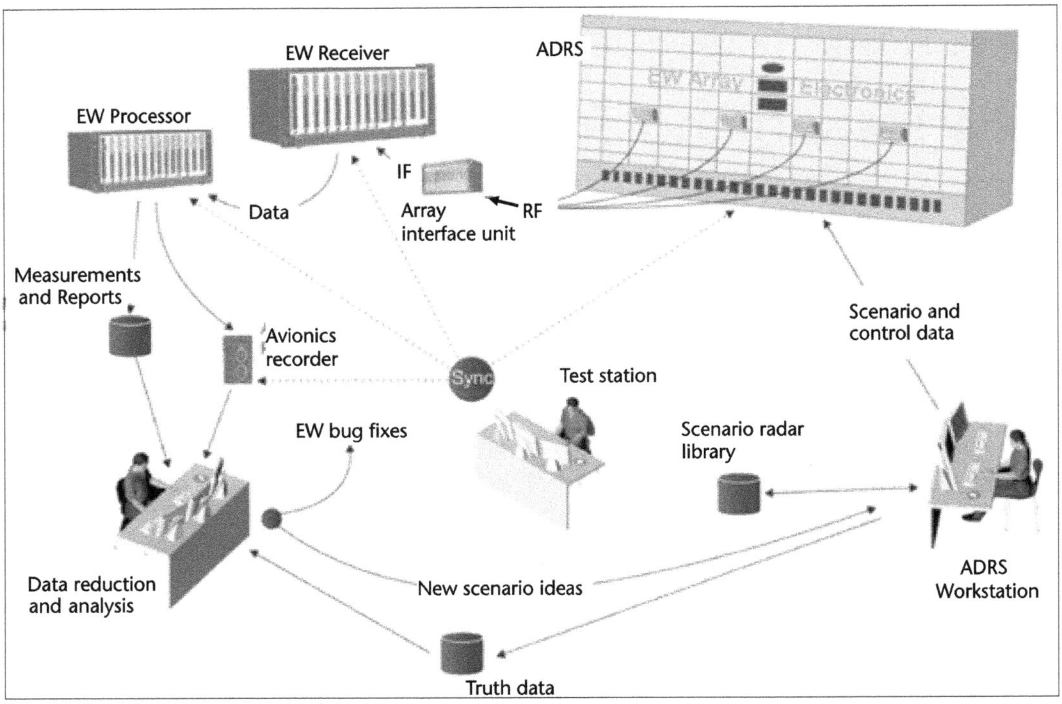

Figure 13.1 Components of a SIL [1]. (Reproduced by kind permission of Mr. Mitch Midgley-Davies, BAE Systems.)

Testing an ESM system in a SIL has an advantage that rapid replays of scenarios can be carried out to test and optimize the ESM radar library. Features of the processing such as the effect of forcing the system to produce so many tracks that the track limit is reached can be studied during testing in the SIL.

Laboratory testing also encompasses testing in an anechoic chamber [2] where several aspects of the system such as establishing antenna patterns can be achieved. There are some facilities that can house an entire aircraft for testing in an anechoic chamber [3].

13.2 Dedicated Test Ranges

Often the first stage of testing the installed performance of the ESM system takes place on an outdoor test or trials range. This is typically located in a very quiet RF environment so that signals from test radars can be received by the ESM and the performance against the specific radar with known parameters and exact ground truth can be evaluated.

Test ranges can usually provide examples of real threat radars and simulations or emulations of threats for use as ground truth for the trials. A radar simulator is a device that generates test conditions approximating actual or operational conditions. They are often emitters with no associated receiver (dumb transmitters). A radar emulator is an accurate imitation of a threat radar system. It transmits and receives and can modify its transmissions in the light of EW system responses [4].

Well-known examples of dedicated test ranges are China Lake in the United States, Spadeadam in the United Kingdom, and Woomera in South Australia.

13.3 The Need for Real-World ESM Testing

Although testing of ESM systems in a laboratory environment gives useful indications of system performance, testing in the real world is needed to determine how the system works in a real RF environment and to aid the optimization of the system.

Factors such as multipath reflections from the surface of the ESM platform and from objects on the ground affect the performance of the system (see Chapters 15, 16, and 17). Features of the ESM installation on an aircraft, such as the location of the antennas, also affect system performance and real-world testing can help to identify anomalous issues.

Common problems with ESM systems such as multitracking for single radars, poor parameter measurement, and errors in direction-finding can be identified and rectified by using test flights in an operational setting.

13.4 Planning for an ESM Test or Trial

There are several factors to be considered in setting up an ESM test or trial. The trial objectives must be decided and a suitable trials area must be identified. The route and altitude (if appropriate) of the ESM platform must be established, taking account of the locations of radars that are to be used as targets of opportunity for the trial. A simulation of the expected pulse profiles for the radars to be detected by the ESM should ideally be carried out to provide a benchmark for performance evaluation.

In the remainder of this chapter, it is assumed that the ESM trial platform is an aircraft. Although the steps in setting up the trial are the same for other platforms such as ships and land-based ESM, the aircraft is the fastest and most agile platform, so more consideration needs to be given to issues such as the trial location and route for airborne ESM platforms.

13.4.1 Setting Test/Trial Objectives

The first step in planning a trial is to set the objectives of the trial. The typical goals for a trial are to find the DOA accuracy of the ESM, to test the deinterleaver or to establish that only one track is being created for each radar (i.e., to check whether there is multitracking by the ESM system). The goals of the trial will influence the choice of trials area and define whether a special target radar is needed or whether radars already present in the environment can be used to provide targets for the ESM during the trial.

13.4.2 The Choice of a Trial Area

The expected radar pulse density in a geographic area will influence whether that environment is suitable for ESM testing purposes. It is advisable to avoid a high pulse density environment unless the objective of the trial is to find the limitations of processing of the ESM system. An example of a high-density RF area is the Straits of Malacca, one of the busiest shipping areas in the world. As described in Chapter 3, there are likely to be 200 ship radars within the detection range of an ESM operating in this environment.

A more suitable environment for an ESM trial is one such as the coastal and littoral region of East Malaysia and Brunei, as shown in Figure 13.2, where there are only 14 land-based radars and a low density of ships radars.

13.4.3 Aircraft Flight Profile

The best flight profile for a trial is one where a repeatable flight route can be flown, for example, a racetrack or box route or circle, as shown in Figure 13.3.

The distance traveled on each straight leg of a route depends on the aircraft speed. Ideally, the aircraft should fly for about 10 minutes on a straight track before turning to allow the ESM picture to stabilize or to build up a new picture if the system is reset.

Table 13.1 shows the distance that the aircraft should fly on each leg of the route at various aircraft speeds.

For a circular trial route, a rate of turn is specified as shown in Table 13.2.

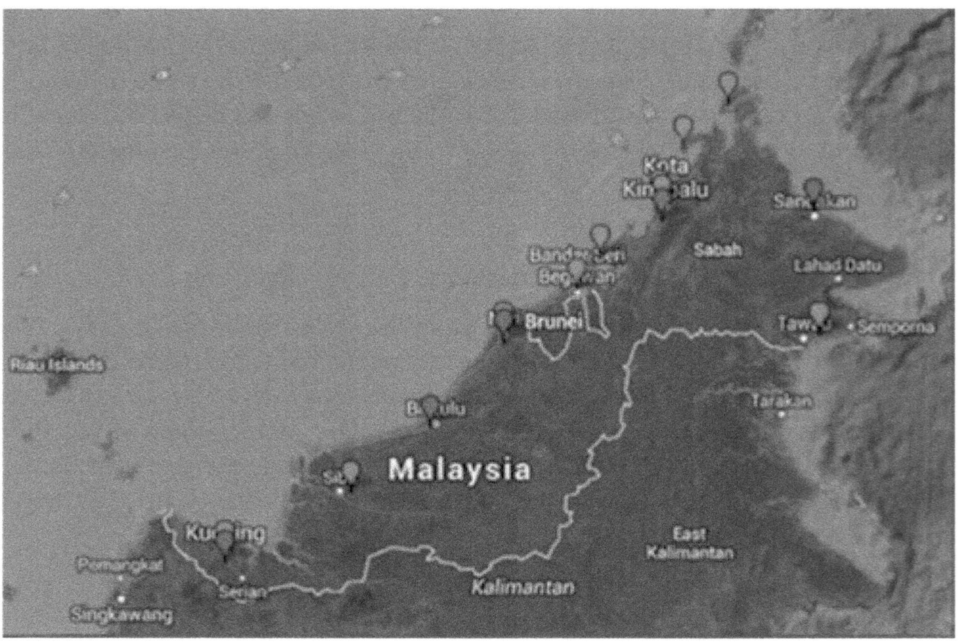

Figure 13.2 Example of a suitable ESM trials environment.

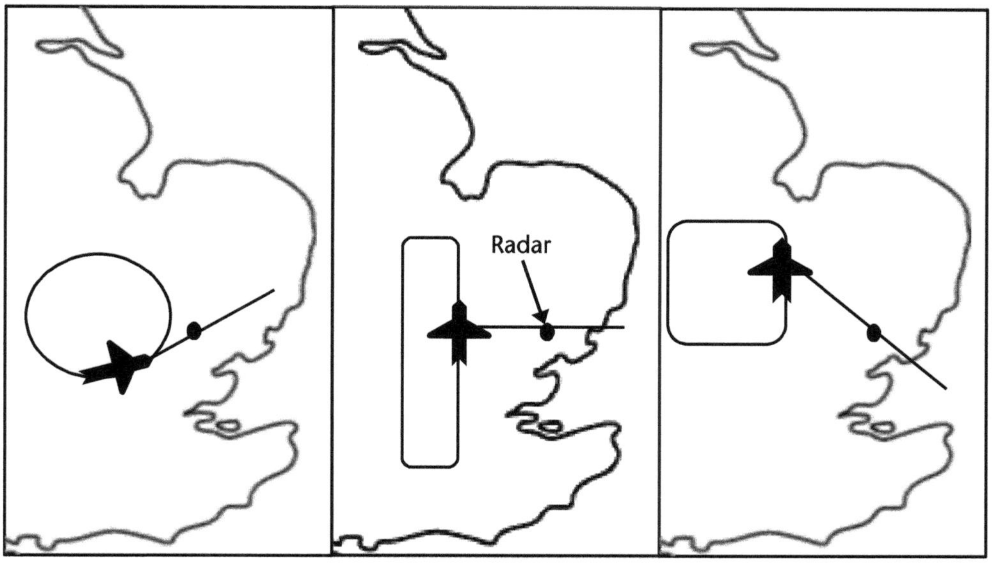

Figure 13.3 Racetrack and box and circle trials/test flight profiles.

Table 13.1 Aircraft Route Leg Length for Various Aircraft Speeds

Aircraft Speed (knots)	*Approximate Distance (km)*
80	25
120	40
200	60
300	90
500	150
1,000	300

Table 13.2 Rates of Turn

Turn	*Degrees/Second*	*Turn Duration for 360°*
Rate 1	3	2 minutes
Rate 2	6	1 minute
Rate 3	9	40 seconds
Rate 4	12	30 seconds

13.4.4 Aircraft Altitude

The altitude at which the trial is flown will determine how far the ESM can see, as described in Section 3.2. For example, a radar with a range of 60 nautical miles could be seen at a range of 3,600 nm by an ESM, if it was not for the curvature of the Earth. The radar horizon increasing with altitude is the real governing factor for the range at which the ESM can detect radars. Table 13.3 shows the radar horizon for various aircraft altitudes.

Table 13.3 Radar Horizon

Aircraft Altitude (ft)	*Radar Horizon (km)*
200	32
1,000	72
2,000	102
5,000	162
10,000	230
20,000	320

13.4.5 Determination of Ground Truth

Having chosen the test area, the exact locations of radars (latitude and longitude) are needed to produce a picture of the ground truth for the testing and to help in the determination of the optimum route for the ESM platform. The range at which an ESM system can detect most radar types (apart from those with very narrow beamwidths) is dependent only on the radar horizon, as discussed in Section 3.2. When planning the exact route that the aircraft is to take, the start and end points of the repeatable flight pattern are set up so that there is a good mix of nearby known radars and those that are at greater range available as the ESM platform performs its flight pattern. This allows several aspects of ESM performance to be tested in the same trial (e.g., probability of intercept and deinterleaving capabilities). The expected density of mobile radars such as those on ships should also be available for designing the trial. Although the precise locations and identities of the ships are not known in advance, the impact of the density of such radars on the trial can be assessed. AIS data can be recorded during the trial to provide the ground truth for shipborne radars.

13.4.6 Preflight Data and Simulation Needed Before Testing Takes Place

Once the trial area has been defined and information has been gathered on radar targets of opportunity, a simulation of the radar pulses that should be received by the ESM in the trial is carried out. The following steps must be accomplished to produce a realistic plot against which the recorded pulse data can be compared in post-flight analysis.

The aircraft route is generated with predicted latitude and longitude at 1-second intervals and the pulse data is simulated using the following process:

```
For each radar
  For each radar scan
    Calculate the range using the known radar location and the aircraft position
    Generate a scan of pulses using the known radar parameters
  Next radar scan time
Next radar
```

The method of simulating the radar beam shape taking account of the radar parameters such as 3-dB beamwidth, first sidelobe level, and ERP is given in

Appendix A. The result of this simulation for a single radar is a set of pulses at each scan time, as shown in Figure 13.4. The amplitude is plotted for these pulses as this is the best way to visualize the individual scans for the radar. The first plot in Figure 13.4 shows the amplitude of the pulses versus time and the second plot shows the amplitude plotted against the pulse number so that the shape of the amplitude profile of the pulses in each scan can be seen. Either way of plotting the data may be used in the simulation of ground truth for a trial and for comparison with the recorded data in post-trial analysis.

13.5 Example of an ESM Test/Trial Preparation

An example of the design and analysis of an airborne ESM test or trial starts by deciding the trial objectives, determining the ground truth, setting the aircraft flight parameters, and simulating the radar pulse data as described next.

13.5.1 Test/Trial Objectives

The objectives of this example trial are:

- To check the creation of an intercept for a radar as it comes into range at the radar horizon (probability of intercept (POI));
- To test the deinterleaver using two of the same type of coastal surveillance (CS) radar;
- To check the extent of multitracking;
- To test the direction finding (DF) or DOA performance.

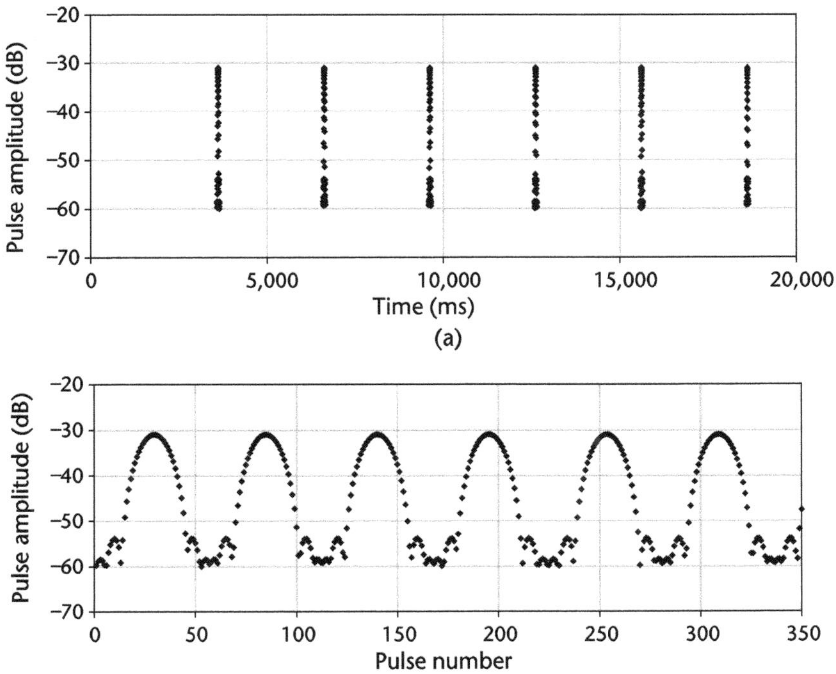

Figure 13.4 Simulated scans for a single radar.

13.5.2 ESM Platform Parameters

In this trial the ESM platform aircraft operates with the following parameters: aircraft speed = 200 knots and maximum altitude = 20,000 ft. For a racetrack route with each of the two legs of duration being 10 minutes, a distance of 60 km is needed for each leg (200 knots = 100 m/s, distance = 100 m/s × 600 seconds = 60 km). The ESM sensitivity is –60 dBmi.

13.6 Preflight Preparations

The first task in setting up a trial is the selection of a suitable trial location. The area shown in Figure 13.5 is chosen as there is relatively low radar pulse density.

The locations of the fixed radar sites in the vicinity of the trial are shown in Figure 13.5 There are six radars in total: two ATC radars, one weather radar, and three CS radars. Although there are six land-based radars in the trial area, it is the three CS radars that are of interest as the analysis of the ESM performance against them will fulfill the stated trial objectives.

13.6.1 Selection of Aircraft Route

One of the objectives of the trial is to test that the ESM can detect a radar as it comes into view at the radar horizon. The choice of the aircraft route and altitude must take this into account. Drawing the radar horizon rings onto the map of the area where the trial is to take place allows the specification of the aircraft route

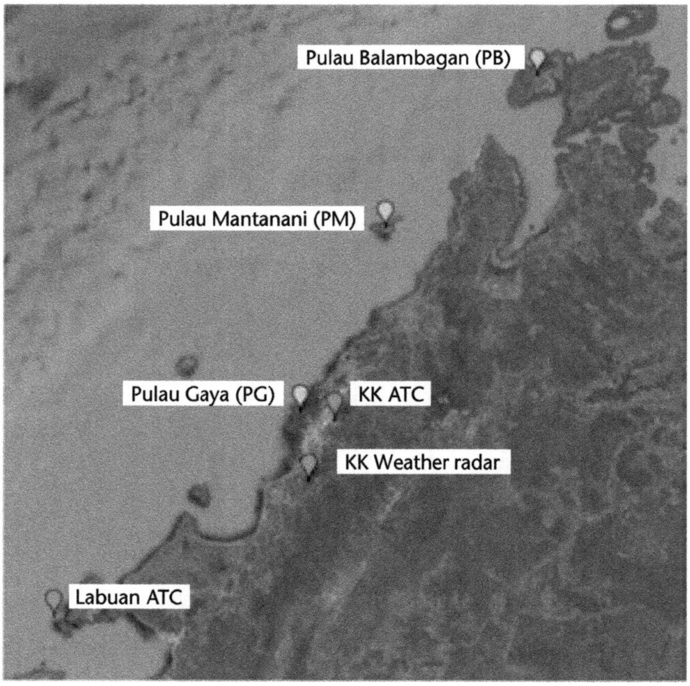

Figure 13.5 A low density radar environment for trials.

and the altitude at which the trial should take place. Figure 13.6 shows the radar horizon rings at various altitudes from 72 km at 1,000 ft to 323 km at 20,000 ft.

If the trial takes place with the aircraft flying a repeatable racetrack pattern off the western coast of East Malaysia at 5,000 ft, the radar at Pulau Balambagan (PB) will come into and out of range as the aircraft flies the racetrack pattern. This will allow the first objective of the trial, the timely creation of an ESM track for a radar, to be tested. The trial route should be set so that two of the radars, Pulau Mantanani (PM) and Pulau Gaya (PG), will be close in the DOA for some of the trial route to test the deinterleaving process and fulfill the second objective of the trial. The position of the aircraft route is not critical for the other two objectives, the test of multitracking and DF. The proposed route consisting of a racetrack with two legs of 10 nautical miles (18.52 km) each is plotted on Figure 13.7. The aircraft is headed in a north northeasterly (NNE) direction for one leg and then south southwest (SSW) for the return leg.

The ranges to the three radars of interest can now be calculated at each end of the aircraft route (i.e., when it is at its most southerly point and also at the aircraft is at the most northerly point of the route). Table 13.4 shows the ranges for the three radars.

13.6.2 Simulation of the RF Environment

A simulation of the pulses from the six land-based radars in the vicinity of the trial is shown in Figure 13.8. This is a plot of the timeline of the pulses for each radar

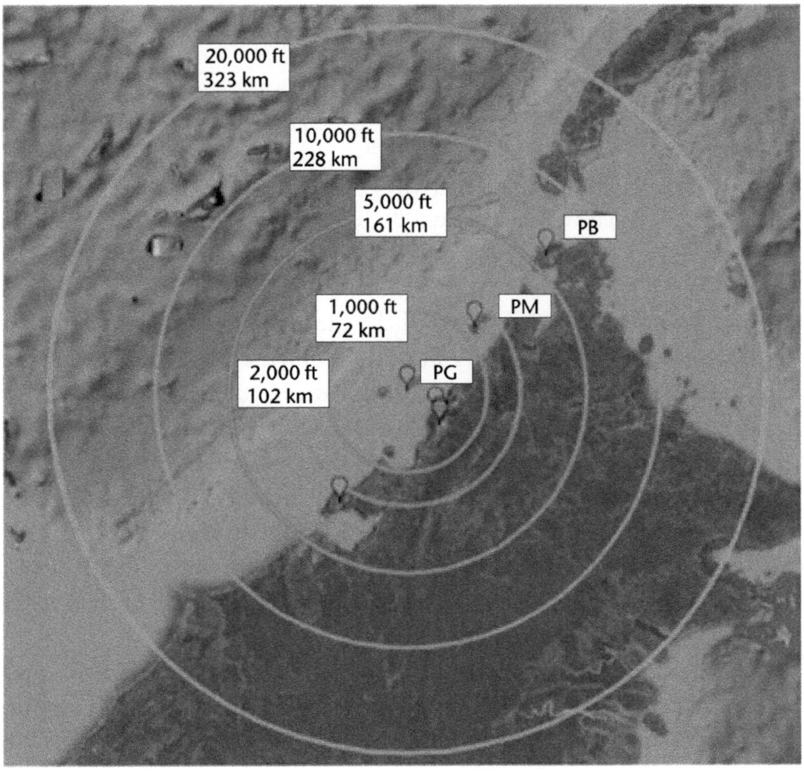

Figure 13.6 Radar horizon rings.

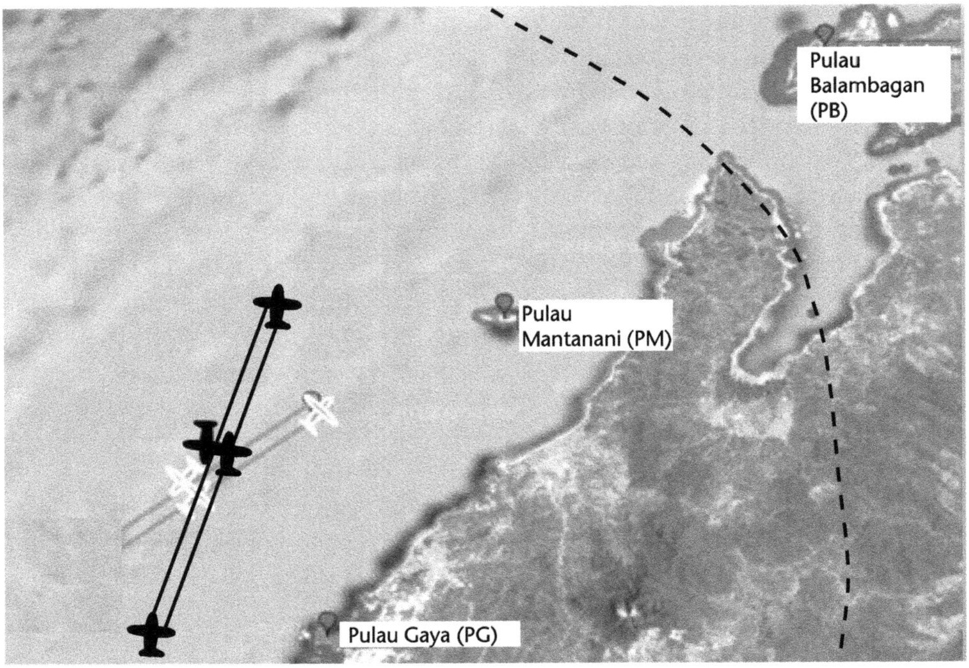

Figure 13.7 Trial route with target radar locations.

Table 13.4 Ranges to the Three Ranges at Each End of the Radar Route

Site Name	*Range at Route Leg Start (km)*	*Range at Route Leg End (km)*
Pulau Gaya (PG)	60	53
Pulau Mantanani (PM)	115	44
Pulau Balambagan (PB)	205	134

for the first half of the NNE leg of the aircraft's route. Each blob on the plot shows the duration for which pulses should be detected for each radar scan. The weather radar has pulses seen for the longest duration, for 350 ms on each scan, whereas the maximum time for the ATC radar scans is for 100 ms. The extent of detection time for each scan depends on the range to the radar. Only the scans that are within the radar horizon should be detected and are shown on Figure 13.8.

The aircraft in this trial will fly at 5,000 ft so the radar horizon is just over 160 km. One of the ATC radars is only detected for part of the way along the NNE route leg. One of the CS radars is not seen until the aircraft has traveled over half of the way along the same leg.

The next step in preparing for the trial is to simulate the amplitude profile of the pulses that are expected to be detected in the trial. The amplitude profiles for all 6 radars for 50 seconds of flight are shown in Figure 13.9. In 1 minute of flight, there are approximately 4,000 pulses received from the six radars.

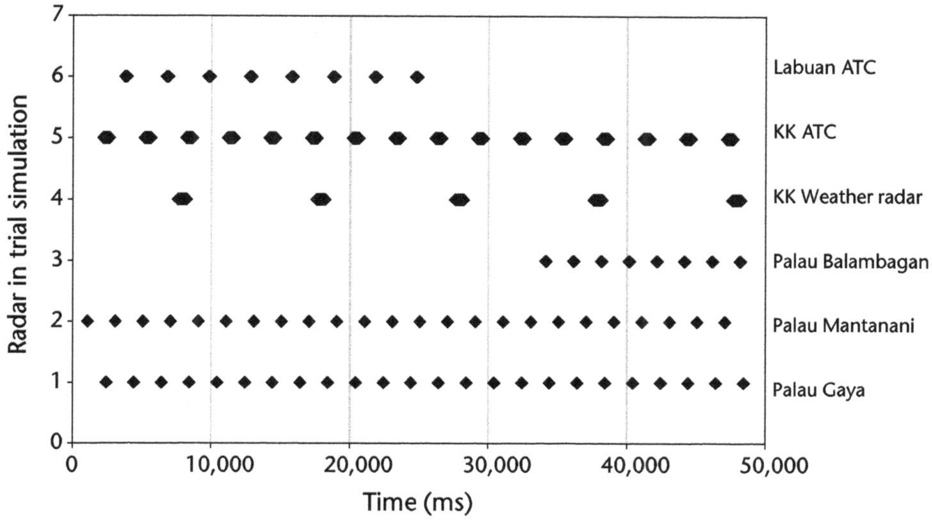

Figure 13.8 Timeline of all radars in the trial.

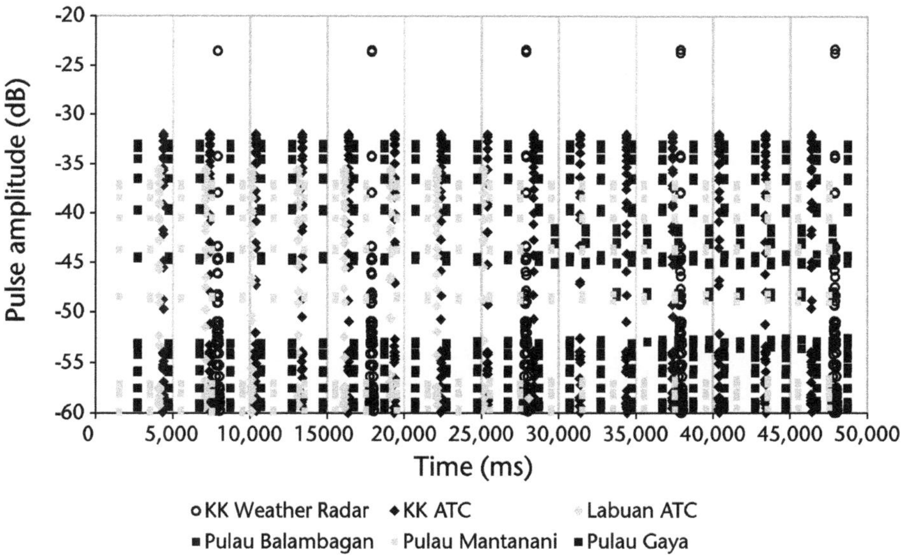

Figure 13.9 Predicted amplitude of all radars showing pulse density.

More details of the amplitude profiles are shown in Figure 13.10 where there is one scan peak from the weather radar and one peak from an ATC radar with two much narrower peaks from two of the CS radars.

13.7 Postflight Data Analysis

13.7.1 Visualization of the Operator's Screen/Replay of Recorded Data

The first stage of the post-flight data analysis is the replay of the recorded data as it seen during the flight on the operator's screen. Figure 13.11 shows the ESM tracks

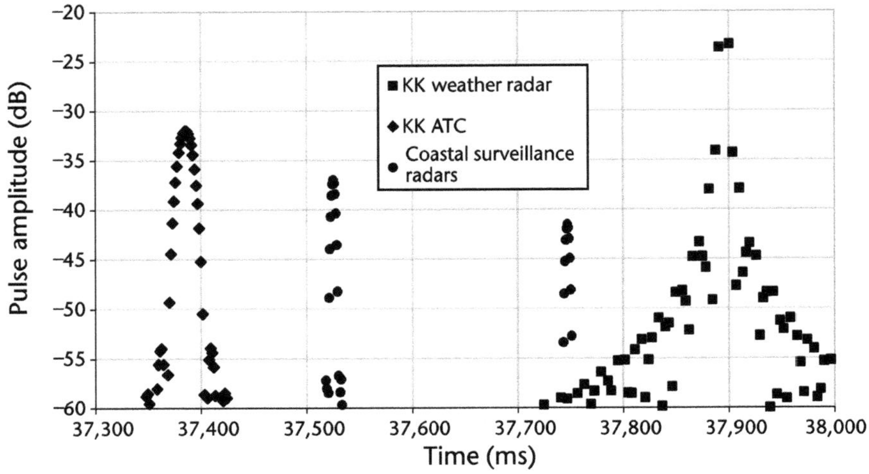

Figure 13.10 Fine detail of predicted amplitude profiles.

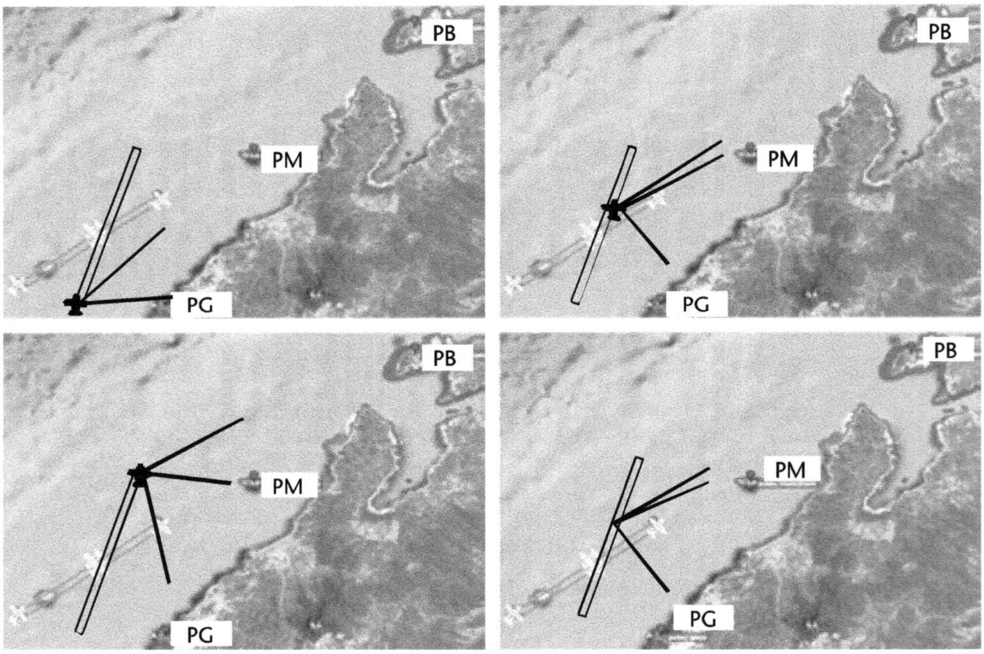

Figure 13.11 Visualization of the trial.

that are seen at various positions along the aircraft route. In this case there is one track for each radar.

13.7.2 Calculation of Ground Truth

A calculation of the correct DOA and range of each radar including when they come into and out of the ESM detection range leads to the graphs of the DOA and range profiles in Figure 13.12. (The methods of calculating the correct DOA and

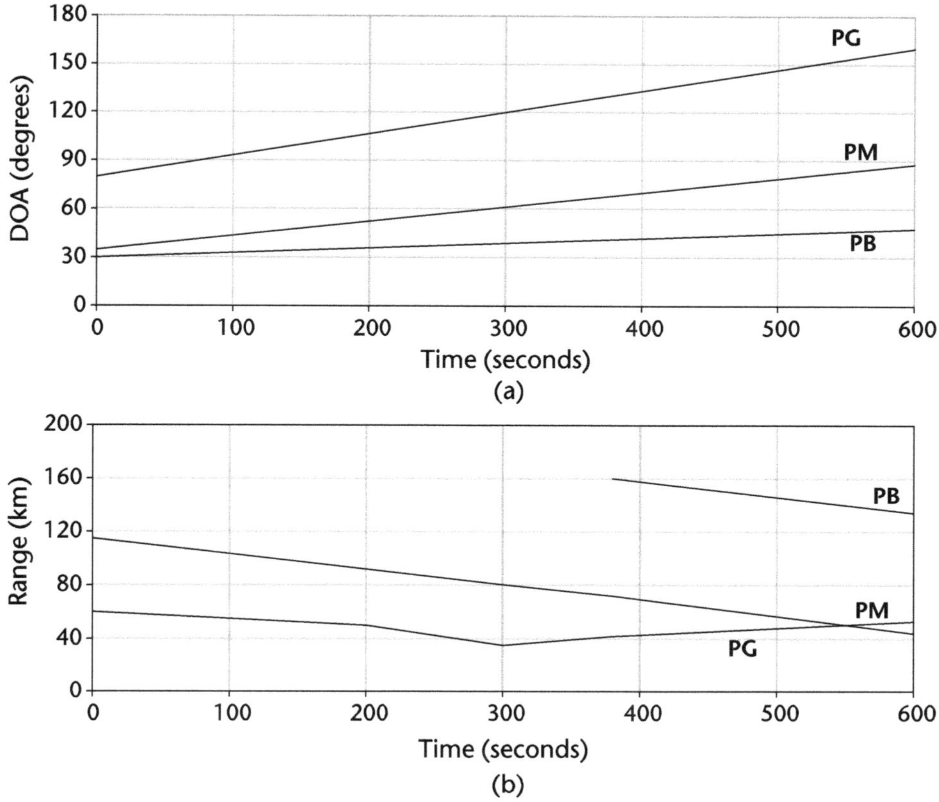

Figure 13.12 DOA and range profiles for trial radars.

range from the ESM platform to the radar are given in Sections 12.3 and 12.10.) The graphs show how the DOA and range change as one leg of the aircraft route (taking 10 minutes with the aircraft traveling NNE) is flown.

Two of the radars are within detection range of the ESM throughout the whole trial, but one of them (PB) is not seen until over halfway through the leg of the trial when the aircraft is traveling NNE; hence, the range profile for this radar does not start until over 300 seconds after the start of the NNE route leg.

13.7.3 Analysis of ESM Track Data

The next stage in the post-flight analysis is to plot the DOA of the ESM track data against the correct DOA of the radars, as shown in Figure 13.13.

There is only one ESM track for the PB radar. This radar is at relatively long range to the aircraft and only comes into the range of the ESM system about halfway through the NNE leg of the route. A track is created at the expected time, in the circle marked A on Figure 13.13 when the aircraft moves to a point where the radar is within the radar horizon. The creation of this track has fulfilled the first objective of the trial to test whether tracks are created on a timely basis when the radar comes into ESM detection range.

The second objective of the trial is to test the deinterleaver by having two radars of the same type with similar DOA profiles in the environment of the trial. It is possible to see by eye which tracks on Figure 13.13 have been created for each

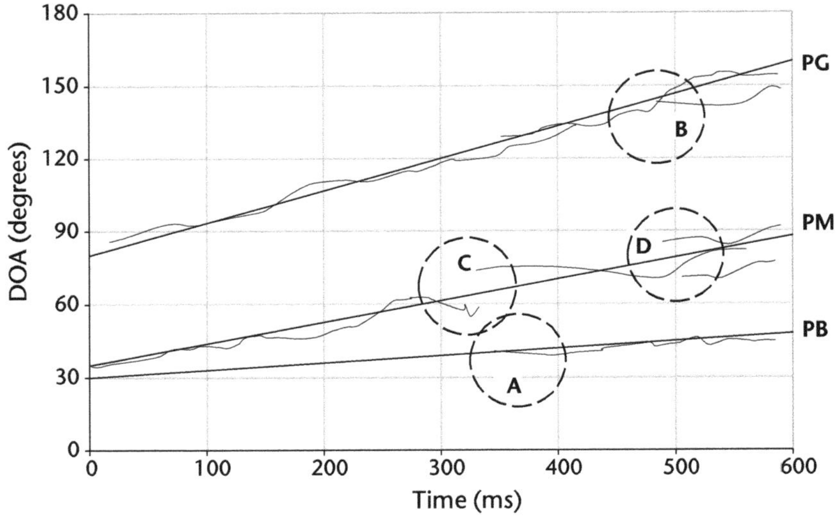

Figure 13.13 ESM track data for the three coastal surveillance radars.

radar, but to be sure that pulses from more than one radar are not contributing to a single track, analysis of the pulse data is necessary.

The third objective of the trial is to check that the extent of multitracking. There are two instances of multitracking in the track data for this example trial. The circle marked B on Figure 13.13 shows two tracks in existence simultaneously for the PG radar. For most of the route leg, there is only one track in existence at any one time for the PM radar. However, towards the end of the leg, when the range to the PM radar is reducing, two extra tracks are created for this radar, shown in the circle marked D. Further investigation of the multitracking issue is required using the recorded pulse data.

The tracks created for all three radars are reasonably close in DOA to the correct DOA, apart from a discontinuity in DOA for the PM radar, shown in the circle marked C.

Further analysis of the recorded pulse data is therefore needed to establish the DOA performance, the fourth objective of the trial.

13.7.4 Analysis of ESM Pulse Data

Plotting the recorded pulse data of the three CS radars against the correct DOA profiles of the radars in Figure 13.14 shows the fundamental DOA performance of the ESM system. Pulses are received from the PG and PM radars for the whole duration of 10-minute route leg, whereas pulses are only received from the PB radar for part of the route leg due to the increased range from this radar.

Figure 13.15 shows more detail of the pulse data for the last 100 seconds of the route leg. The PB radar has the smallest spread in DOA as it is at relatively long range from the ESM platform. Only one track is created for the PB radar and it is within 5° of the correct DOA at all times so the DOA performance for this radar is deemed satisfactory. The PG and PM radars have the greater DOA spreads for their individual scans as they are much nearer to the aircraft during the time period shown in Figure 13.15.

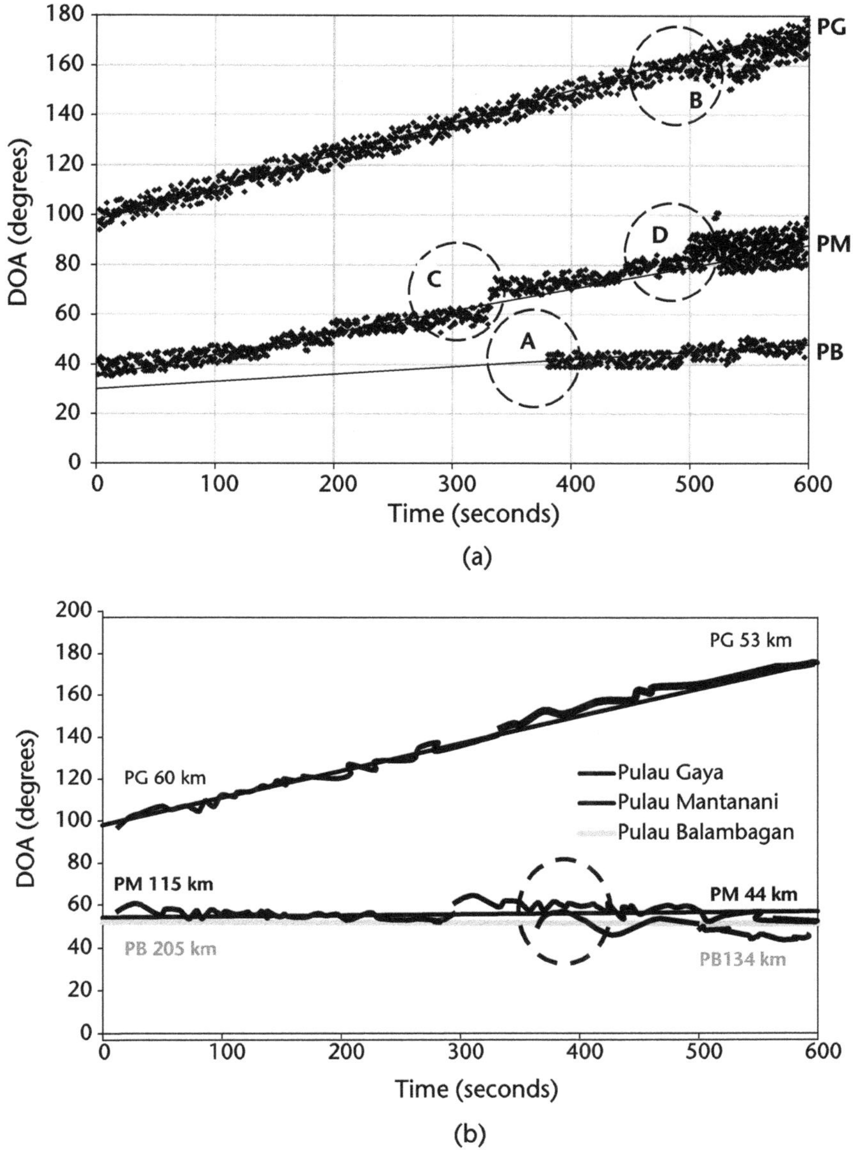

Figure 13.14 Radar pulse data from the trial.

Pulse DOA errors are likely to be greater for pulses that are not near the radar scan peak and for pulses that are detected from the sidelobes of the radar (see Chapters 5 and 15). The reason for the multitracking for the PM and PB radars is therefore likely to be because of DOA errors. Analysis of individual radar scans confirms that there is a spread in DOA seen across single scans and also indicates that the average DOA from scan to scan is different as might be expected for radars at relatively short range to the aircraft (see Chapter 16). Figure 13.16 shows the DOA profiles of consecutive scans for the PM radar recorded near the end of the route leg when the range to this radar is at a minimum. The DOA change from scan to scan leads to the creation and maintenance of two separate tracks for this radar in this case.

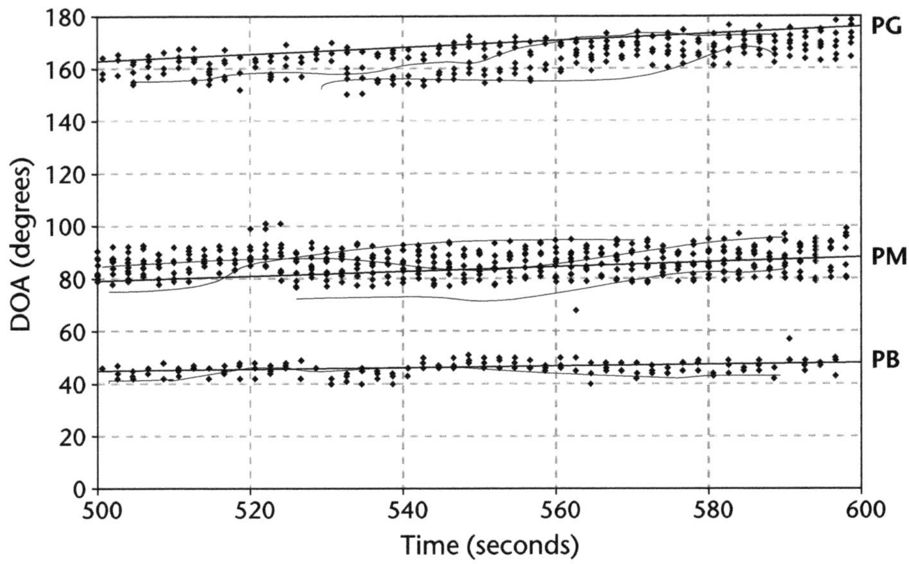

Figure 13.15 Detail of recorded radar pulse data.

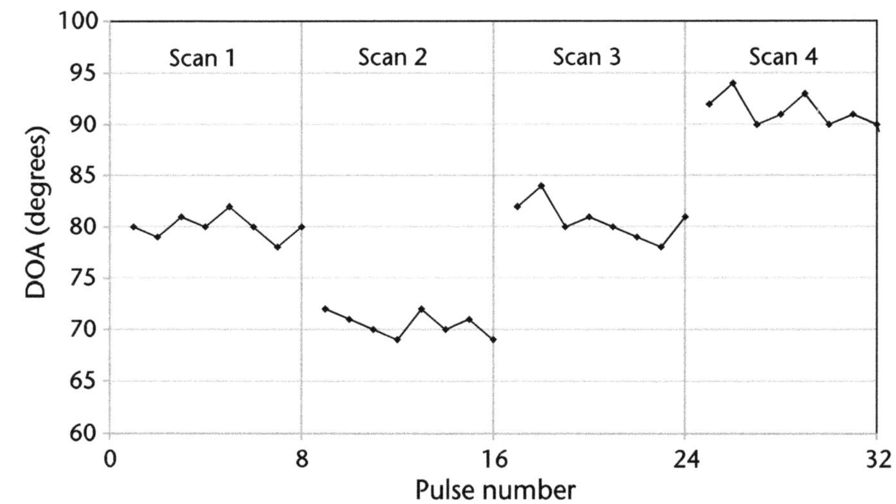

Figure 13.16 DOA profiles for consecutive scans for the PM radar.

A final task in the analysis of the pulse data is to compare amplitude profiles for individual scans with those generated using a simulation of the radar beamshape. Figure 13.17 shows an example where a theoretical scan peak is compared to a recorded scan peak. In this case, there is good agreement between the two graphs. However, examination of the recorded pulse data in this way is often needed to start to diagnose and remedy issues with ESM systems.

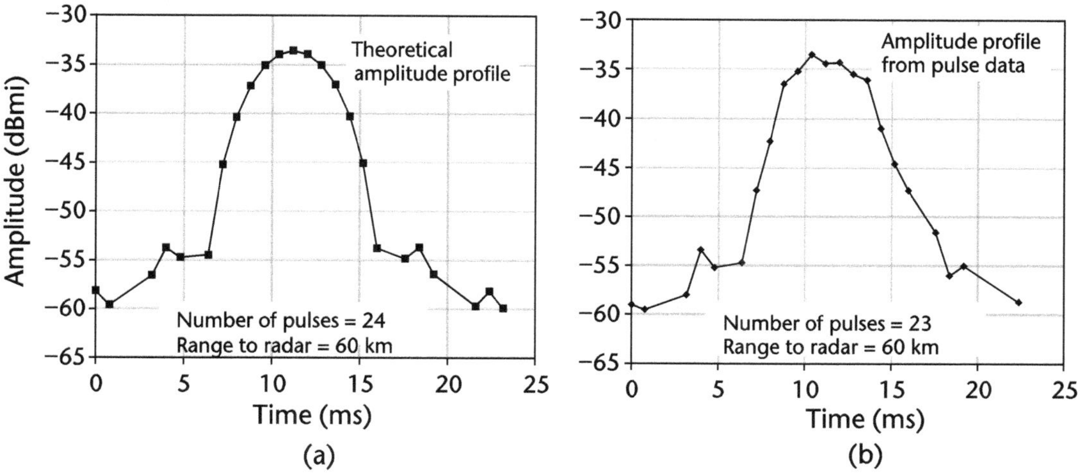

Figure 13.17 Individual radar scan comparison.

References

[1] Midgely-Davies, M., http://tangentlink.com/wp-content/uploads/2014/07/2.-Electronic-Warfare-Test-Evaluation-Mitch-Midgley-Davies.pdf, *Proc. EW Asia Conference*, Kuala Lumpur, 2014.

[2] Dash, G., http://glendash.com/Dash_of_EMC/Anechoic_Chambers/Anechoic_Chambers.pdf, 2005.

[3] Hehs, E., http://www.codeonemagazine.com/article.html?item_id=57, October 2010.

[4] Davies, W., http://tangentlink.com/wp-content/uploads/2014/03/6.-Integrated-Air-Electronic-Warfare-Ranges-Wynne-Davies.pdf, *Proc. EW Asia Conference*, Kuala Lumpur 2014.

CHAPTER 14

Multitracking

Multitracking is one of the most common adverse issues in the performance of ESM systems. The creation and update of more than one track for each radar in the RF environment is a major source of confusion for the ESM operator. Although experienced ESM operators recognize that multitracking is present and compensate for the extra tracks using their specialist knowledge, it is useful to look at some of the reasons for multitracking and to explore some measures for reducing the effect.

There is another important reason for multitracking to be minimized as most ESM systems have a limit on the number of tracks that can be in existence at any one time. Once this limit is reached there can be no new tracks created until some of the existing tracks have been deleted, usually timing out through a lack of updates. This means that new radars appearing within the detection range of the ESM are not seen or recorded once the ESM track limit has been reached. An ESM system being operated in a tactical role cannot be relied upon as a defensive aid if it is likely to be blind to new threat radars.

The opposite problem of overmerging of pulses from more than one radar into a single intercept or the merging of intercepts from more than one radar into a single ESM track is also an issue that needs to be addressed to ensure that the ESM can produce a reliable picture of the RF environment. This is discussed further in Chapter 9.

14.1 Causes of Multitracking

There are many causes of multitracking, mostly due to mismeasurement and miscalculation of radar parameters, considered in this chapter. However, it should be noted that correlation failures, between intercepts and tracks, may also cause multitracking. These issues are considered in Chapter 9. Parameter-based causes of multitracking are:

- DOA errors due to antenna separation;
- DOA errors due to multipath interference;
- PRI calculation errors;
- PRI errors due to missing pulses;
- PRI errors due to complex PRI sequences;

- PRI errors due to PW measurement errors;
- PW measurement errors;
- RF agility.

An example of the tracks shown on the ESM operator's screen when there are multiple tracks created for a single radar is shown in Figure 14.1. There are four tracks in Figure 14.1: one of them is at the correct DOA, and five of them are at incorrect DOA. In this case, all the extra tracks are due to mismeasurement of the DOA. In reality, there may be other tracks overlaid with these tracks that have the same DOA, but have been created due to the miscalculation of other parameters, such as PRI.

14.2 DOA Errors Due to Antenna Separation

In an ESM system with separated antennas, there are DOA measurement errors due to different input pulse amplitudes at each antenna. This effect is present at ranges up to 100 km and is a commonly seen phenomenon, described in detail in Chapter 5. The main effect of antenna separation is a slope in the DOA across the main beam of a scanning radar, as shown in Figure 14.2. The spread in DOA shown on this graph leads to the creation of at least two separate intercepts, one at about 60° and the other for pulses near the peak of the radar scan at 43°. These two intercepts either correlate to separate existing tracks or cause the creation of new extra tracks for the radar.

14.3 DOA Errors Due to Multipath Interference

The effects of multipath on the input pulse amplitude lead to a variety of the DOA profiles for intrascan DOA profiles. At short range, the DOA can vary considerably

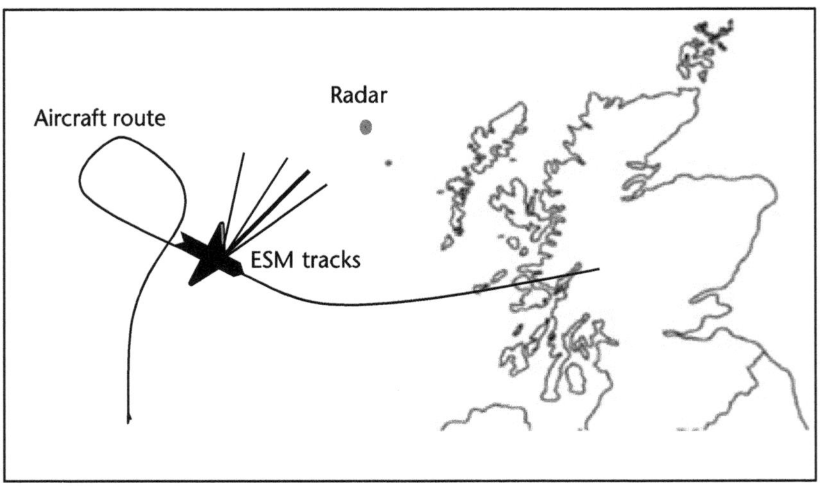

Figure 14.1 Example of a map with multiple tracks for a single radar.

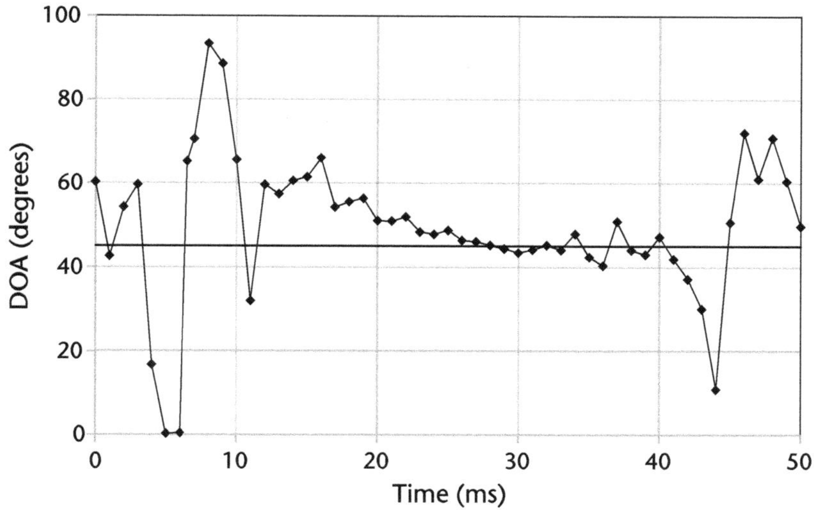

Figure 14.2 DOA profile of a radar scan received by an ESM with separated antennas.

from pulse to pulse (see Chapter 16), and there are several categories of the DOA profile that are commonly present in recorded pulse data (see Chapter 17). An example of one of the DOA profiles for a single scan of radar pulses affected by multipath is shown in Figure 14.3.

Figure 14.3 is a typical DOA profile with errors due to multipath effects starting part of the way through the radar scan. The pulses from this scan have two intercepts created for them, one for pulses at the correct DOA near the start of the scan and the other at a DOA error of nearly 20°.

14.4 PRI Calculation Errors Due to DOA Errors

PRI is the most important parameter in the specification of ESM library entries. There is an assumption in the design of ESM systems that a complete set of pulses will be received from each radar so that PRI can be correctly calculated.

As described in Chapter 8, the first part of the deinterleaving process is the separation of pulses into pulse chains (intercepts) for each radar (i.e., deinterleaving). Deinterleavers usually work by clustering radar pulses according to RF and DOA or DTOA (for TDOA ESM systems). In the ideal case, one cluster is created for each radar. In the example shown in Figure 14.4, there are pulses from three radars on an RF/DOA plot.

In addition to the clusters of pulses from the three radars at the correct DOA, there is a second RF/DOA cluster for pulses from radar 3 having incorrect DOA. There is no way for the ESM to recognize that the extra cluster belongs to radar 3 at this stage of the processing. The existence of this extra cluster will lead to a gap in the PRI sequence of the cluster at the correct DOA for radar 3 leading to a miscalculation of the PRI.

There are lots of reasons for DOA errors, from the configuration of the antennas (Chapter 5) to multipath effects (Chapter 15). To illustrate the effect of the

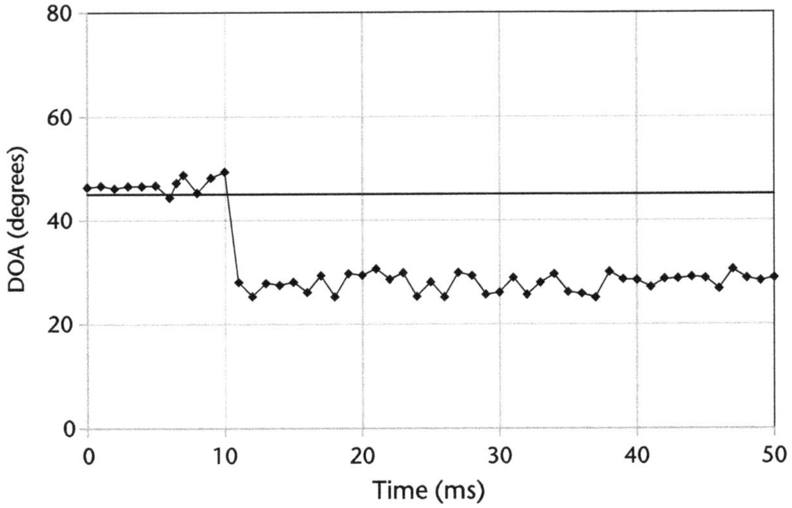

Figure 14.3 DOA profiles of a single scan of a radar affected by multipath.

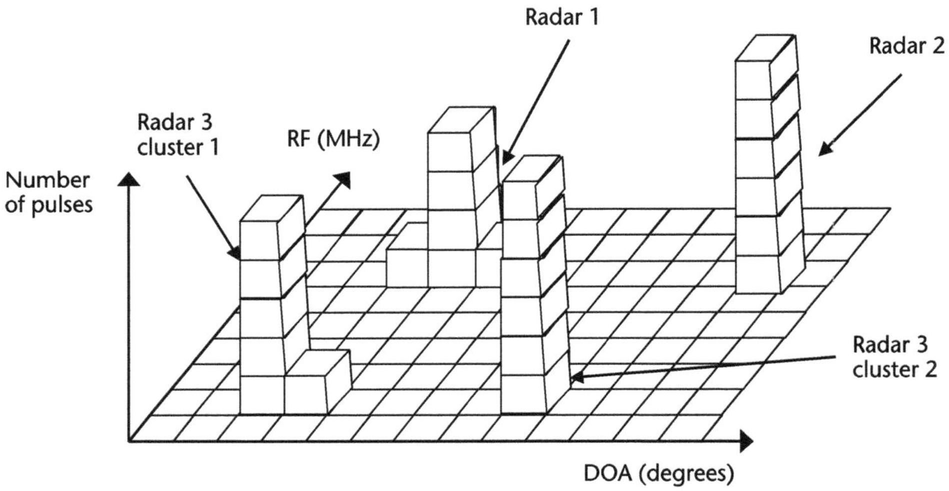

Figure 14.4 RF/DOA cluster diagram for a buffer of pulses received by an ESM system.

DOA error on PRI calculation, Figure 14.5 shows a DOA profile of a radar with simple PRI of 1m but with DOA errors for some of the pulses.

In Figure 14.5, there are 16 pulses with DOA ~ 20° and 14 pulses with DOA ~ 30°, so two RF/DOA clusters are formed.

If all the pulses are in the same RF/DOA cluster, the PRI is correctly calculated in a single TOADH, as shown in Figure 14.6. This shows the TOADH that would have been formed if there had been a single RF/DOA cluster with the first peak at 1 ms and the second peak at the first harmonic of 2 ms. There are at least half the number of pulses in the peak at 2 ms as there are in the peak at 1 ms, leading to the correct PRI calculation.

Returning to the DOA profile of the pulses with the two separate DOA clusters, the clusters at 20° and 30° are outlined in Figure 14.7.

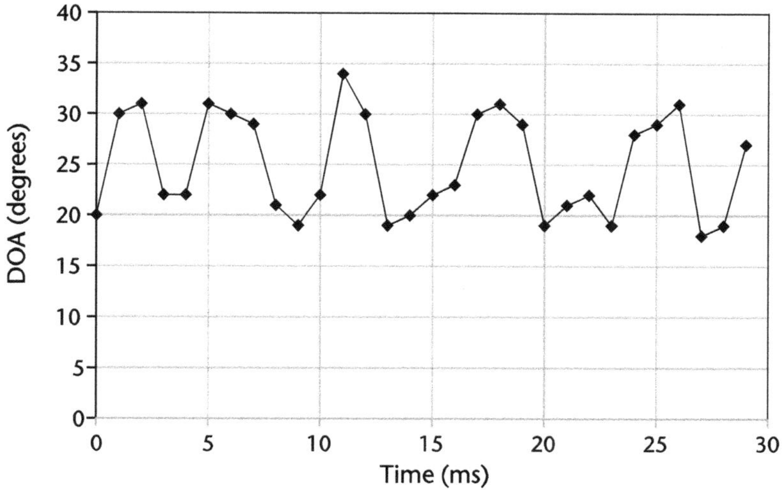

Figure 14.5 DOA of pulses received from a single radar scan.

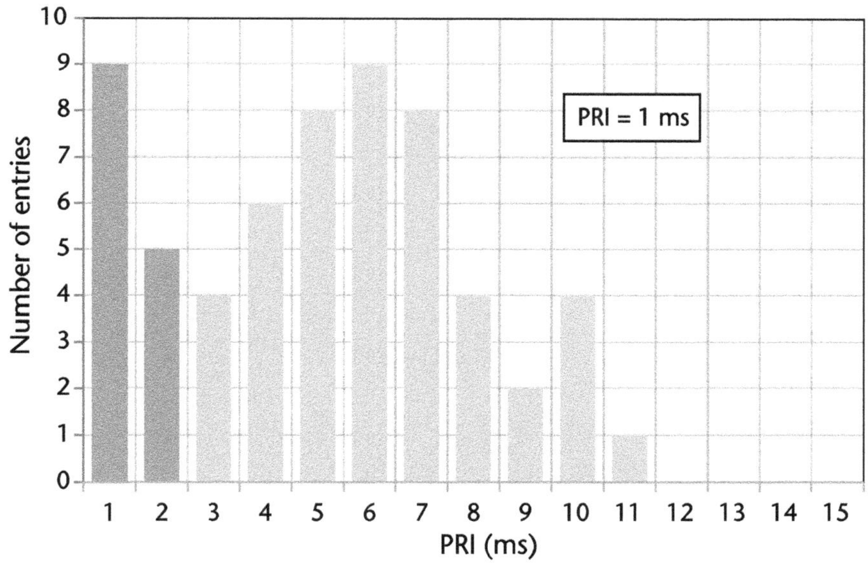

Figure 14.6 TOAD histogram for correct PRI of 1 ms.

Two TOADHs are formed from the two DOA clusters shown on Figure 14.7. Figure 14.8 shows a TOADH with the correct PRI of 1 ms, but from the same set of pulses a second TOADH is formed with an incorrect PRI of 5 ms, also shown in Figure 14.8. Two intercepts are therefore formed for the pulses from the single radar.

14.5 PRI Errors Due to Missing Pulses

There are almost always pulses missing from a pulse sequence received from a radar by an ESM system. To measure DOA, a pulse must be seen on more than one ESM

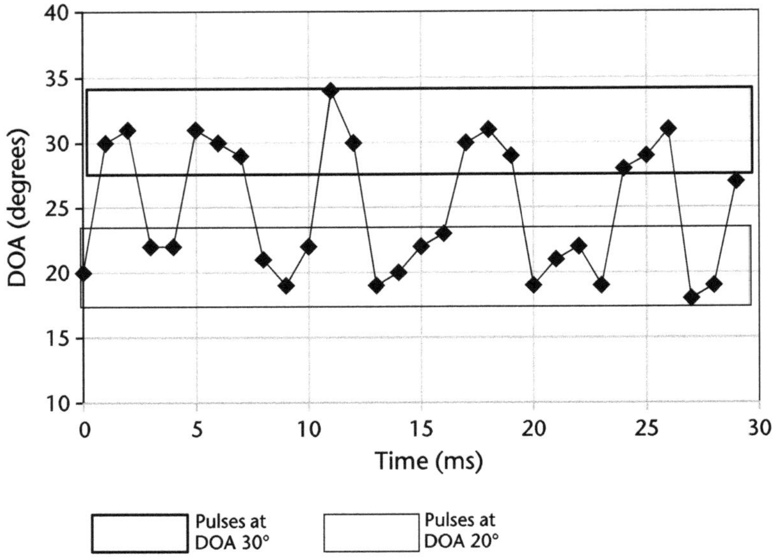

Figure 14.7 DOA of pulses from a single radar scan with two separate DOA clusters.

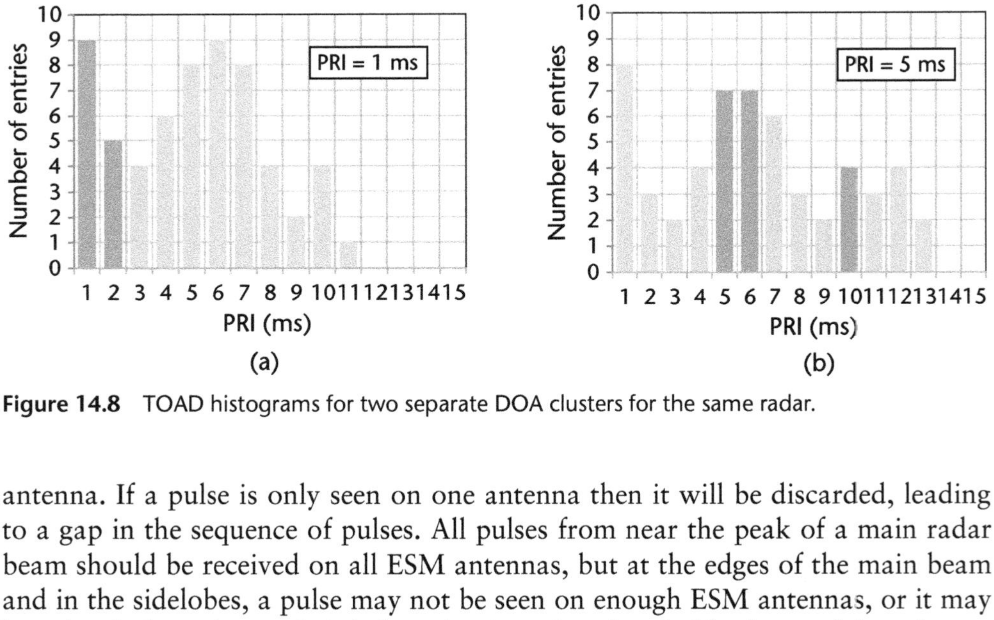

Figure 14.8 TOAD histograms for two separate DOA clusters for the same radar.

antenna. If a pulse is only seen on one antenna then it will be discarded, leading to a gap in the sequence of pulses. All pulses from near the peak of a main radar beam should be received on all ESM antennas, but at the edges of the main beam and in the sidelobes, a pulse may not be seen on enough ESM antennas, or it may be missed altogether as it is below the detection threshold of one of the relevant ESM antennas.

It is the shape of the radar beam that determines the number of missing pulses. Figure 14.9 shows a radar beam from a typical ATC radar at a range of 25 km from an ESM system with a detection threshold of –50 dBmi.

There are 40 pulses received from this radar above the ESM threshold. These pulses are from the main beam and the first sidelobes. However, there are seven pulses missing from the pulse sequence as they are at amplitudes below the ESM threshold. The fact that there are missing pulses means that the PRI may be miscalculated. In practice, ESM PRI processing should be set up to cope with some

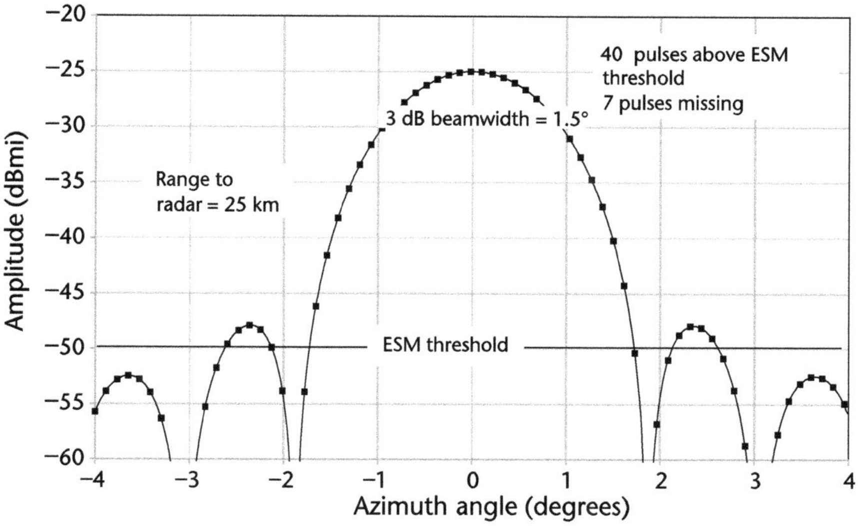

Figure 14.9 Amplitude profile of pulses from a single scan of a typical ATC radar.

missing pulses, but usually such measures are only effective if the radar has a fixed PRI or a very simple stagger sequence, as described in Chapter 8.

Another example of a radar where pulses are even more likely to be missing is shown in Figure 14.10. In this plot a radar beam for an HS radar with a very small 3-dB beamwidth is shown.

This time, there are nine pulses above the ESM threshold and three pulses missing as pulses are seen from the main beam and first and second sidelobes, but not those from the troughs between the main beam and the first sidelobes and the first and second sidelobes. With 25% of the pulses missing and so few pulses detected from this radar, the PRI calculation is seriously compromised.

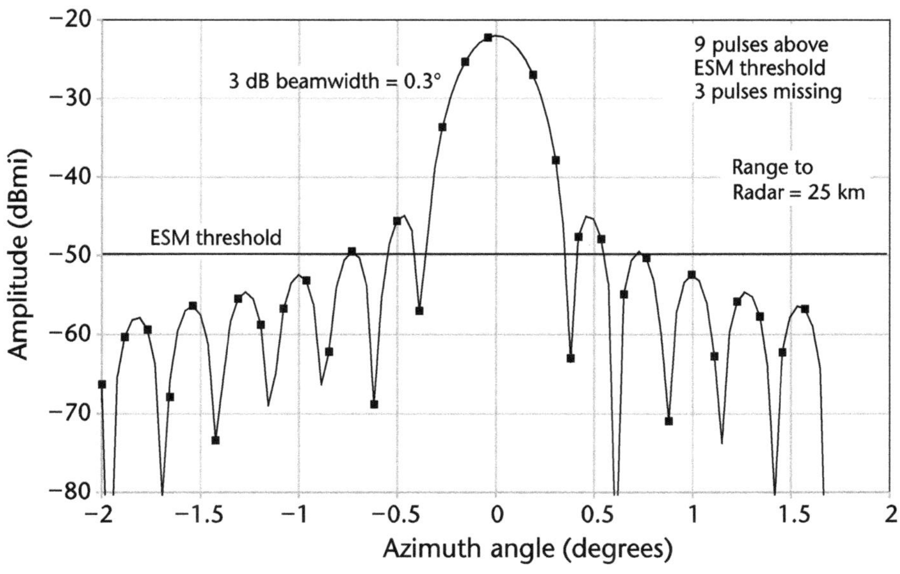

Figure 14.10 Amplitude profile of pulses from a single scan of a typical HS radar.

14.6 PRI Errors Due to Complex PRI Sequences

Most ESM systems have difficulty in detecting radars emitting multi-element PRI sequences. Radars that have complicated PRI sequences include some ATC radars and military radars with target acquisition and search functions. An example of a typical PRI sequence for an ATC radar is given in Table 14.1. There are six elements in this sequence, but the order of the PRI elements is varied to give a 12-position stagger.

If there are no other radars contributing to a TOADH apart from the one in Table 14.1, then the PRI elements can be extracted, as long as there are sufficient pulses available in the pulse sequence. However, when these pulses are interleaved with those of just one other radar, the ESM cannot usually calculate individual PRI elements for a PRI sequence of six or more elements and may not even be able to calculate the pulse group repetition interval (PGRI).

TOADH algorithms are often designed assuming that there will be as many pulses received from a radar as are necessary for the correct operation of the algorithm (i.e., usually at least twice the number of pulses as in the PRI sequence). This is unrealistic as the shape of the radar beam means that, unless the radar is located at short range, there will be too few pulses received for the criteria for validating the TOADH peaks to be met.

To scope the problem of there being too few pulses available for the TOADH to work on, the number of pulses received from a typical ATC radar (ERP of 80 dBW, 3-dB beamwidth = 1.5°) and also from an HS radar (ERP of 73 dBW, 3-dB beamwidth = 0.3°) by an ESM system with sensitivity –55 dBmi at various ranges are shown in Table 14.2.

There are many more pulses received from the ATC radar, with 22 pulses being received for each scan at the long range of 300 km. However, far fewer pulses are seen from the HS radar. At the relatively short range of 30 km, only eight pulses are detected from this type of radar from each scan. By the time the range to the radar

Table 14.1 Example of a Complex PRI Sequence

Pulse No.	*PRI (μs)*	*Pulse No.*	*PRI (μs)*	*Pulse No.*	*PRI (μs)*
1	980	5	910	9	950
2	930	6	890	10	810
3	950	7	980	11	930
4	870	8	870	12	890

Table 14.2 Number of Pulses Seen from ATC and HS Radars at Various Ranges

Range (km)	*ATC Pulses*	*HS Pulses*
15	180	16
30	120	8
75	44	4
150	30	4
300	22	3

is 75 km, only four pulses are seen, not enough for the ESM system to be able to construct an intercept.

Most ESM systems create intercepts labeled as complex PRI if there are sufficient pulses at a specific RF and DOA and a PRI value cannot be determined for the pulse cluster. In practice, the number of pulses for complex intercept creation is set to be large, with 30 pulses being a typical value for this criterion. The use of intercepts labeled as complex PRI causes problems for the identification process of the ESM system and inevitably causes multitracking. Some correlation processes are unable to link complex PRI intercepts to existing identified tracks leading to the creation of extra tracks for complex PRI intercepts.

14.7 PRI Errors Due to PW Measurement Errors

PRI errors may occur because there are errors in PW measurement due to intentional or unintentional modulation of the pulse. Pulses with no intentional modulation may have their amplitude and phase profiles altered by multipath, as shown in the example in Figure 14.11.

In the example shown in Figure 14.11, the PW measurement is shortened as the ESM deems that the pulse has ended due to the drop in amplitude. If PW is not used as a deinterleaving parameter, then the shortening of the pulse like the one in Figure 14.11 does not lead to the creation of extra tracks. However, the situation is very different for pulses that have intentional modulation (i.e., chirped pulses). These pulses are usually relatively long, of width at least 10 μs, and they are susceptible to adverse effects when multipath reflections are present. The effect of multipath reflections on a chirped radar pulse is to cause oscillations in the amplitude profile of the pulse. This leads to the possibility of the pulse being measured as several shorter pulses. Figure 14.12 shows an amplitude profile of a chirped pulse that has been subject to multipath.

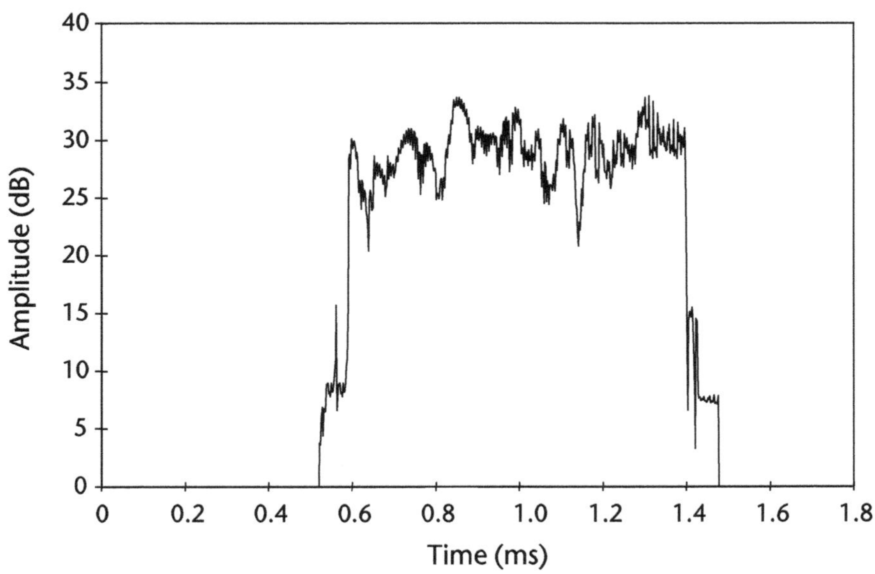

Figure 14.11 Amplitude profile of a pulse affected by multipath (unintentional modulation).

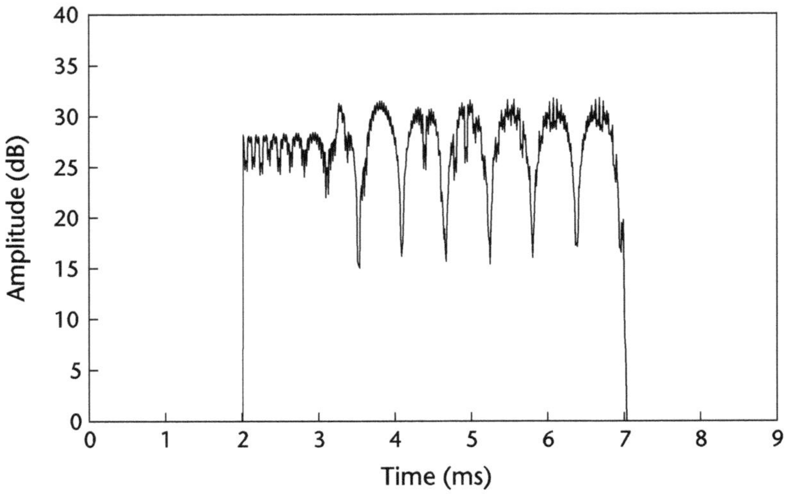

Figure 14.12 Amplitude profile of a chirped pulse affected by multipath.

To illustrate the effect of pulse splitting on the creation of extra intercepts, consider Table 14.3, an example of a pulse sequence from a radar with a simple PRI of 100 μs, but relatively long PW of 10 μs. Initially, all the pulses are measured correctly and only one intercept is formed with the correct values of PRI and PW.

However, if the same pulses are subject to multipath, the pulse sequence appears as shown in Table 14.4. The perfectly measured pulses from Table 14.3 are each split into several shorter pulses so that there now appear to be 24 pulses from the radar.

The PRI sequence is altered significantly and there are now two intercepts created for the radar, one with a PRI 93 of ms, near the correct PRI and one with a PRI of just 3 ms. Both intercepts have an incorrect PW of 2 μs.

An intercept with a PRI as short as 3 ms is likely to cause serious identification issues for the ESM as such intercepts may be mistaken for threat radars such as missile guidance signals using pulse position modulation with PRIs of the order of a few milliseconds.

Table 14.3 Perfect Pulse Sequence with No PW Measurement Errors

Pulse	*TOA (ms)*	*PRI (μs)*	*PW (μs)*
1	0.1	100	10
2	0.2	100	10
3	0.3	100	10
4	0.4	100	10
5	0.5	100	10
6	0.6	100	10
7	0.7	100	10
8	0.8	100	10

Table 14.4 Pulse Sequence Affected by Multipath

Pulse	*TOA (ms)*	*PRI (μs)*	*PW (μs)*	*Pulse*	*TOA (ms)*	*PRI (μs)*	*PW (μs)*
1	0.100	100	1	13	0.405	3	3
2	0.102	2	2	14	0.500	95	6
3	0.105	3	1	15	0.503	3	2
4	0.200	95	3	16	0.505	2	1
5	0.203	3	2	17	0.508	3	1
6	0.206	2	2	18	0.600	92	1
7	0.208	3	1	19	0.602	2	1
8	0.300	92	2	20	0.604	2	1
9	0.303	3	3	21	0.700	94	2
10	0.307	4	2	22	0.703	3	2
11	0.400	93	1	23	0.706	3	2
12	0.402	2	1	24	0.800	94	3

14.8 PW Measurement Errors

In the previous section, a description of how PW measurement errors can lead to PRI errors was given. The mismeasurement of PW can itself lead to multitracking if PW is used as a deinterleaving parameter. As already stated, radars with long, chirped pulses are particularly susceptible to multitracking.

The effect of multipath reflections on a chirped radar pulse is to cause oscillations in the amplitude profile of the pulse leading to the possibility of the pulse being measured as several shorter pulses. An example of the amplitude profile of a chirped pulse that has been affected by multipath appears in Figure 14.13.

The rippling effects typically cause a dip in amplitude by several decibels, followed by a rise again several nanoseconds later.

To determine the extent of the PW measurement problem, pulses from a typical ATC radar are simulated. The radar emits equal numbers of pulses at two widely

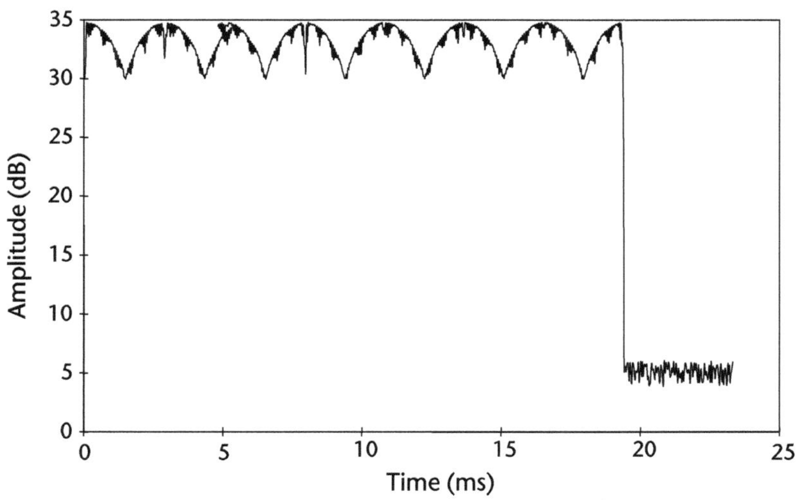

Figure 14.13 Simulation of an amplitude profile of a chirped pulse affected by multipath.

spaced PWs. Half the pulses are long, of width 15 μs, and half are short pulses of a width less than 1 μs. The long pulses are chirped at 4 MHz over the PW and they are therefore subject to interference due to multipath reflections. This level of chirp is within an RF resolution cell of the ESM system, so no pulse splitting should take place due to frequency changes through the pulse lifetime. The multipath interference takes the form of a rippling effect on the amplitude profile of the pulse, which results in a false PW measurement.

When the amplitude of a pulse drops by 3 dB, the ESM system deems that a pulse has ended and measures its PW to that point. Sometimes the long radar pulses are split up into a series of shorter pulses by the ESM pulse measurement system with a PRI of few microseconds, as shown in the example in Section 14.7. However, much more often, the ends of the pulses are lost altogether. These groups of pulses, an example of which is shown in Table 14.5, have the correct PRIs and are often all the same, wrong, length.

Figure 14.14 shows histograms of the PWs of three sets of pulses simulated in 1-minute time slots. Ninety-three percent of the short pulses are correctly measured, but only 33% of the long pulses are correctly measured (i.e., with PWs ~15 μs). The range of the aircraft to the radar was short (<15 km) for the generation of the 1-minute data sets.

Most of the short pulses (i.e., < 1 μs) fall into the 0.5 → 0.6 μs bin, which is as expected for this radar. However, many of the long pulses are wrongly measured, as can be seen in Figure 14.14. The possible consequence of this problem is shown in Figure 14.15 where a plot of the correct DOA for a single radar shows that the

Table 14.5 Examples of Shortened Chirped ATC Pulses

Pulse No.	*Expected PW*	*Measured PW*
1	Five short pulses at 0.9 μs	0.90
2		0.90
3		0.85
4		0.90
5		0.85
6	Five long pulses at 15 μs	11.60
7		11.55
8		11.55
9		11.55
10		11.55
11	Five short pulses at 0.9 μs	0.50
12		0.45
13		0.45
14		0.45
15		0.50
16		0.50
17	Five long pulses at 15 μs	2.40
18		2.55
19		2.30
20		2.30

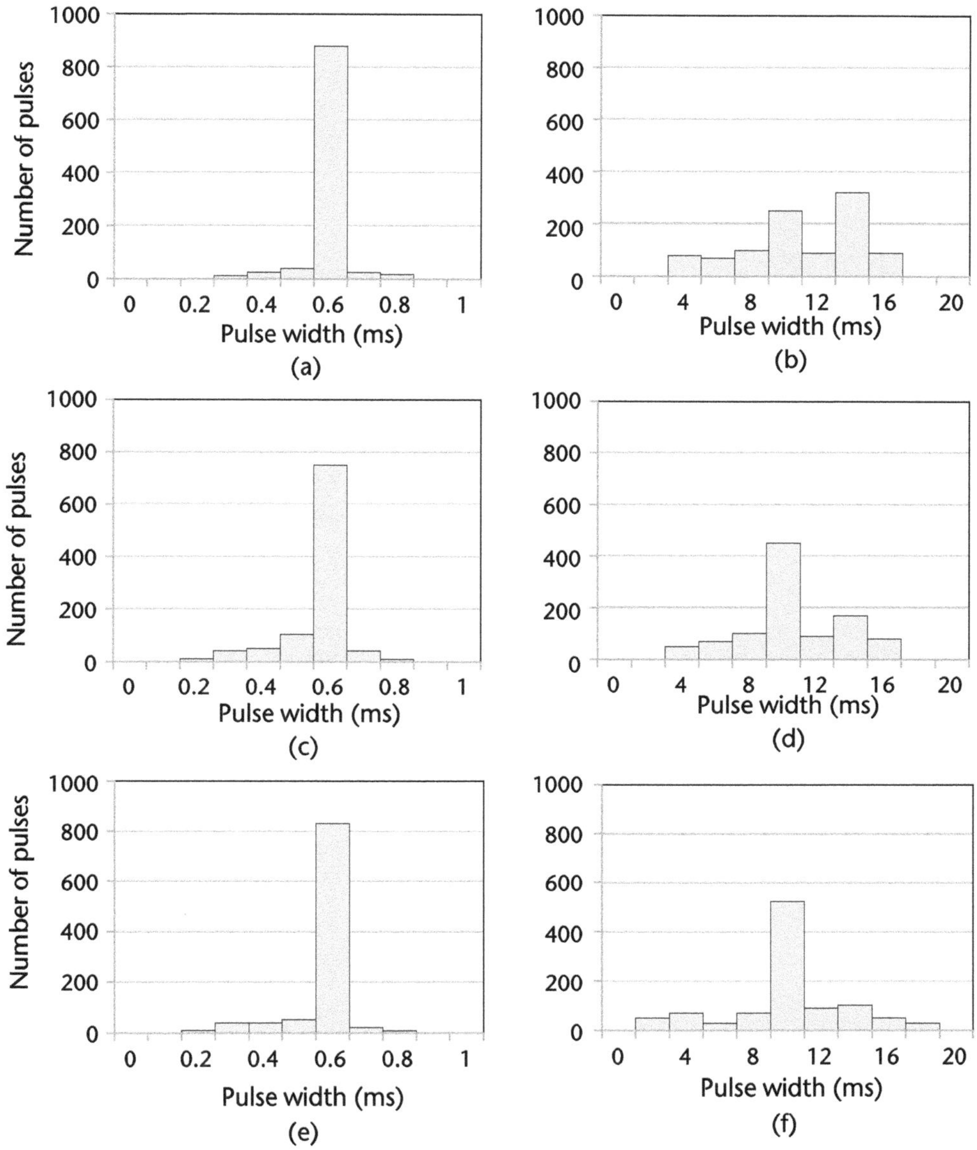

Figure 14.14 Histograms from simulations of ATC PWs for a radar with long and short PWs.

DOA is not significantly in error for the intercepts, but that there are many different tracks, each one with a different PW. Most of the tracks created with the different PWs are never updated and their presence in the track table and on the operator's screen leads to serious limitations in the capacity and usefulness of the ESM system.

14.9 RF Agility

Most ESM systems attempt to create intercepts for RF agile radars by removing the requirement for clustering in RF once pulses from fixed RF radars have been extracted from the pulse buffer for TOADH processing. Figure 14.16 shows a very

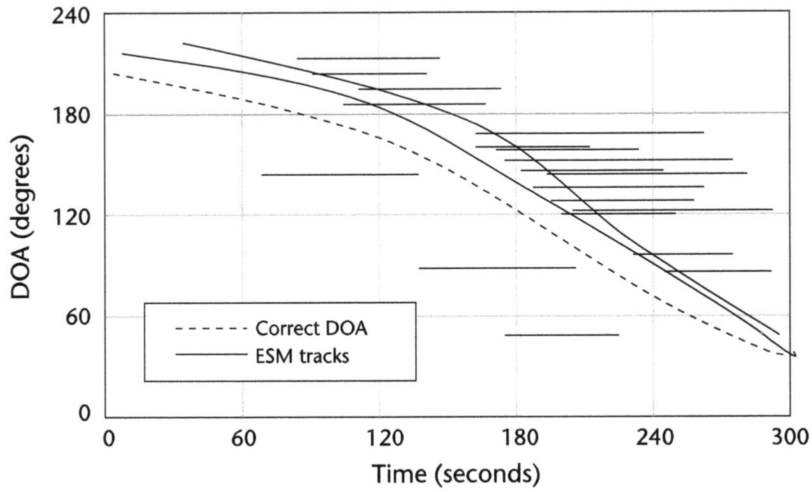

Figure 14.15 Multiple track creation for a single radar with PW as a deinterleaving parameter.

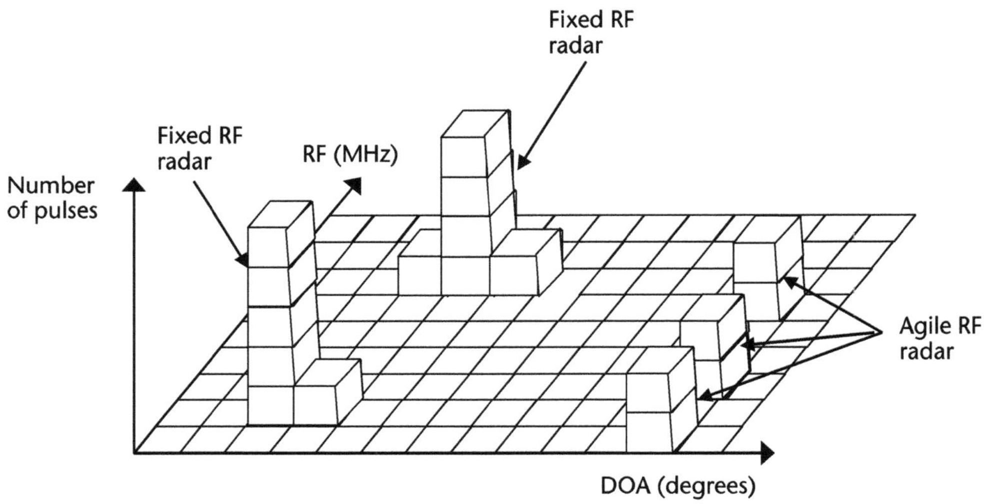

Figure 14.16 RF/DOA cluster diagram with fixed RF and agile RF radars.

simple example of an RF/DOA cluster diagram containing pulses from both fixed RF and agile RF radars.

If there are pulses received from only one RF agile radar in a buffer of pulses, then the removal of the RF criterion for clustering should allow for the creation of an intercept of pulses from that radar, once the pulses from fixed RF radars have been removed. However, problems with the calculation of DOA mean that it is unlikely that all the pulses from a single RF agile radar are present in a single DOA (or TDOA) cluster. It is also likely that in a single DOA cluster there are pulses from other radars whose DOA has been incorrectly calculated. Combining these pulses with those of the RF agile radar leads to incorrect PRI calculation and subsequent problems with radar identification.

The removal of the RF criterion for clustering causes particular problems when there are several radars with identical PRI sequences and the ESM combines pulses from more than one radar into an intercept. For example, most nations use the same types of ATC radar within their country. There may be several tens of a particular type of ATC radar, many having the same complex PRI sequence within range of the ESM system. In this case, it is common for the ESM to form intercepts, tracks, and location fixes made up of pulses from more than one radar of the same type.

14.10 Methods to Reduce Multitracking

There are some measures that can be taken to improve multitracking:

- Improve the DOA measurement;
- Ignore intercepts with poor DOA;
- Design the PRI calculation algorithm to allow for missing pulses;
- Define a PRI quality measure;
- Use a realistic number of pulses for a complex PRI intercept;
- Do not use PW as a deinterleaving parameter;
- Time out single intercept tracks and do not display on the operator's screen.

14.10.1 Improve DOA Measurement

DOA errors are undoubtedly the biggest cause of multitracking. Methods for improvements in DOA measurements are discussed in the chapters on the DOA measurement by the different types of ESM and also in Chapter 17 where multipath mitigation techniques are explored.

14.10.2 Ignore Intercepts with Poor DOA

Although it is difficult to correct poor DOA measurements, it is not difficult to determine that a pulse is in error in DOA. For amplitude comparison systems, using a method of estimating the incident pulse amplitude at each antenna can give an indication that the DOA is in error (see Sections 5.7.3 and 17.5). In TDOA systems, the width of a histogram of the DTOA of the pulses on each baseline can help to determine that the DOA of an intercept is likely to be in error (see Section 6.7). For phase difference systems, taking samples of phase difference for several antenna pairs and finding that there are differences in the calculated DOA (see Chapter 7 and Section 17.7) indicates that the DOA may be incorrect.

14.10.3 Design the PRI Calculation Algorithm to Allow for Missing Pulses

There are often pulses missing from the pulse sequence from a radar, due to DOA errors or the radar beamshape (see Section 14.5). This may lead to the TOADH process producing more than one intercept for the radar. A method of linking two or more intercepts with the same PRI sequences from a single pulse buffer should be adopted. Then, if pulses are missed, causing a break in the pulse sequence, the consequent extra intercepts can be merged.

14.10.4 Define a PRI Quality Measure

A measure of the quality of PRI can be determined according to the number of times that particular PRI elements are seen in a sequence of pulses. This could be defined as in Table 14.6. When attempts are made to correlate the intercept into an existing track, wider correlation tolerances could be used for poorer quality intercepts. This should reduce multitracking. In addition, poor-quality intercepts are allowed to create tracks that are only kept in existence for a very short time so that they do not take up valuable space in the limited track table available.

14.10.5 Use a Realistic Number of Pulses for a Complex PRI Intercept

A complex PRI intercept is created when the TOADH fails to determine a PRI pattern in a cluster of pulses. This often happens for multistagger PRI sequences. Typically, this type of intercept will not be reported unless there are 30 or more pulses in the intercept.

Reducing the number of pulses for the creation of complex PRI intercepts to 15 or 20 pulses will increase the number of complex intercepts in existence. There is no significant increase in the possibility that an intercept will be created from random pulses in the pulse buffer if this change in criterion is made because complex PRI intercepts are created from pulses within a tight RF range. Once a complex intercept is created with a reduced number of pulses, it should only be allowed to exist for a short time (perhaps 20 seconds) unless it can be correlated into an existing track.

The exact mechanism for the correlation of complex PRI intercepts to existing tracks depends on the order in which parameters are compared in the correlation process. In the case of an ESM system that identifies an intercept and then attempts to correlate it into an existing track, a complex PRI intercept will only be correlated into a track if the track identity has been specified in the ESM library to have

Table 14.6 Definition of PRI Sequence Quality

PRI Type	*PRI Sequence Quality*	*Definition*
Fixed	Good	The PRI element is repeated 8 or more times in the PRI sequence
	Medium	The PRI element is repeated at least 4 times in the PRI sequence
	Poor	The PRI element is repeated less than 4 times in the PRI sequence
Staggered	Good	4 or more elements are repeated in the PRI sequence
	Medium	1 to 3 elements are repeated in the PRI sequence
	Poor	None of the elements in the PRI sequence are repeated
Jittered	Good	At least 2 sets of 3 jittered PRI elements in the PRI sequence
	Medium	One set of 3 jittered PRI elements in the PRI sequence
Switch and dwell	Good	At least 3 sets of 3 fixed PRI elements in the PRI sequence
	Medium	1 or 2 sets of 3 fixed PRI elements in the PRI sequence

a complex mode line. More details on the specification of mode lines in the ESM library are given in Chapter 10.

14.10.6 Do Not Use PW as a Deinterleaving Parameter

PW should not be used as a deinterleaving parameter. The range of possible PWs for most types of radar is small, so there is not enough discrimination between different radars for this parameter to valid for deinterleaving. For radars that have longer PWs, and especially those that are chirped, the possibilities for multitracking are immense due to the severe impact that multipath can have on these types of pulse. However, PW may be used as an identification parameter if the radar has a relatively unique PW.

14.10.7 Time Out Single Intercept Tracks and Do Not Display on the Operator's Screen

When a track is created from an intercept made up of pulses from a particular radar, it is assumed that another intercept will be available to update the track the next time that pulses from that radar are detected. If the radar has a simple circular scan, it should be assumed that this will occur on the next scan cycle, typically of the order of a few seconds. If the radar is far away from the ESM platform and/or it has a beamshape such that only a few pulses are likely to be received on each scan, then it is probably reasonable to allow a time period of a few scan cycles before deleting the track. However, it is rarely useful to keep tracks that have only one intercept associated with them in existence for more than 30 seconds.

The time for keeping tracks that have not been updated alive is sometimes specified in the ESM library entry for the radar type. Tracks for radars that are identified as threats should be kept in existence for longer whereas tracks that have the identities of civilian radars should be allowed to time out more quickly.

Most ESM systems have an option to prevent single intercept tracks from appearing on the operator's screen. However, provision should be made in the design of the system for single intercept tracks for radars identified as threats to override this option.

CHAPTER 15

Reflections and Multipath

Most ESM systems suffer from errors in the measurement of DOA of radar pulses, sometimes to such an extent that their use as a defensive aid is seriously compromised. If pulses from individual radars are detected with widely varying DOA, the ESM system will conclude that there are more radars in the environment than there really are and will create extra tracks. The presence of large DOA errors will also significantly degrade geolocation accuracy. It is therefore important to explain the cause of the DOA errors and to explore ways of obtaining more accurate estimates of this parameter.

The most important contribution to DOA error (after antenna separation effects have been taken into consideration; see Chapter 5) is multipath interference from objects and surfaces in the environment of the radar and the platform on which the ESM is mounted. All types of ESM DOA measurement (amplitude comparison, TDOA, and phase comparison) are affected by multipath interference.

Amplitude comparison and TDOA systems make a major assumption that the pulse amplitude incident at each ESM antenna is equal. The effect of multipath is to change the incident amplitude of the pulse, due to constructive or destructive interference of the waveform. Multipath also changes the phase of the pulse, causing problems for phase comparison systems.

15.1 Types of Reflections Affecting ESM Systems

There are three categories of reflections that affect radar pulse detection by an ESM system:

- Reflections of the main radar beam from objects distant from the radar and from the ESM platform, typically with a difference from the direct path of hundreds of meters, may be detected a few microseconds after the direct path pulse. These reflections are seen as separate pulses. These types of reflection are usually seen at wide angle differences to the main beam pulse that has just been detected and they do not fit the amplitude profile of pulses being detected from the direct path. They often have an amplitude more than 20 dB lower than the direct path pulses.

- Reflections of the main beam from objects near to the radar cause more complicated effects in the pulse stream. These reflections are not interleaved with direct path pulses and usually a whole set of reflected pulses are detected. The reflected pulses are offset in time by a few milliseconds from the direct path main beam with an amplitude profile mirroring that of the direct path, but having lower peak amplitude. This type of reflection is only seen for scanning radars as the extra pulses are received when the main beam is scanning past the reflection object either before or after the radar scans past the ESM platform.
- Reflections from objects very near either to the radar or to the receiver, at distances of a few meters or less, cause the most serious interference effects. Extra pulses are not seen in this case, but the amplitude and phase of the direct path pulse are affected. Different fluctuations in amplitude are seen at different receiving antennas causing errors in DOA measurement.

Examples of these three reflection types are shown in Figures 15.1 to 15.3.

Figure 15.1 shows a single scan from a radar that has an interleaved direct path and reflection pulses. The reflection object is at a DOA of about 30° from the radar and the reflected pulses have much lower amplitude, 30 dB below the direct path pulses. The time difference between the direct path and reflected pulses in the example shown here suggests that the reflected path length is less than 1 km in excess of the direct path length. The decrease in amplitude of the reflected pulses is largely due to the coefficient of reflection (see Chapter 16).

Figure 15.2 shows a single scan from a radar with a reflection of the whole main beam detected a few milliseconds later. The x-axis scale on this graph shows pulse number and does not indicate the time difference between the direct path and reflected beams. The radar in this example is too far from the ESM receiver for sidelobes to be detected, so there is no doubt that the second peak that is seen is due to a reflection of the main beam. The DOA of both the direct main beam and that of the reflection are accurately measured, in that the spread in DOA for both sets of pulses is only ±5°. The DOA of the reflection is seen at a DOA difference of 20°, which is consistent with the reflected main beam having been received 150 ms after the main beam, since the scan period of this radar in this example is 2.5 seconds.

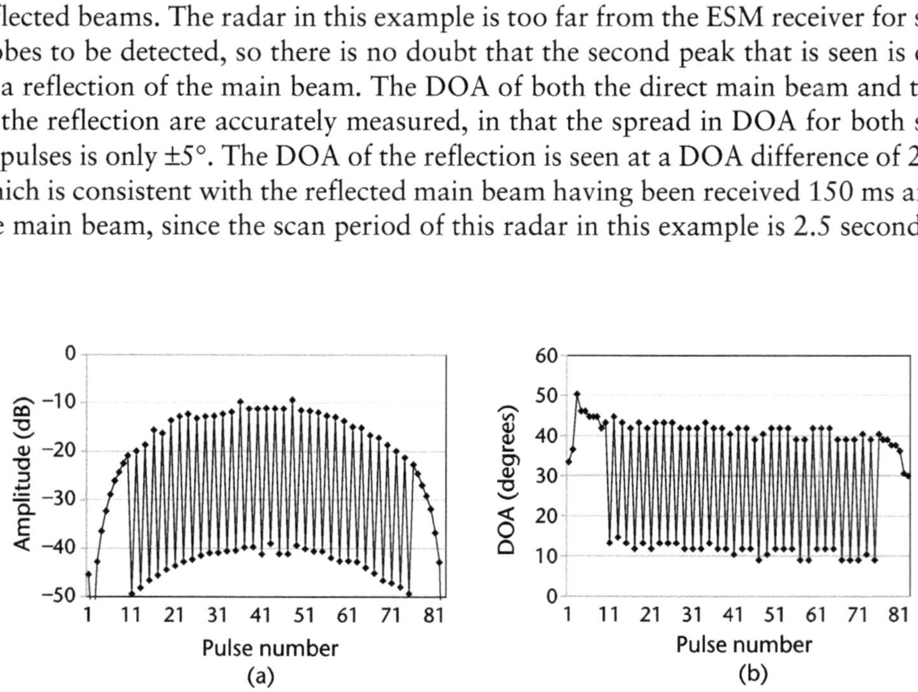

Figure 15.1 Example of amplitude and DOA profiles for a simple reflection.

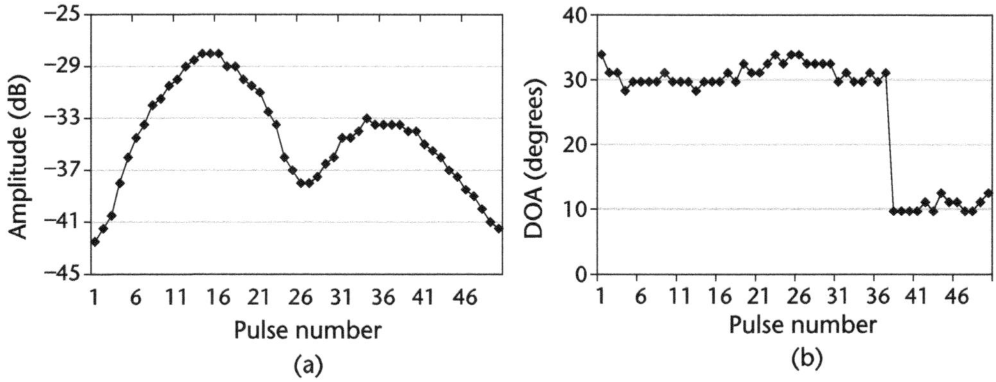

Figure 15.2 Example of amplitude and DOA profiles for a whole main beam reflection.

This example shows what a difficult job an ESM system has in recognizing that what appear to be two perfectly good main beam peaks are actually a main beam and a reflection of the same radar.

Figure 15.3 shows two consecutive main beam scans that have been subject to multipath. Most of the pulses in the first scan are not significantly affected, but one of the pulses near the peak has suffered destructive interference to the extent that its amplitude is 10 dB below that expected. The corresponding DOA profile shows that the DOA of the affected pulse is in error by 15° as a result.

In the second scan, detected only 2.5 seconds after the first scan, all the pulses are affected by multipath. The amplitudes of the pulses towards the peak of the scan are all decreased and the corresponding DOA profile shows that all the pulses in the scan have DOA errors of about 20°.

The first type of reflection (Figure 15.1) is easy to identify in the data, and should be filtered out. The second type (Figure 15.2) is more difficult to deal with as there may be some doubt about whether the extra pulses are due to a reflection effect or whether they are due to a separate radar. However, the third category of reflections (Figure 15.3) has the most serious effect on the performance of ESM systems and it is with this type of reflection that the remainder of this chapter is concerned.

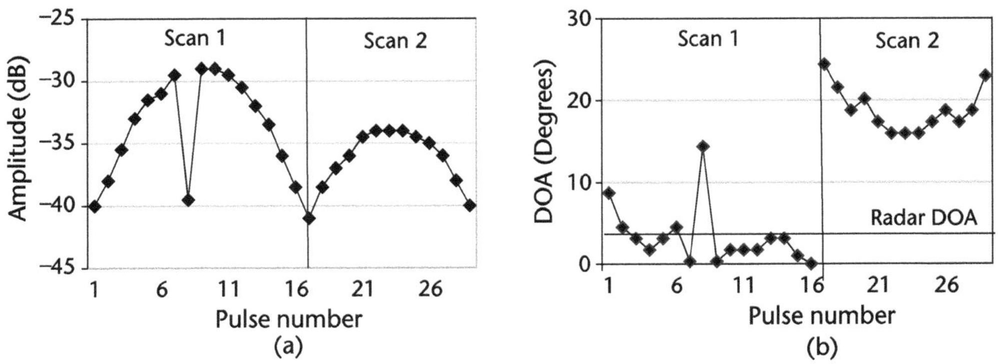

Figure 15.3 Example of multipath effect on amplitude and DOA profiles.

15.2 Multipath

The environment in which the ESM system must operate contains a wealth of surfaces and objects from which radar signals can reflect. There are two main sources of reflection surface. One is objects in the environment of the radar, such as: the ground, buildings, chimney stacks, the sea, and hills. The other source of reflection surfaces is the ESM platform itself near to a receiving antenna. It is expected that the second case is generally more serious as only one antenna may be affected by a particular multipath reflection and the reflection coefficient of an aircraft surface is likely to be significantly higher than the reflection coefficient from a reflection object in the environment of the radar. Exceptions are always possible; a truck passing very near to the radar may have similar reflection properties to a platform-based reflector.

The radar pulse seen by an ESM system usually comprises signals that have arrived via several different paths. However, it is useful to begin by considering the effect of a single reflection on pulse amplitude, showing the extent of interference possible for a direct path and a reflected pulse of equal amplitude. Figure 15.4 is the effect of phase difference on the amplitude of a pulse consisting of a direct path and a single reflection.

For a random phase difference between two signals, two-thirds of the cases will result in constructive interference. In one-third of the cases, the amplitude of the combined signal will be lowered by destructive interference.

Figure 15.5 shows the amplitude in decibels resulting from the interference of two signals of equal amplitude. The effects of total destructive interference are worse than for the case of total constructive interference for a single reflection as the maximum amplitude resulting from the constructive interference of two equal amplitude signals is an increase of 3 dB. Therefore, the magnitude of the phase differences between the direct and reflected paths is a less important factor in the

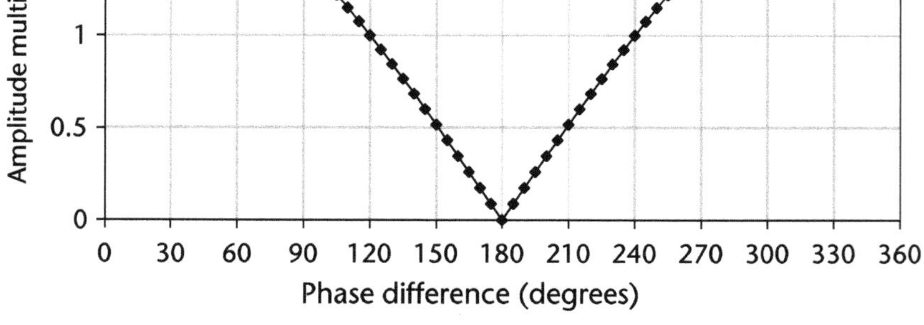

Figure 15.4 Amplitude multiplication factor for the combination of two equal amplitude signals.

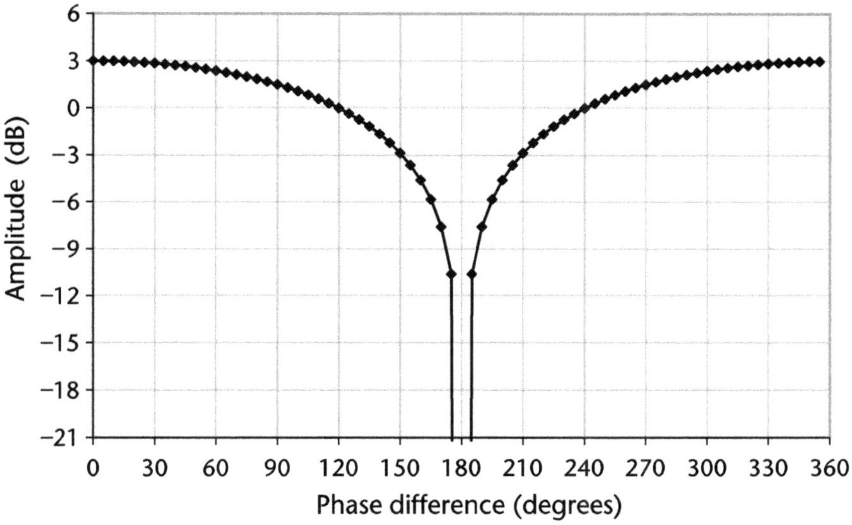

Figure 15.5 Amplitude in decibels for the combination of two signals of equal amplitude.

size of the DOA error than the part of the curve in Figure 15.5 on which the phase differences lie.

For phase differences near 180° the reduction in amplitude can be severe, as seen in Figure 15.6 which shows the amplitude in decibels resulting from the interference of two signals of equal amplitude with phase differences from 175° to 185°.

In amplitude comparison systems, the algorithm for the calculation of DOA relies on log amplitude measurements. The reduction of −145 dB, seen in Figure 15.6, for a phase difference very near 180°, would have a more damaging effect on the calculation of the DOA than the worst case of constructive interference. This

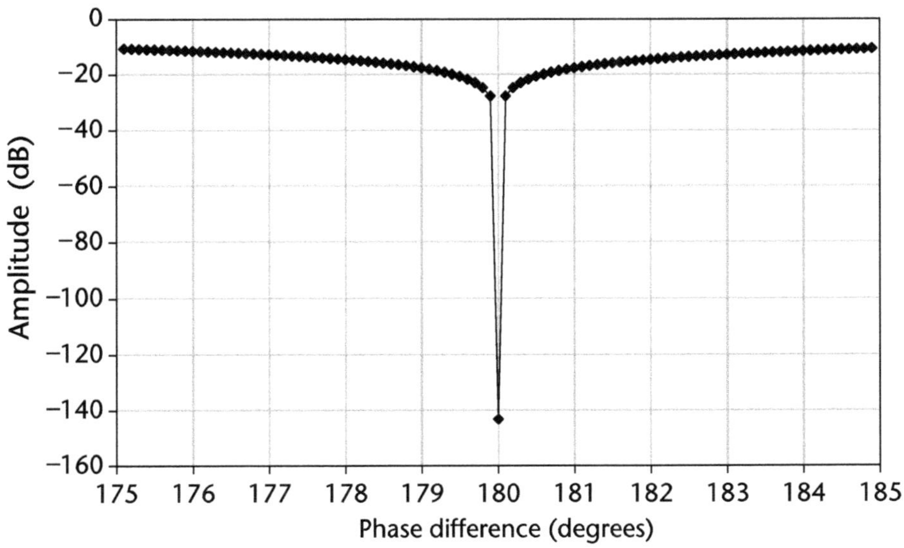

Figure 15.6 Amplitude in decibels for two signals of equal amplitude with phase difference near 180°.

would only have a 3-dB difference between a pulse with no reflection and one that had been subject to complete constructive interference.

There are many factors that govern the way that multipath affects the DOA profile of a scanning radar. The range to the aircraft, the ESM antenna configuration, radar parameters such as beamshape, and the nature of the reflection surface all have to be considered. Chapter 16 discusses these effects in detail.

15.3 Multipath in Amplitude Comparison Systems

The basis of operation of an amplitude comparison ESM system is that the emitted pulse arrives with equal amplitude at all receiving antennas, but that the detected amplitude at each antenna depends on the receiver antenna beamshape. The DOA of the pulse can therefore be determined from the difference in amplitude of the receiving antennas.

The effect of multipath is to change the effective emitted amplitude of the pulse, due to constructive or destructive interference of the waveform. Multipath changes the emitted amplitude during the transmission of the pulse. It is always the case that the direct path pulse arrives at the receiver before the reflected pulse does, but often there is such a short time difference between the two paths that the receiver cannot be activated to measure the direct path alone.

An example of an intrapulse amplitude profile is given in Figure 15.7. Figure 15.7 is a very much simplified pulse profile; in reality, all pulses are subject to noise and amplitude fluctuations, especially at the rising and trailing edges.

Some amplitude comparison systems measure the pulse amplitude at a fixed time after the initial detection of the pulse, typically at 30 or 50 ns, and some sample the pulse amplitude throughout the lifetime of the pulse to find the peak amplitude. Either method can lead to problems.

In the first case, multipath may have already affected the pulse amplitude by the time the amplitude measurement is taken. Also, the time taken for the reflected path to reach the different receiver antennas may be sufficiently different for some

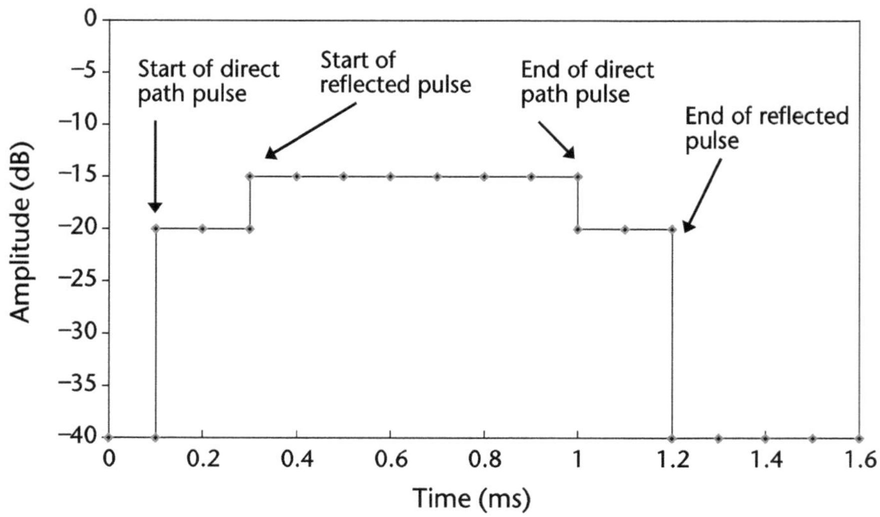

Figure 15.7 Example of an intrapulse amplitude profile.

of them to be affected by multipath and some of them to have only received the direct path pulse.

In the second case, the path difference between the direct and reflected paths is different for each of the antennas; hence, the phase difference is different and the amplitude at each antenna is affected to a greater or lesser extent by the multipath. Since the input pulse amplitude at each antenna is different, DOA errors result. If all the antennas are affected identically by multipath, then there are no DOA errors due to multipath.

15.4 Multipath in TDOA Systems

TDOA systems are typically used on large aircraft often with several tens of meters separating the receiver antennas. Multipath affects the amplitude of the pulse, often with different levels of interference at each antenna. There is therefore great potential for DOA errors and for many pulses to be missed altogether from the detection process.

A more subtle effect of multipath interference in TDOA systems is the change in the time of arrival that is induced due to the inevitable slope in the rise time of the pulse. Figure 15.8 shows a diagram of a direct path pulse of width 1 μs with a 200-ns rise time. A pulse with such a long relative rise time is unlikely in reality, but is used here for clearer illustration of multipath effects.

The two methods of measuring the pulse arrival time in TDOA systems (amplitude threshold and amplitude reduction) are described in detail in Chapter 6. With no multipath, the pulse TOA is measured either when the amplitude of the pulse reaches some amplitude threshold level or when the amplitude is reduced by 3 dB from the peak amplitude of the pulse. In the example in Figure 15.8, both these measurements take place 100 ns after the start of the pulse.

Figure 15.9 shows what happens when a single reflection arrives at the ESM antenna with a 50-ns time delay to the direct path pulse. When using the amplitude

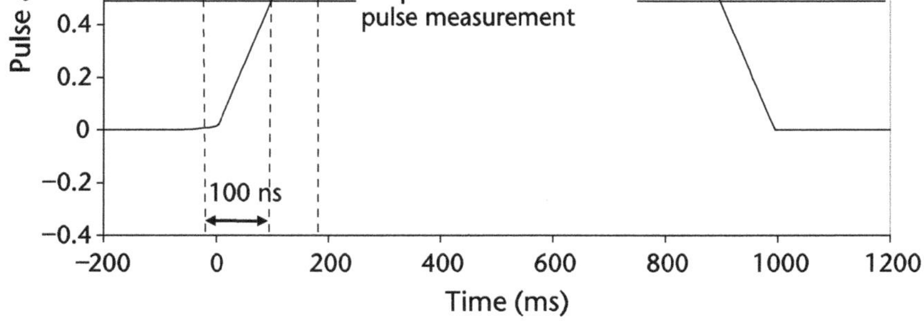

Figure 15.8 Measurement of time of arrival of a pulse.

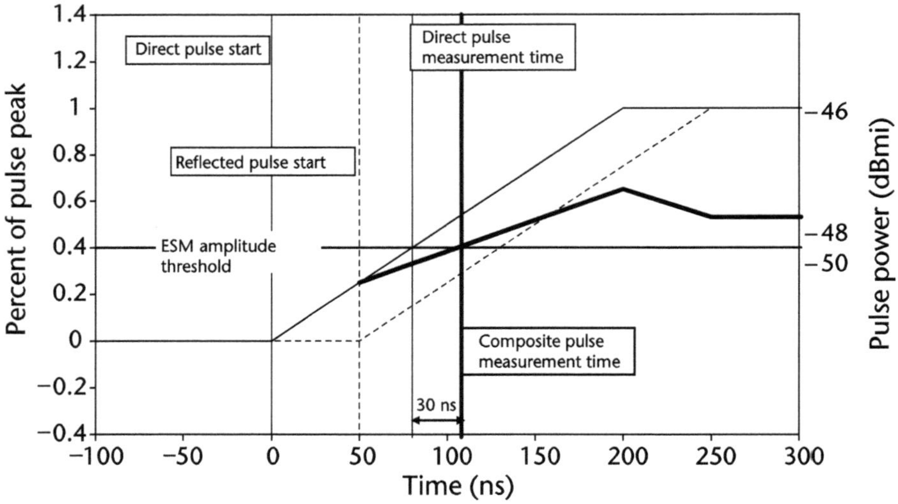

Figure 15.9 Delay in pulse measurement time due to multipath (amplitude threshold method).

threshold method, the amplitude of the composite signal does not exceed the measurement threshold until 30 ns after the direct path pulse would have been measured. This delay in measurement leads to an error of over 15° in the DOA calculated for this pulse for a TDOA system with antenna separation of 30m.

A TDOA system using the amplitude reduction method does not detect the pulse shown in Figure 15.9. The peak amplitude of the composite pulse is reduced from –46 dBmi (the peak amplitude of the direct path pulse) to –48 dBmi, only 2 dB above the ESM detection threshold of –50 dBmi.

Another example of multipath interference is shown in Figure 15.10. This time, the peak amplitude of the direct path pulse is 10 dB above the ESM threshold. The reflected pulse begins 50 ns after the direct path.

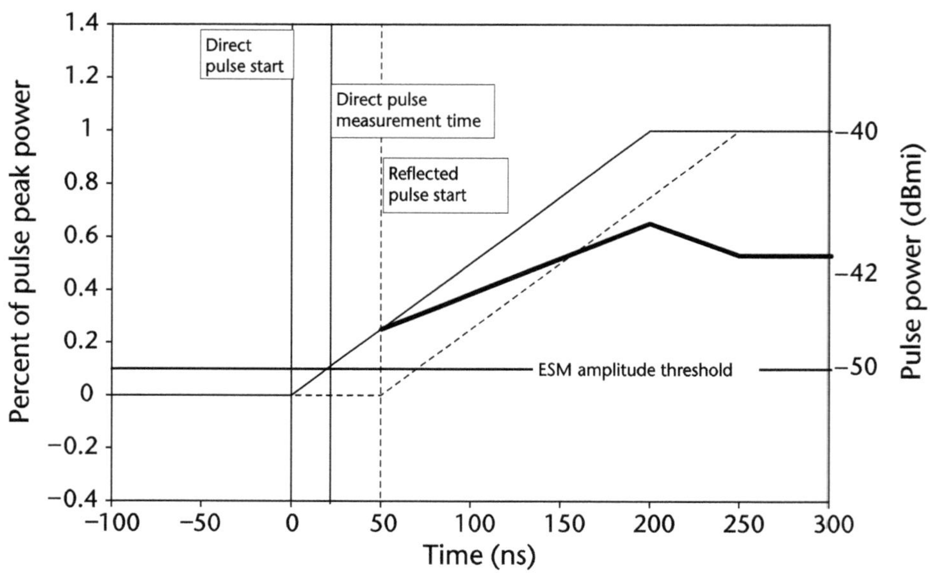

Figure 15.10 Delay in pulse measurement time due to multipath (amplitude threshold method).

Using the amplitude threshold method the time of arrival of the pulse is measured at 20 ns after its start, before the reflection has had any effect on the pulse amplitude or on the slope of its rise.

Figure 15.11 shows exactly the same pulse being affected by the same reflection as in Figure 15.10. The composite peak amplitude of –42 dBmi is measured and the time when the amplitude was 3 dB lower than the peak is recorded as the pulse time of arrival. This is measured as 80 ns and is 20 ns earlier than the time that would have been measured if the reflection had not been present.

As can be seen in Figures 15.9 to 15.11, multipath interference has the effect of changing the slope in the rise of the pulse during the rise of the pulse.

The fact that each ESM antenna receives pulses with different multipath effects, especially in the case of widely spaced antenna installations means that time differences due to multipath are likely to cause DOA errors for TDOA systems using either method of time measurement. However, ESM systems using the amplitude reduction method suffer more from multipath than those using the amplitude threshold method. The pulses have been in existence for longer when their TOA is measured using the amplitude reduction method, so there is more opportunity for multipath interference to have occurred.

The pulse rise time is an important factor in considering the effect of multipath in TDOA systems. Most pulses have short rise times, with more than 50% of current radars having rise times of 50 ns or less. Another 25% of radars have rise times that are greater than 50 ns, but less than 100 ns.

There are few pulses that have rise times of more than 300 ns and, as can be seen in Figure 15.12, the rise times of current radars are usually less than 10% of the total PW.

Figure 15.12 is limited to pulses of width less than 5 μs to allow the detail of the plot to be seen. There are many radars with pulses of a width more than 5 μs, but their rise times are usually less than 500 ns (i.e., less than 10% of their width).

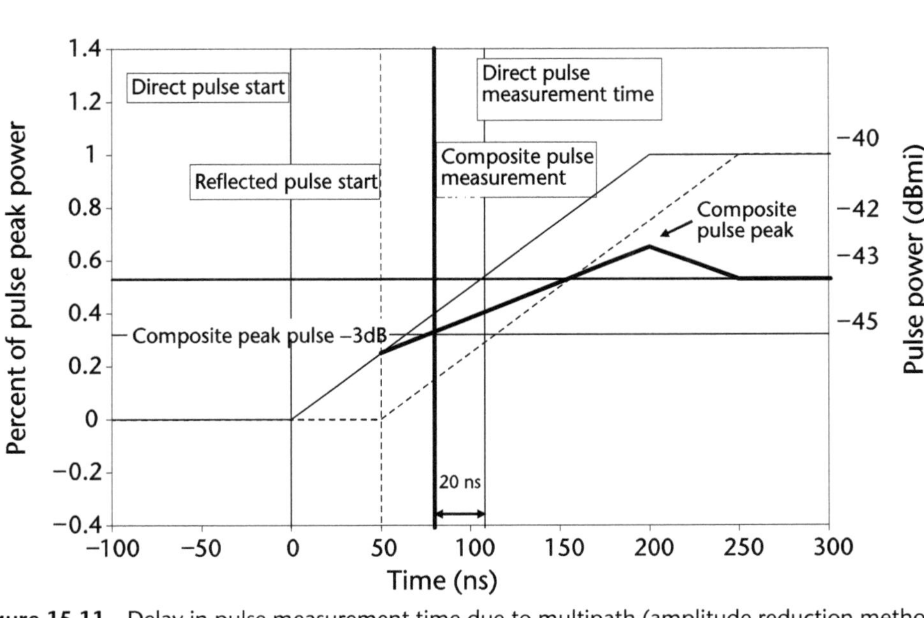

Figure 15.11 Delay in pulse measurement time due to multipath (amplitude reduction method).

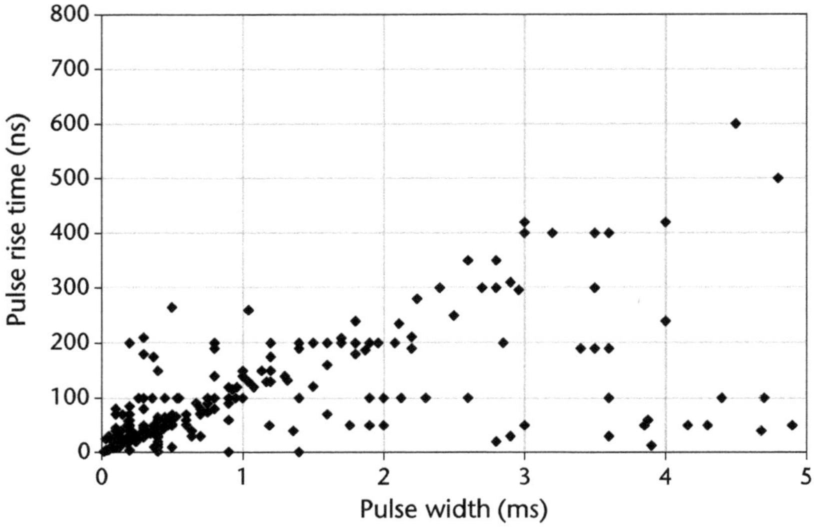

Figure 15.12 Pulse rise time against PW for current radars.

15.5 Multipath in Phase Comparison Systems

Because multipath alters the phase of the pulse, phase comparison systems are bound to be susceptible to its effects. The effect of multipath on the phase profile of a pulse is that there are jumps in phase at the points where the reflected pulse starts and also where the direct pulse ends. The fact that the amplitude profiles are also affected may deceive the receiver into believing that the pulse has ended and therefore limit the amount of information available for phase detection. Figure 15.13 shows the effects of a single reflection on the phase profile of a pulse. The single reflection arriving 1 ns after the direct path pulse causes a reversal in phase and significantly reduces the pulse amplitude.

The distance between the antennas in a phase comparison system is small, typically of the order of 5 to 15 cm so an example of the phase profiles of a pulse arriving at two antennas separated by 15 cm is shown in Figure 15.14. The phase profiles of a 3-GHz pulse at an angle of arrival of 30° to the linear array show that phase difference between the antennas in this case is 90°.

For a pulse at 3 GHz, the phase cycle time is 0.33 ns and this does not change when multipath effects are present. However, the phase of the signal does change and this is what causes problems for phase comparison systems. The pulse phase is typically sampled at each antenna with an interval of between 1 ns and 5 ns between samples. If the sampling interval is 5 ns, then for a PW of 1 μs, 200 phase samples are taken. Figure 15.15 shows the phase profiles at the two antennas when multipath has taken effect. In this case, antenna 1 has received only the direct path pulse at the time of sampling the phase, but antenna 2 receives the composite pulse after the phase change has happened. The phase difference is now 40°, leading to a 17° error in the DOA.

The simple, one reflection, example presented here is made much more complicated by the multiple reflections that are present in reality. Figure 15.16 shows how there are likely to be many phase and amplitude changes throughout the lifetime of the pulse, leading to the possibility of a DOA error.

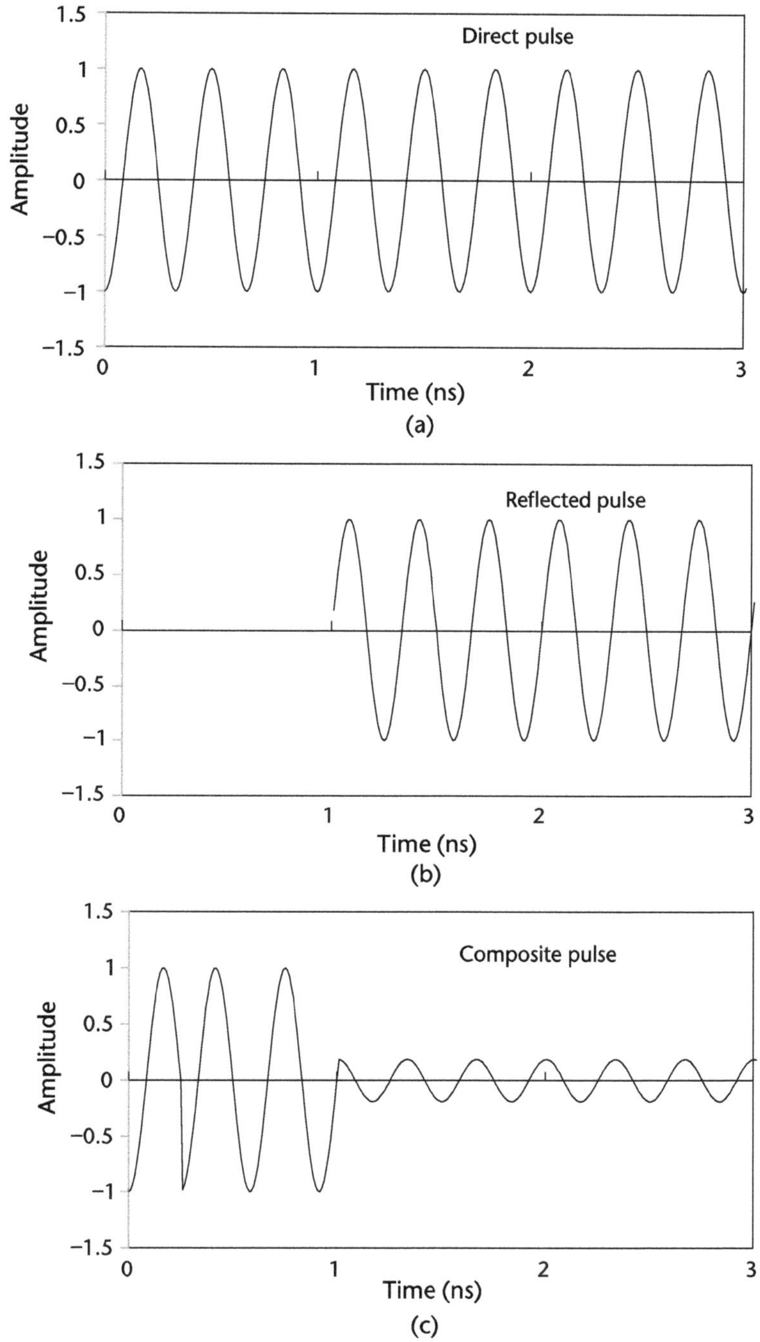

Figure 15.13 Effect of a single reflection on the phase profile of a pulse.

Phase comparison ESM systems are usually quoted as having very accurate DOA performance. However, in practice, multipath can cause severe corruption in the measurement of the DOA.

When phase changes occur throughout the lifetime of the pulse, some of the phase difference samples lead to incorrect DOA (i.e., those where one only antenna has been affected by multipath at the time of taking the phase sample). If

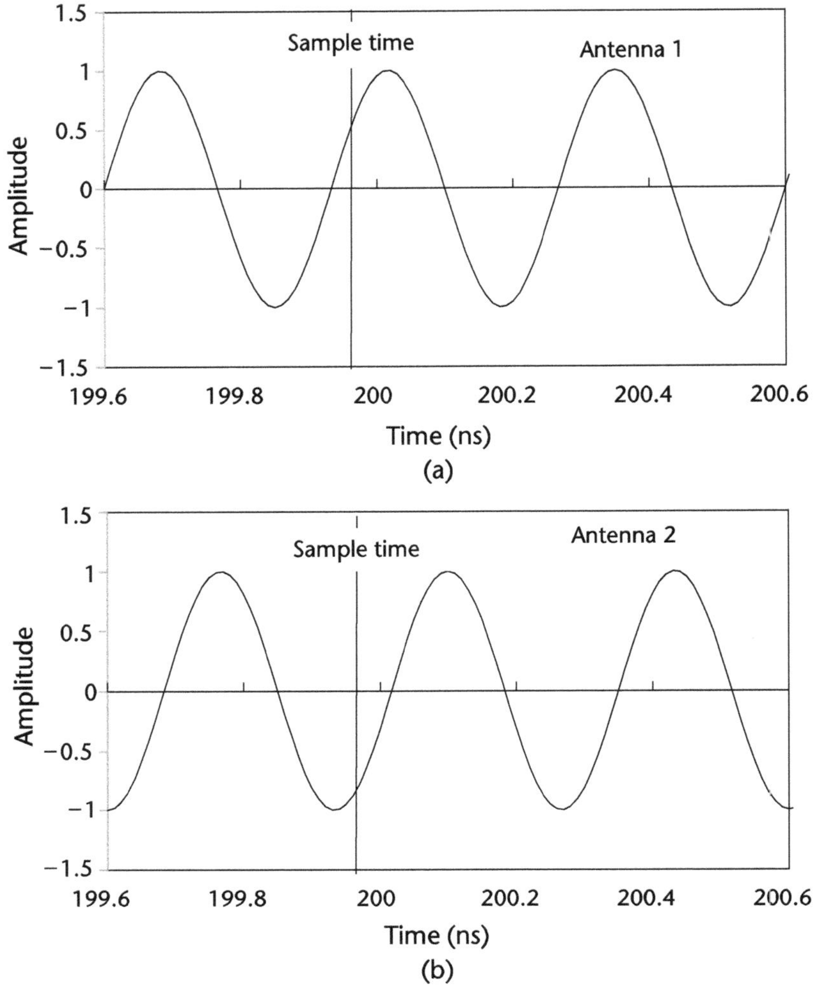

Figure 15.14 Phase profiles for a 3-GHz pulse arriving at two antennas separated by 15 cm.

both antennas are affected, then no DOA error should be seen. A simulation of thousands of pulses of PW of 100 ns with 5-ns sampling time showed that 35% of pulses had incorrect DOA. It should be possible to reduce the effect of DOA errors by taking the mode of the phase differences for a pulse, but it seems that this level of processing is not carried out in current ESM systems.

15.6 Evidence for Multipath

There have been several reports of multipath effects in recent years. One experiment carried out in Switzerland in 2016 showed that multipath interferences in radar images can be caused by the superposition of direct waves and indirect waves that have been reflected at a surface between radar and observation area. Especially affected are scenarios where the observation areas and the radar are separated by a highly reflective flat surface, such as water, bare soil, short grass, or wet snow [1].

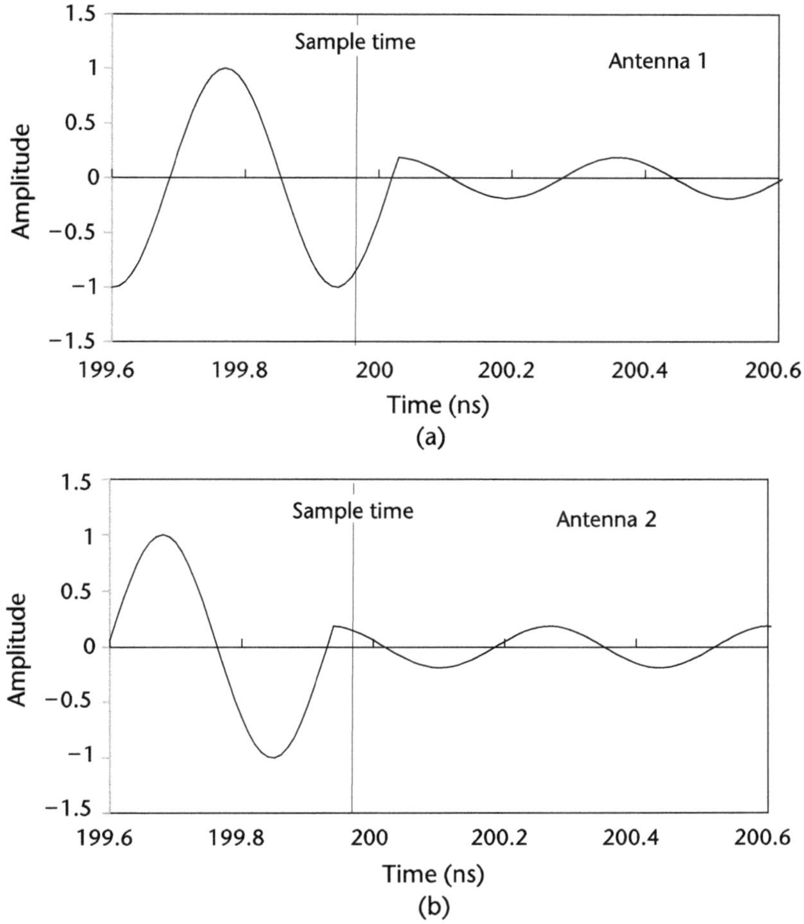

Figure 15.15 Phase profiles for a 3-GHz pulse arriving at two antennas separated by 15 cm with multipath effects.

A study comparing predicted multipath effects with data collected from a wide-band radar installation was carried out using two identical X-band radars sited on a coastline with a tethered sphere a few miles offshore. One of the radars tracked the sphere while the other radar antenna pointed downwards and observed the sea surface. The downward antenna only received radar signals so that direct observation of the one-way multipath could be made [2].

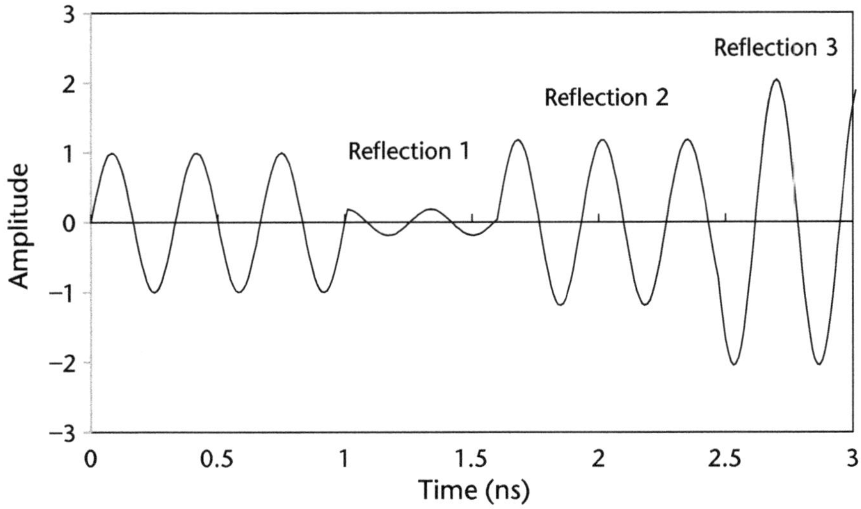

Figure 15.16 Phase profile of a pulse showing the effect of multipath reflections.

References

[1] Lewis, C., et al., "Multipath Interferences in Ground-Based Radar Data: A Case Study," *Remote Sensing*, Vol. 9, No. 12, 2017.

[2] Haspert, K., and M. Tuley, "Comparison of Predicted and Measured Multipath Impulse Responses," *IEEE Transactions on Aerospace and Electronic Systems*, 2010.

CHAPTER 16

Factors Affecting Multipath

Multipath interference from objects and surfaces in the environment of the radar and the platform on which the ESM system is mounted are a significant factor in the calculation of the DOA in ESM systems. If pulses from a particular radar are detected with widely varying DOA values, the ESM system will conclude that there are many more radars in the environment than there really are and will create extra tracks, obscuring the correct data. The effect of multipath interference is to change the incident amplitude and phase of the pulse, due to constructive or destructive interference of the waveform.

To explain how various factors such as ESM antenna separation, radar beamwidth, range to the radar, and reflection geometry affect the calculation of DOA, the results of running a simulation program are presented in this chapter. This multipath model allows the simulation of any one of six different radar types and the specification of ESM system parameters such as the sensitivity, dynamic range, and pulse measurement time. The number and location of the ESM antennas can be specified. The geometry of a single reflection is described in terms of the distance from the radar and the angle of the reflection object/surface relative to a line drawn from the aircraft to the radar. Figure 16.1 shows the basic reflection geometry used in the simulation program.

16.1 ESM Antenna Configuration

At short range to a radar, an aircraft subtends a relatively large angle to the radar beam and there may be a large differences in amplitude seen at the various receiver antennas, even if there is no multipath. Figure 16.2 shows the amplitude profiles of six consecutive scans of a radar at an initial range of 5 km as seen by an ESM system with antennas separated by a maximum of 20m. At this range and a heading of 0°, an aircraft of length 20m subtends an angle of 0.12° to a radar at a DOA of 90°.

To generate the graphs in Figure 16.2, a simulation was carried out using a reflection object located so that the angle between the direct and reflected paths is small, <1°. The path difference between the direct and reflected pulses is only about 10m, giving a time difference of arrival between the direct and reflected signals of less than 3 ns. A perfect reflector is assumed. The amplitude values shown on the graphs are calculated using the following formula:

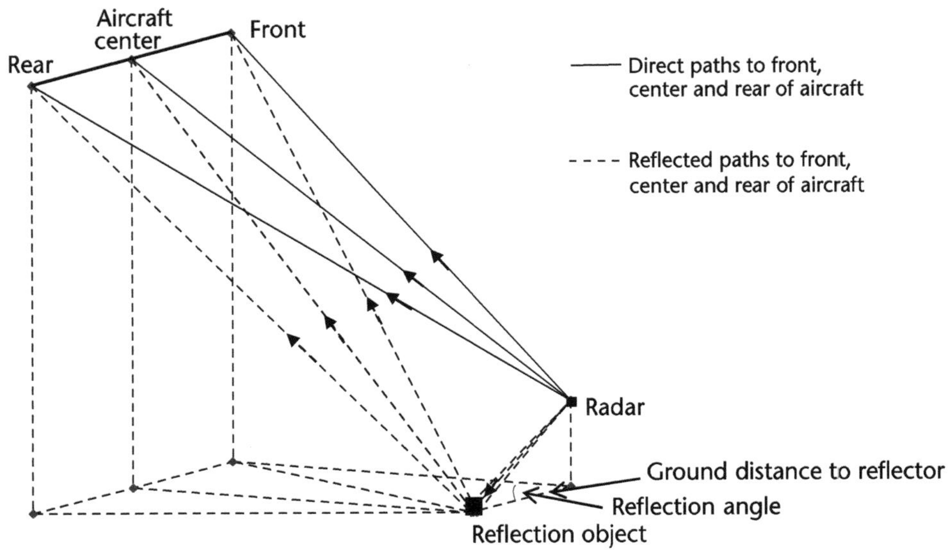

Figure 16.1 Simplified diagram of geometry used in multipath simulation.

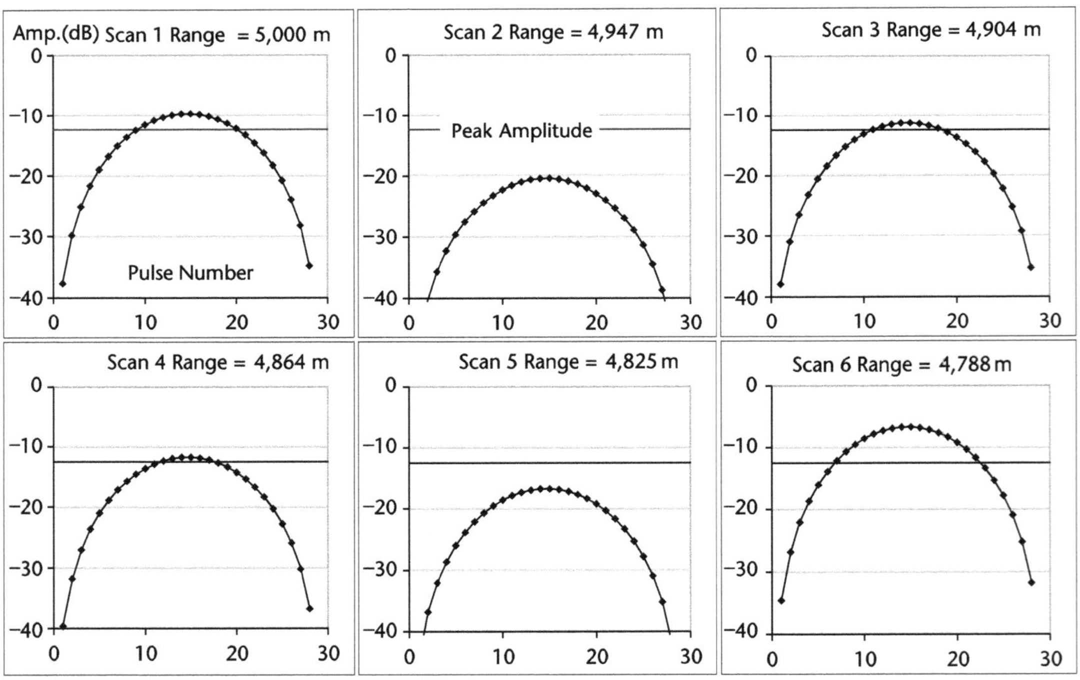

Figure 16.2 Amplitude profiles of consecutive scans of a radar affected by multipath seen by an ESM with separated antennas.

$$Amplitude = 0.66A_{max} + 0.33A_{max-1} \tag{16.1}$$

where A_{max} is the amplitude of the antenna detecting the largest amplitude for the pulse and $A_{max\text{-}1}$ is the amplitude of the antenna detecting the next largest amplitude.

Although the change in aircraft to radar range due to the aircraft flight from scan to scan is small, less than 100m, there are large differences seen in the peak amplitudes for each scan. The corresponding DOA profiles are shown in Figure 16.3, calculated using an amplitude comparison algorithm using three ESM antenna activation levels. The effect of this single reflection object on the DOA profiles is different for each scan, with the largest errors being seen for scan 2 where the peak amplitude is at least 6 dB below that expected.

The phase differences between direct and reflected paths at each of 3 ESM antennas (forward, aft, and center of the aircraft) for the peak pulse in each of the first three scans are shown in Table 16.1. The phase differences for all the relevant antennas in scan 2 are between 179° and 199° so destructive interference will occur to lower the amplitude seen at all three antennas.

In scan 2, the phase difference for antenna 2 is only 3.2° away from the total destructive interference expected for a 180° phase difference. The 183° phase

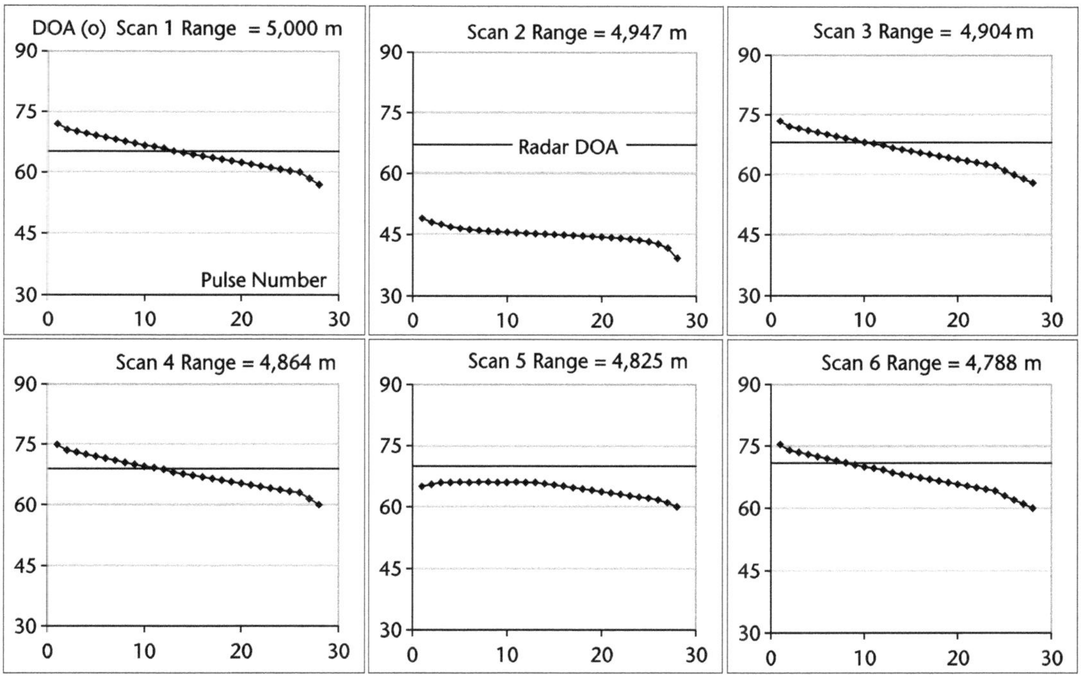

Figure 16.3 DOA profiles of consecutive scans of a radar affected by multipath as seen by an ESM with separated antennas.

Table 16.1 Phase Differences for Individual ESM Antennas

	Peak Pulse Phase Difference Between Direct Path and Reflection		
Antenna	1	2	3
Scan 1	106.9	81.3	47.2
Scan 2	198.9	183.2	179.5
Scan 3	329.6	323.9	310.6

difference causes a reduction in input amplitude of 12.5 dB for antenna 2. However, the most destructive interference occurs for antenna 3, which has its boresight towards the rear of the aircraft. The phase difference in this case is only 0.48° less than 180° and lowers the amplitude input to this antenna by over 20 dB. Antenna 1 suffers only a 4.5-dB decrease in amplitude. The DOA calculated for the peak pulse of scan 2 is therefore drawn towards the front of the aircraft by the input amplitude difference between the three relevant antennas. It is 15° less than it should be due to the more extreme destructive interference experienced by antenna 3 at the rear of the aircraft.

Running the same simulation again but this time with the ESM antennas colocated produces the set amplitude profiles in Figure 16.4. Similar excursions from the expected peak amplitude values are seen as for the separated ESM antenna configuration in Figure 16.2.

However, the DOA profiles shown in Figure 16.5 do not exhibit the errors seen for the separated antenna case. The small constant error seen for the colocated antennas is due to the elevation effect discussed in Chapter 5.

The first conclusion from these results is that the separation of the antennas causes problems in the measurement of DOA of pulses that are subject to multipath interference, which are not seen when the ESM antennas are colocated.

Table 16.2 shows the phase differences for the three antennas for the peak pulse for each of the first three scans for the colocated antenna case. The second scan has phase differences of 187.87° for the three antennas whose outputs are used in the DOA calculation. All the antennas have the same phase difference between the direct and reflected path, as expected due to their colocation.

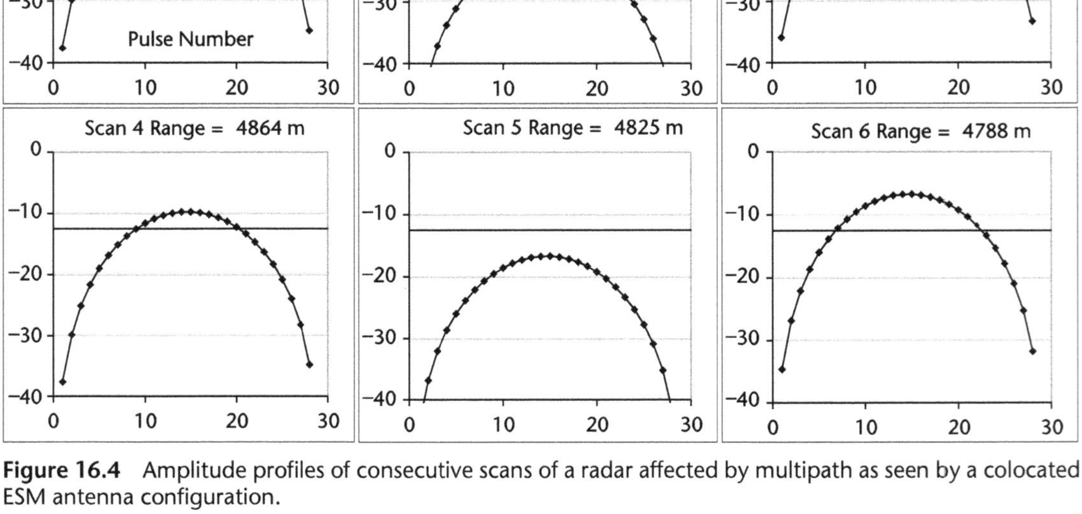

Figure 16.4 Amplitude profiles of consecutive scans of a radar affected by multipath as seen by a colocated ESM antenna configuration.

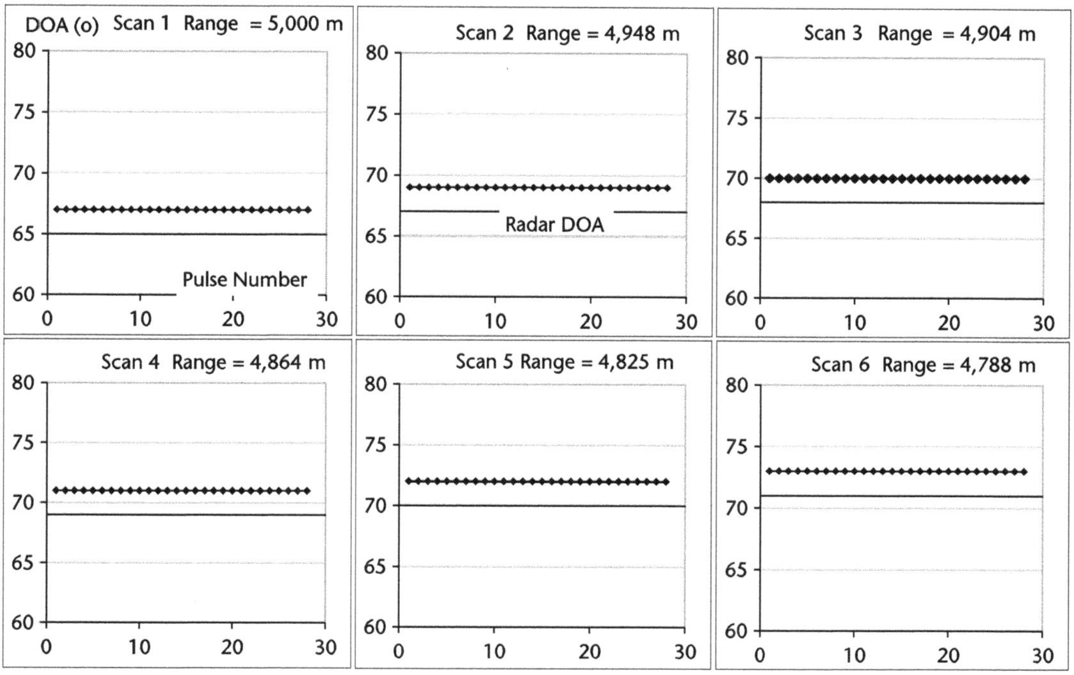

Figure 16.5 DOA profiles of consecutive scans of a radar affected by multipath as seen by a colocated ESM antenna configuration.

Table 16.2 Phase Differences for Individual Antennas in a Colocated ESM Antenna Configuration

	Peak Pulse Phase Difference Between Direct Path and Reflection		
Antenna	1	2	3
Scan 1	56.3	56.3	56.3
Scan 2	187.8	187.8	187.8
Scan 3	318.4	318.4	318.4

A second conclusion drawn from these results is that it is not the phase difference which matters to the size of the DOA error; it is the difference in phase difference seen at the individual ESM antennas that is the important factor.

A feature of all the DOA profiles seen in Figure 16.3 is that there is a trend in the DOA error from pulse to pulse across the main radar beam scan. The result can be explained by looking at the change in phase difference between direct and reflected path from pulse to pulse. Table 16.3 shows the phase difference for the three antennas used to measure the DOA for all pulses in a single scan of a radar at range 5 km from an aircraft with a single perfect reflection causing the phase differences shown at each antenna.

The third conclusion from these simple results is that a trend is likely to exist in the DOA error across a single radar scan due to multipath. This is important in that most deinterleaving algorithms currently use DOA as a deinterleaving parameter. A

Table 16.3 Phase Differences for Individual Pulses in a Single Scan Seen by an ESM System with Separated Antennas

	Phase Difference Between Direct Path and Reflection									
Pulse	1	2	3	4	5	6	7	8	9	10
Antenna 1	271.8	271.9	271.9	271.9	271.5	272.0	272.0	272.0	272.1	272.1
Antenna 2	157.1	157.1	157.1	157.1	157.1	157.2	157.2	157.2	157.2	157.2
Antenna 3	80.1	80.1	80.1	80.1	80.1	80.1	80.2	80.2	80.2	80.2
Pulse	11	12	13	14	15	16	17	18	19	20
Antenna 1	272.1	272.1	272.2	272.2	272.2	272.3	272.3	272.3	272.3	272.4
Antenna 2	157.3	157.3	157.3	157.3	157.3	157.4	157.4	157.4	157.4	157.5
Antenna 3	80.2	80.2	80.2	80.3	80.3	80.3	80.3	80.3	80.3	80.3

whole set of pulses (as few as six in most ESM systems) with an incorrect DOA are likely lead to the creation of an extra track.

To demonstrate the effect of ESM antenna separation on the DOA of pulses subject to multipath, the results of simulations with various antenna separations from 0.1m to 20m are shown next.

Figures 16.6 and 16.7 show the amplitude and DOA profiles, respectively, for simulations of a single perfect reflection interfering with pulses from a radar with a wide beamwidth of 5°. The range to the radar is 25 km and the radar beamwidth is large so the slope in the DOA across the main beam is not perceptible even for the case with 20-m distance between the front and rear ESM antennas.

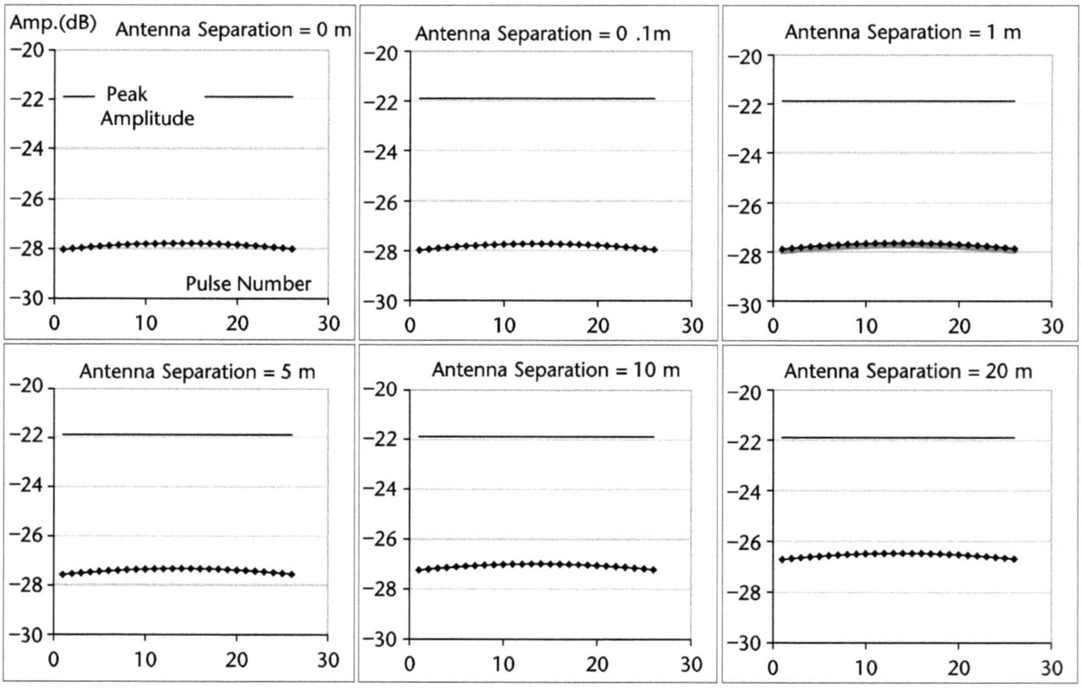

Figure 16.6 Effect of a single perfect reflection on amplitude for various ESM antenna separations.

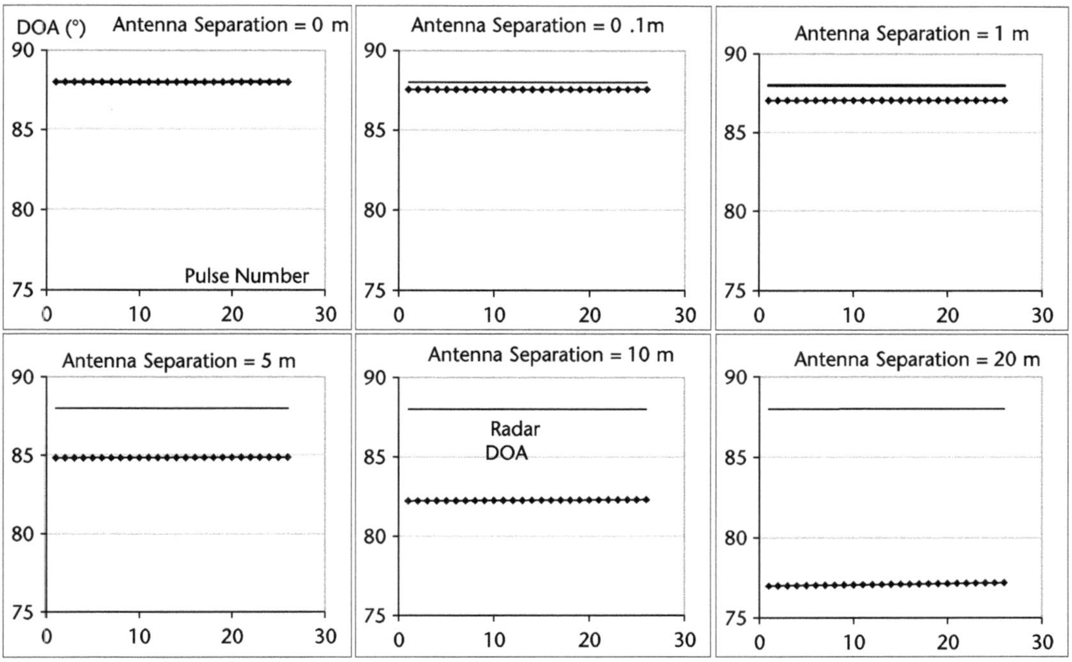

Figure 16.7 Effect of a single perfect reflection on DOA for various ESM antenna separations.

Figure 16.6 shows that for all ESM antenna separations the main beam amplitude has been significantly decreased due to multipath. This is because the phase difference between direct and reflected paths for each of the antennas is in the region 145° to 165° for all these examples, resulting in some destructive interference. Table 16.4 shows the phase differences and amplitude differences for the front and rear antennas for the various antenna separations.

The DOA profiles in Figure 16.7 show an increase in DOA error with the increase in antenna separation and the increase in phase difference between the front and rear antennas. The worst case shown is for a separation of 20m, which has errors of 13° for all pulses in the scan. For the 20-m antenna separation case, the difference in phase difference between front and rear antennas was approximately 20°.

Table 16.4 Phase and Amplitude Differences for Various ESM Antenna Separations (for a Correct DOA of 88°)

Separation (m)	*Front ΔPhase (°)*	*Rear ΔPhase (°)*	*Phase Diff (°)*	*Front Amp. (dB)*	*Rear Amp. (dB)*	*Amp. Diff.*	*Calc. DOA (°)*	*DOA Error (°)*
0	156.3	156.3	0.0	−25.8	−25.8	0.0	88.0	0.0
0.1	155.9	156.7	0.8	−25.7	−25.9	0.1	87.6	0.4
1	155.4	157.2	1.7	−25.7	−26.0	0.3	87.0	1.0
5	153.4	159.4	5.9	−25.4	−26.4	1.0	84.9	3.1
10	151.1	162.3	11.2	−25.0	−27.0	1.9	82.2	5.8
20	147.1	168.8	21.7	−24.5	−28.6	4.0	77.1	10.9

16.2 Radar Beamwidth

The separation of the ESM antennas causes a slope in the DOA profile across the main beam of a scanning radar, even when there is no multipath present (see Chapter 5). In the previous example, the radar 3-dB beamwidth was chosen to be relatively wide, 5.0°, to isolate the effects of ESM antenna separation and radar beamwidth.

Now the effect of radar beamshape on the DOA is considered for an ESM system that has antennas separated by 20m. The case with no multipath is presented in Figure 16.8, which shows the main beam amplitude profiles of six different 3-dB beamwidths from 5.0° to 0.3°. Apart from the variation in beamwidth, all other parameters in the generation of these plots are identical. The simulated radar has an RF of 3 GHz, a 4-second scan period, and a PRI of 1 ms. The aircraft is heading due North, the range to the radar is 15 km, and the DOA of the radar is 65°. A different number of pulses appear in each plot, as only the main beam of the radar is being considered.

The corresponding DOA profiles are shown in Figure 16.9. The slope in the DOA increases with the decrease in beamwidth to the most extreme case of the 0.3° beamwidth where the difference in DOA from the first pulse of the main beam to the last pulse is nearly 40°. The initial conclusion here is that even in the absence of multipath current amplitude comparison systems would have great difficulty in correctly determining the DOA of radars with narrow azimuth beamwidth at the detection range of 15 km that has been chosen for this example.

Figure 16.10 shows the effects on the same radar amplitude profiles of a perfect reflection at a small angle (0.5°) to the direct path. In all cases constructive

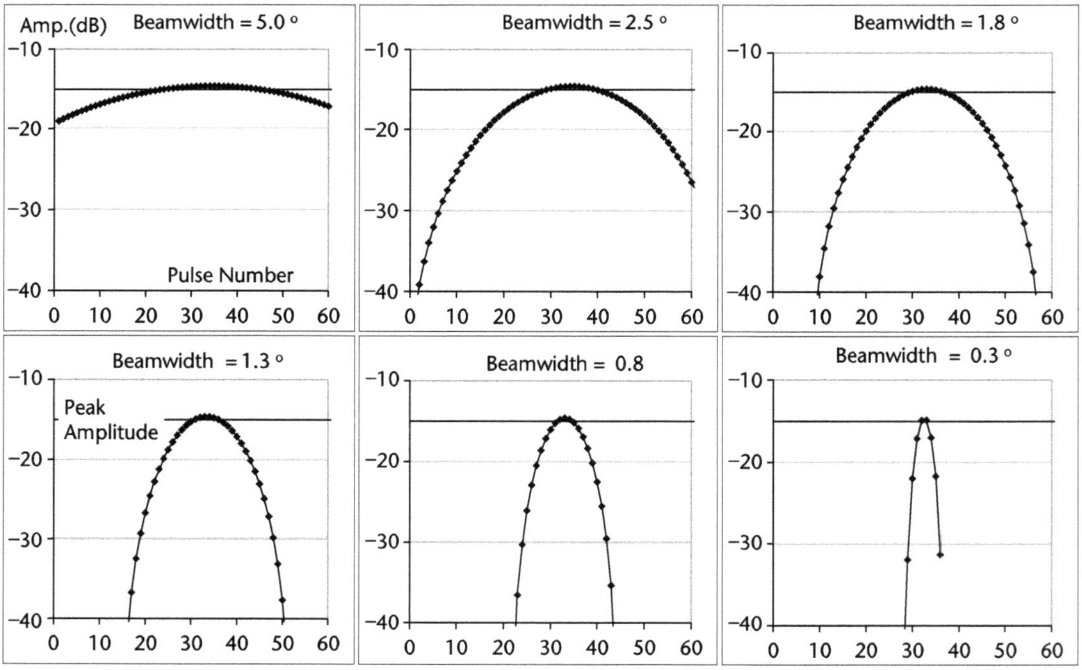

Figure 16.8 Amplitude profiles of the main beam of a scanning radar with various 3-dB beamwidths (no multipath).

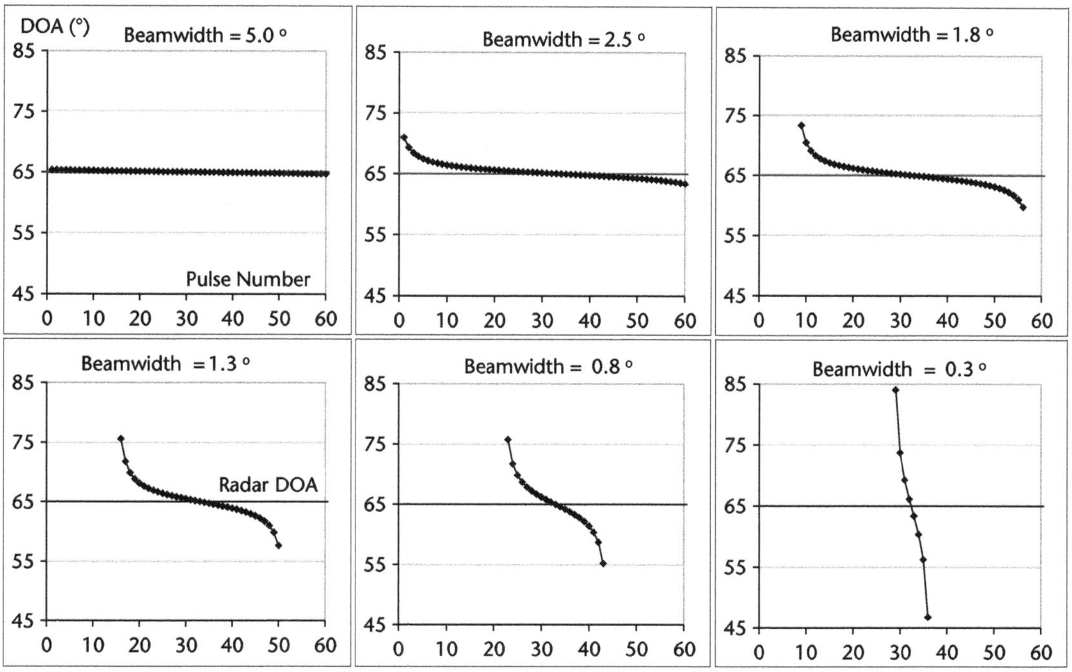

Figure 16.9 DOA profiles of the main beam of a scanning radar with various 3-dB beamwidths (no multipath).

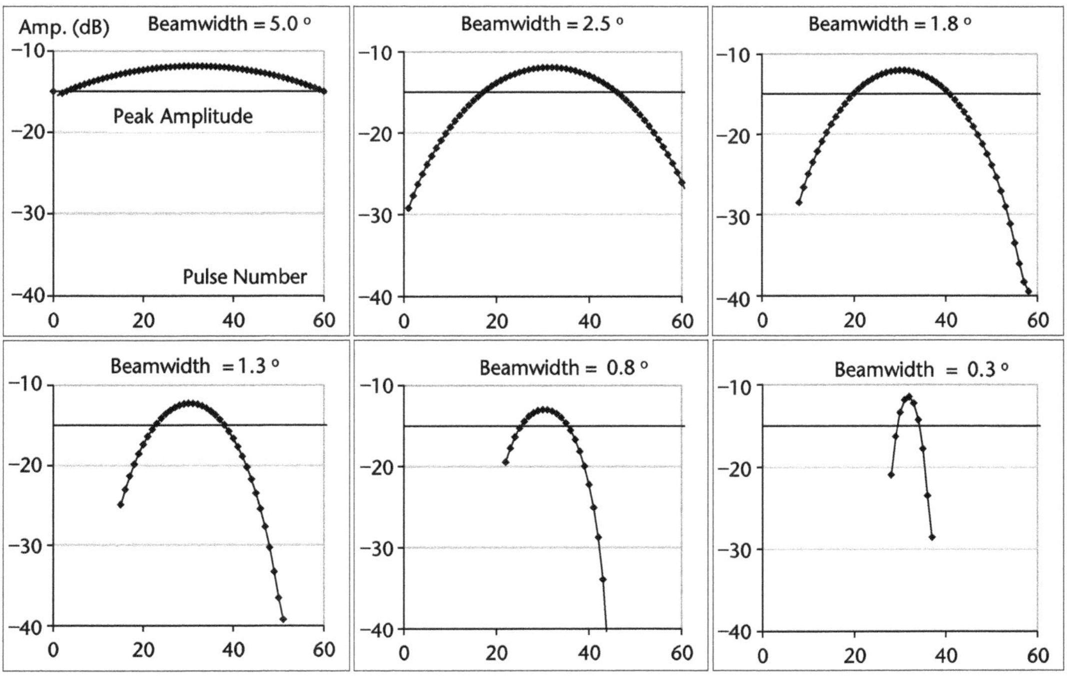

Figure 16.10 Amplitude profiles of the main beam of a scanning radar with a single reflection for various radar beamwidths.

interference increases the amplitude levels of the main beam. Figure 16.11 shows the corresponding DOA plots. All the plots show the same trend in DOA, but with a maximum DOA error of 10° for the first four plots. The maximum DOA errors are no worse than those for the case with no multipath for the remaining two plots with beamwidths of 0.8° and 0.3°. For the narrow radar beamwidths, the DOA profiles do not follow the predicted slope across the main beam seen when there is no multipath and pulses near the peak of the radar beam have smaller DOA errors for the narrow radar beams when there is no multipath.

Figure 16.11 illustrates that multipath interference does not always have a detrimental effect on the calculation of DOA. Sometimes it serves to counter the effects of antenna separation by boosting the amplitude of pulses that would otherwise be undetected by one or more antennas.

So far, the effects of multipath on main beam scans only have been considered. Figure 16.12 shows the amplitude and DOA profiles of a scanning radar detected by an ESM with separated antennas. The radar is at 15-km range from the aircraft and the first two sidelobes are detected.

The slope across the main beam can be seen in the DOA profile and there are also slopes across the sidelobes as expected. There are DOA errors of up to 30° between the main beam and sidelobes and between the first and second sidelobes due to the difference in amplitude seen at the various antennas. Figure 16.13 shows the amplitude input to each of the three ESM antennas for the two pulses between the main beam and first sidelobe and for one pulse between the first and second sidelobes.

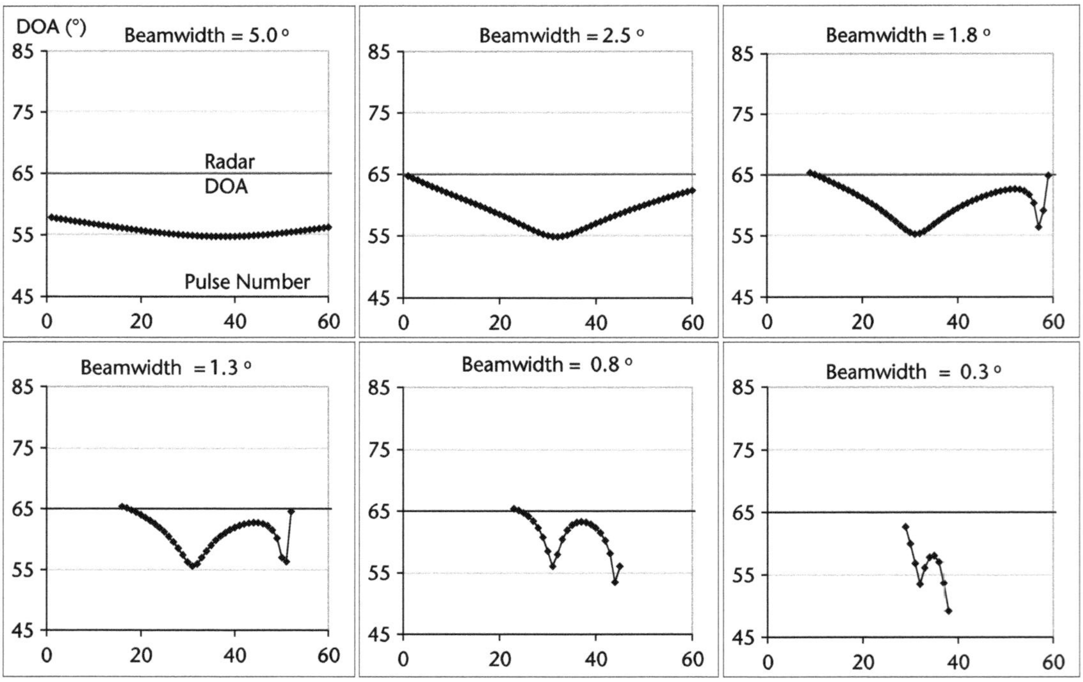

Figure 16.11 DOA profiles of the main beam of a scanning radar with a single reflection for various radar beamwidths.

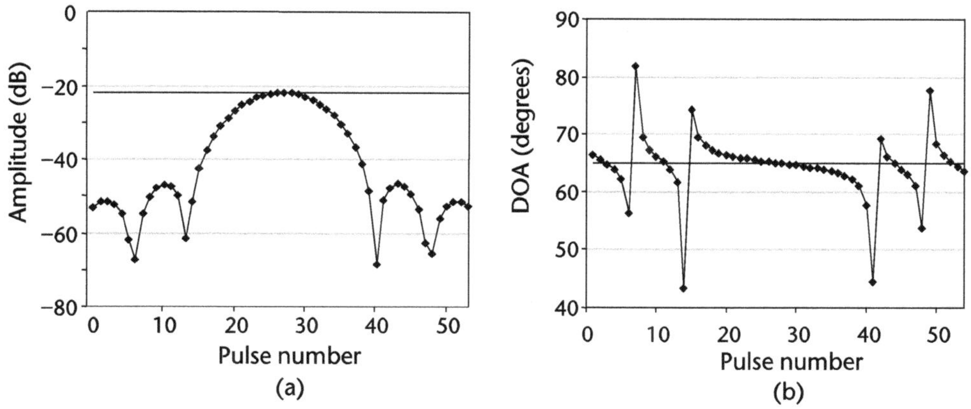

Figure 16.12 Detection of sidelobes in amplitude and DOA profiles of a scanning radar.

The amplitude input to each of the three relevant ESM antennas for these three pulses is shown in Table 16.5. For pulse 13, which lies between the main beam and the first sidelobe, there is a difference of 21 dB between the input amplitudes for antennas 1 and 3, leading to a 21° error in DOA. The other two pulses also suffer DOA errors, one of 16° and the other of –20°.

Figures 16.14 and 16.15 show three examples of the effects of a perfect reflector on the amplitude and DOA profiles of radars with various beamwidths in which the first sidelobes are detected in addition to the main beam. The distance to the reflection object is 0.5 km and the maximum path difference between direct and reflected signals is 10m.

Multipath changes the DOA profile in all the examples shown in Figure 16.15. It even improves the spread in DOA for the 1.8° and 1° beamwidths in that the

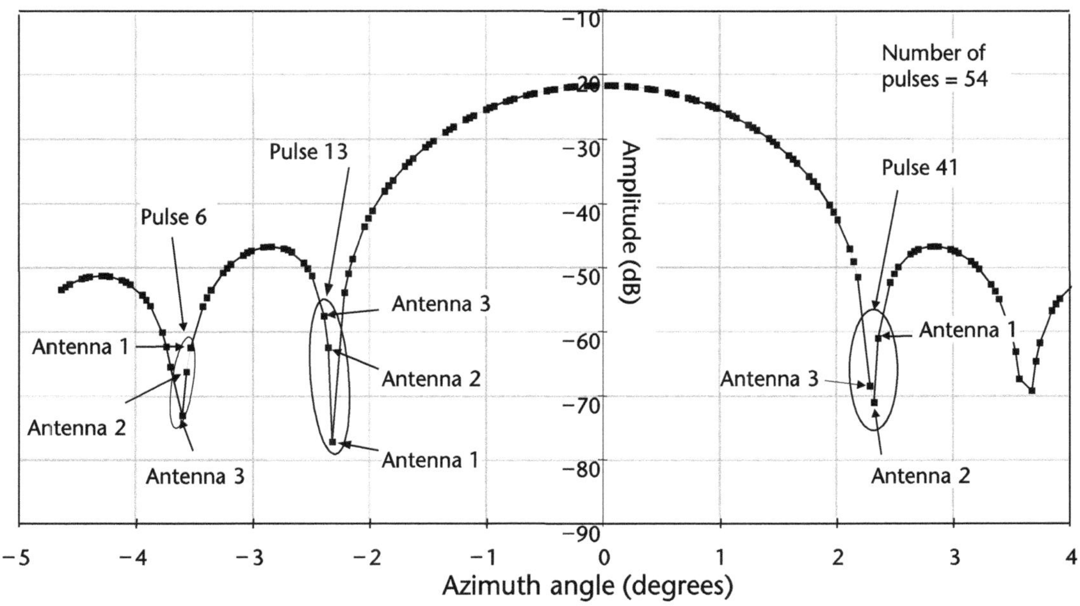

Figure 16.13 Amplitude input to individual antennas for selected pulses.

Table 16.5 Amplitudes and DOA for Pulses Between a Main Beam and Sidelobes (No Multipath)

Pulse Number	*6*	*13*	*41*
Antenna 1 amplitude in	–63	–77	–61
Antenna 2 amplitude in	–66	–63	–71
Antenna 3 amplitude in	–74	–56	–68
Calculated DOA	81	44	45
DOA error	+16	–21	–20

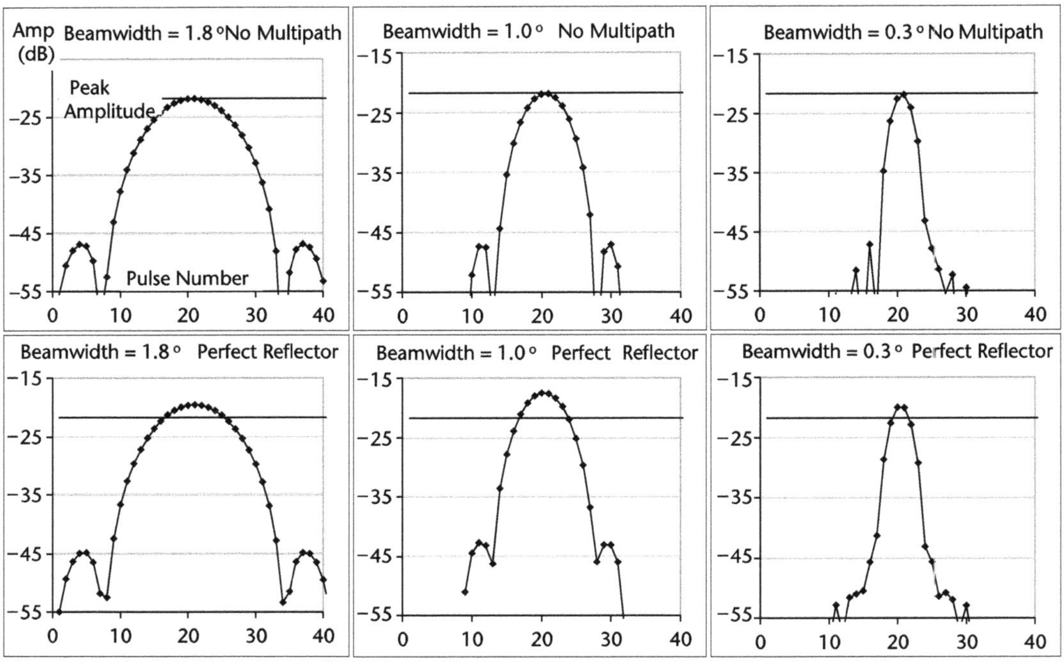

Figure 16.14 Effect of a single perfect reflection on amplitude profiles for main beams and sidelobes for various beamwidths.

spikes seen in the case with no multipath are reduced. The slopes in DOA across the sidelobes are in evidence only for the beamwidth of 1.8°.

Table 16.6 shows the values of amplitude and phase difference for the individual antennas for the pulse between the trailing edge of the first sidelobe and the main beam for each of the cases shown in Figure 16.15. The values of amplitude that would be seen with no multipath are also given. The DOA error is actually reduced in all these cases by multipath.

16.3 Reflection Geometry

For intrapulse reflections to occur, the difference between the direct path and reflected paths must be less than the PW, which for many radars is of the order of 1 μs. For a 1 μs PW, the path difference between direct and reflected paths must therefore be less than 300m.

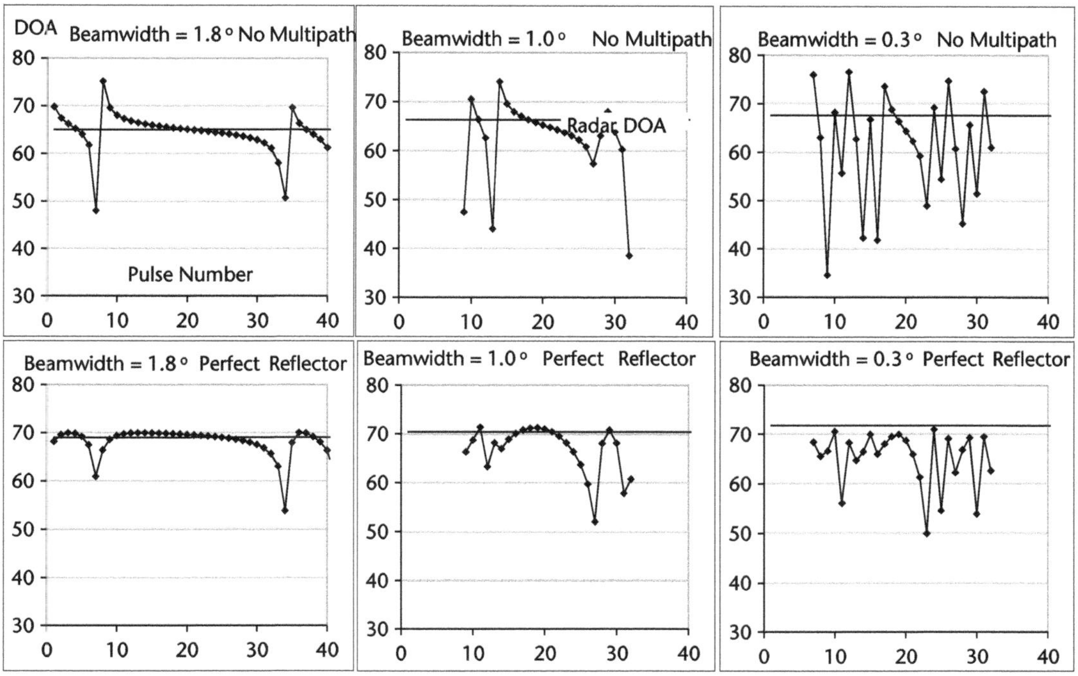

Figure 16.15 Effect of a single perfect reflection on DOA profiles for main beams and sidelobes for various beamwidths.

Table 16.6 Amplitudes and Phase Differences for Individual ESM Antennas for a Pulse Between the First Sidelobe and the Main Beam

	With Multipath		*No Multipath*			
Beamwidth	1.8	1	0.3	1.8	1	0.3
Antenna 1 phase difference	11.9	11.9	11.9			
Antenna 1 amplitude in	–42	–46	-45	–56	–53	53
Antenna 2 phase difference	16.0	16.0	16.0			
Antenna 2 amplitude in	–42	–47	-44	–59	–56	57
Antenna 3 phase difference	16.3	16.5	16.3			
Antenna 3 amplitude in	–45	–48	-46	–76	–73	69
DOA Error	–10	-7	3	–17	25	–24

The angle between the direct and reflected paths affects the relative amplitudes of the direct and reflected parts of the signal. A reflection angle that is very small (~0.1° for a radar with a 1.5° 3-dB beamwidth) produces a reflected amplitude near to that of the direct path for each pulse in a scan. The composite pulse amplitude is therefore largely determined by the phase difference between the direct and reflected pulses as discussed in Chapter 15.

As the reflection angle increases to about 1° or 2°, the reflection causes a broadening of the amplitude profile of the main beam. Further increases of the reflection angle cause a second main beam peak to appear very near the real main beam, but diminished in amplitude due to the reflection coefficient as discussed in

Section 16.5. In this case, the reflection of the main beam is interfering with direct path sidelobe pulses and the appearance is of a whole main beam reflection.

Figures 16.16 and 16.17 illustrate these points by showing the effect of the reflection angles of 0.1° to 5° on the amplitude and DOA profiles of a scanning radar of 3-dB beamwidth 1.8° at a range of about 25 km. The geometry of the reflection was chosen such that the phase difference between direct and reflected paths at antenna 3 is about 180°.

There are different numbers of pulses in each of the plots in Figures 16.16 and 16.17 as the angular extent of the reflection path must be taken into account in each case. For example, the main beam of the 10° reflection case is unaffected by multipath, but pulses in the trailing edge sidelobes suffer interference by the reflected main beam. This significantly enhances their amplitude and gives the appearance of another main beam received very close in time to the real main beam.

A perfect reflection has been assumed in all these cases. However, the reflection is of lower amplitude than the real main beam in the 10° reflection case due to the phase shift of nearly 180° causing destructive interference to the composite pulse amplitudes in the reflection peak. These pulses are reduced in amplitude from the direct path main beam level by nearly 10 dB.

16.4 Range to the Radar

Following the description of the effects of various reflection geometries on the extent of DOA corruption due to multipath, limits are placed on the type of reflection used to determine whether multipath effects depend on the range to the radar. The

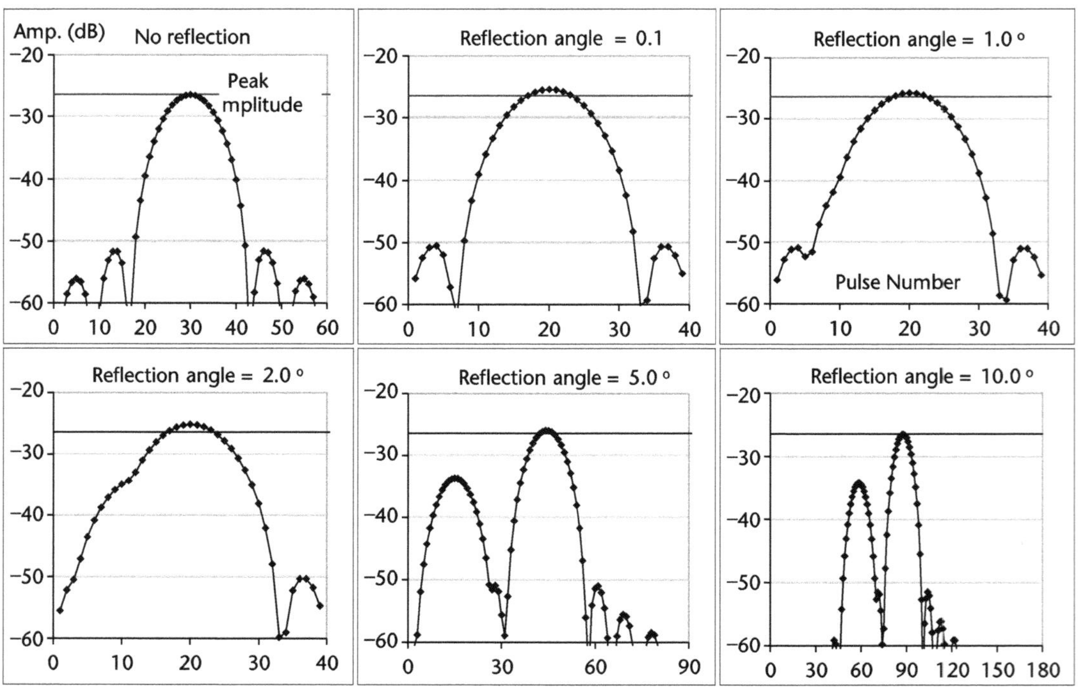

Figure 16.16 Effect of reflection angle on the amplitude profile of a scanning radar.

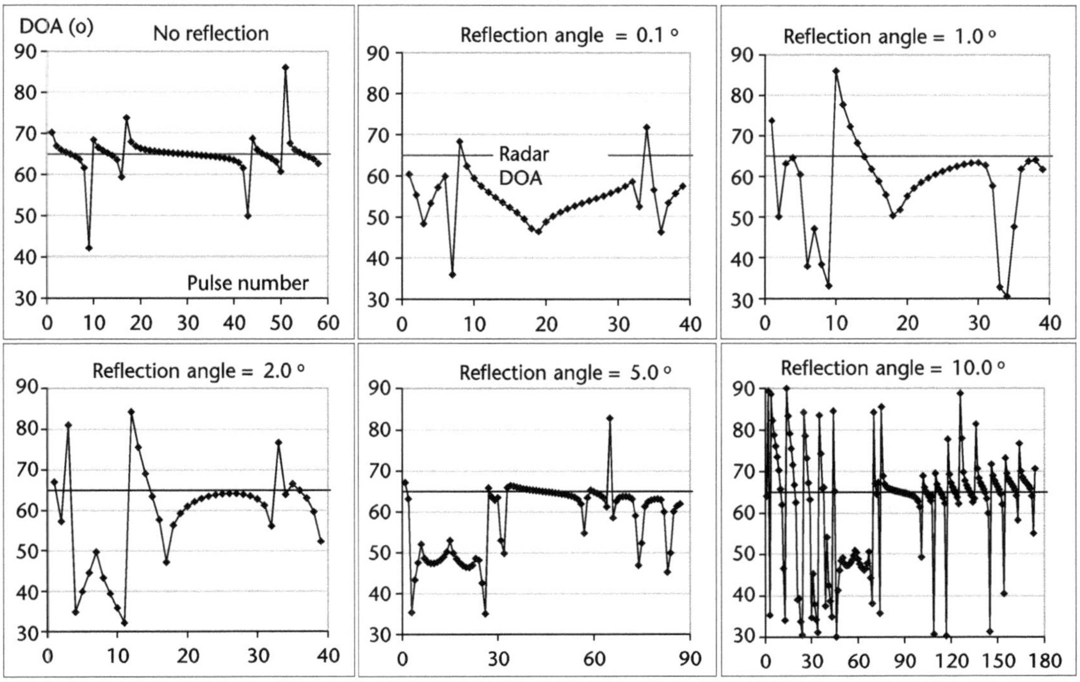

Figure 16.17 Effect of reflection angle on the DOA profile of a scanning radar.

only type of reflection considered here are those expected to have the most impact on the amplitude levels of the composite pulse. Small reflection angles are chosen so that the amplitudes between the direct and reflected pulses are similar.

The effects of a single reflection at 0.5° to the direct path on the detection of single scans from a radar broadside on to the aircraft at various ranges from the aircraft are shown in Figures 16.18 and 16.19.

The DOA plot for the closest range shown in Figure 16.19 has a reversed slope in the DOA for the first half of the scan and an error of 20° for the peak pulse. The same profile is seen for all the ranges, but the DOA error of the peak pulse reduces as the range increases.

The amplitude profiles of Figure 16.18 show that the peak amplitude is reduced for all ranges, for this reflection geometry. The amount of amplitude reduction is the same at all ranges due to the phase difference seen at antenna 2, which should receive the strongest signal for a radar DOA of 90°.

The signal arriving at antenna 2 is subject to a phase difference of 194° for all the ranges in Figure 16.18. The phase differences for all three relevant antennas are shown in Table 16.7, together with the Δ phase difference (difference in phase difference between antennas 1 and 2 or antennas 2 and 3). The Δ phase difference decreases as the range increases which accounts for the decrease in DOA error experienced at longer range.

For the 135-km example, all the antennas are subject to approximately the same amount of destructive interference and at this long range the antenna separation is negligible compared with radar beamwidth so no serious DOA errors result for any of the pulses in the scan.

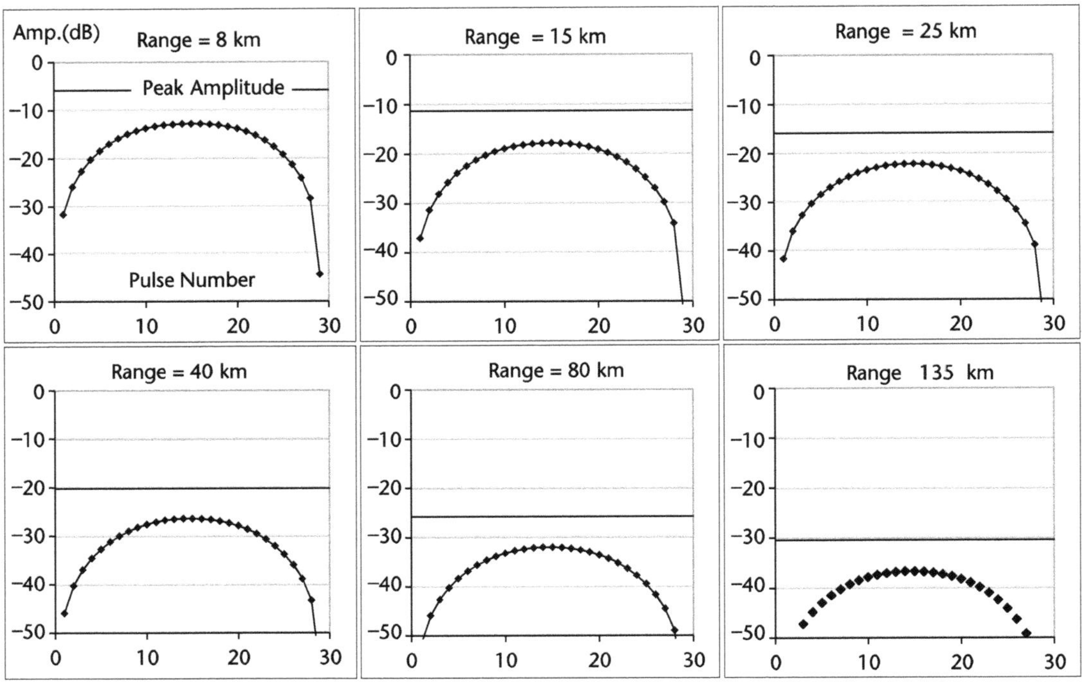

Figure 16.18 Amplitude profiles of a scanning radar at various ranges with a perfect reflection at 0.5° to the direct path.

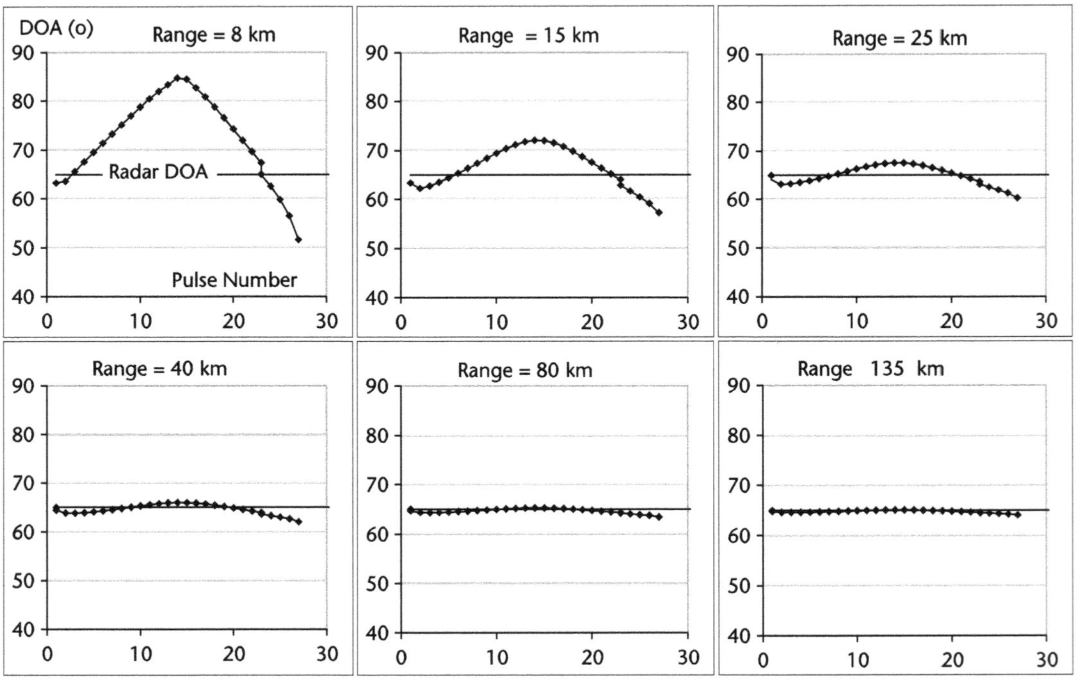

Figure 16.19 DOA profiles of a scanning radar at various ranges with a perfect reflection at 0.5° to the direct path.

Table 16.7 Phase Differences Seen at ESM Antennas for Various Ranges

Range (km)	*Antenna*	*Peak Pulse Phase Difference*	*Δ Phase Difference*
5	1	162.48	
	2	194.98	32.50
	3	274.76	77.43
8	1	180.75	
	2	194.07	13.33
	3	225.79	31.71
15	1	189.88	
	2	194.07	3.88
	3	225.79	9.24
25	1	193.67	
	2	195.02	1.38
	3	198.31	3.27
40	1	193.51	
	2	194.02	0.51
	3	195.22	1.22
80	1	193.91	
	2	194.05	0.14
	3	194.38	0.33
135	1	193.91	
	2	193.96	0.05
	3	194.08	0.11

The first conclusion on the effect of range on the extent of the multipath problem is that, when all antennas are affected, the DOA errors that result decrease with increasing range. This is because the phase differences between the antennas converge at long range.

Two cases of multipath interference are considered now. The first case is the effect of multipath on consecutive scans of a radar at short range (6.5 km) to the ESM platform, and the second case is the effect of multipath on a radar at long range (80 km).

First, Figures 16.20 and 16.21 show amplitude and DOA profiles for a set of six consecutive scans that were generated for a radar of beamwidth 1.5° at a short range of about 6.5 km from the aircraft.

Each of the six scans has a different DOA profile, with the phase difference from pulse to pulse for each antenna following a trend during the individual scans. However, there is no trend from scan to scan, as shown in Table 16.8, which gives the phase difference for the three relevant antennas for the peak pulse in the six scans.

Although there is no trend in the DOA profile from scan to scan, Table 16.8 shows that there are some scans that have similar phase differences for the three antennas. For example, scans 2 and 5 have similar values for the phase differences for the three antennas. Both these scans have similar DOA profiles as shown in Figure 16.21. Scans 3 and 6 in Table 16.8, which immediately follow the two similar scans 2 and 5, do not have DOA profiles that are similar to each other.

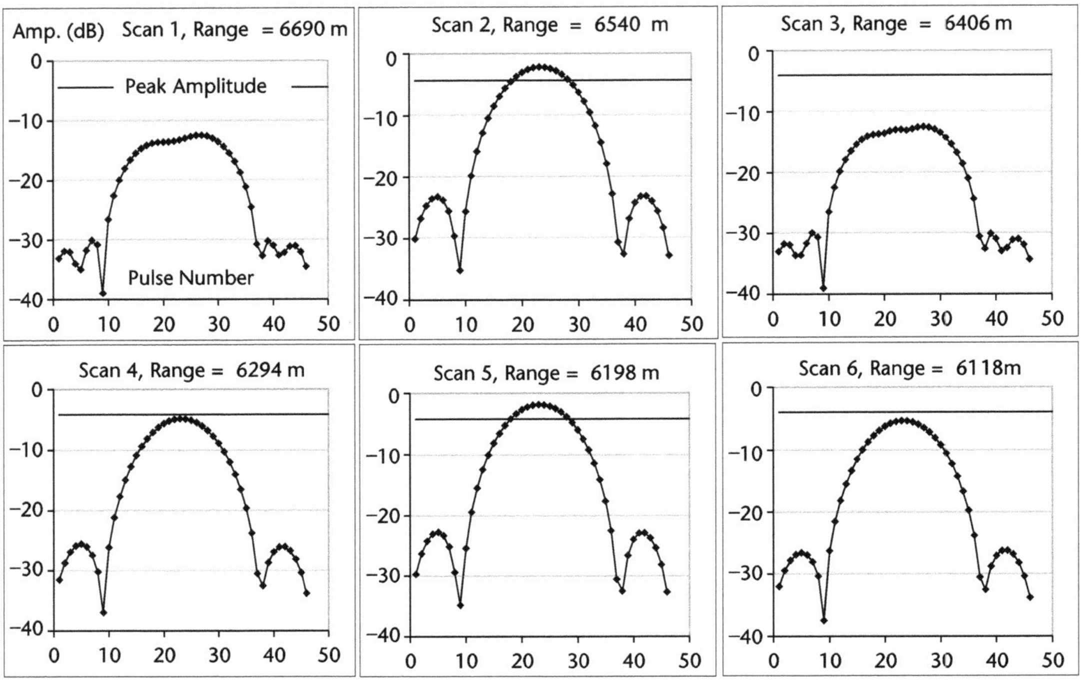

Figure 16.20 Consecutive single-scan amplitude profiles for a radar at short range with a single reflection at 0.5° to the direct path.

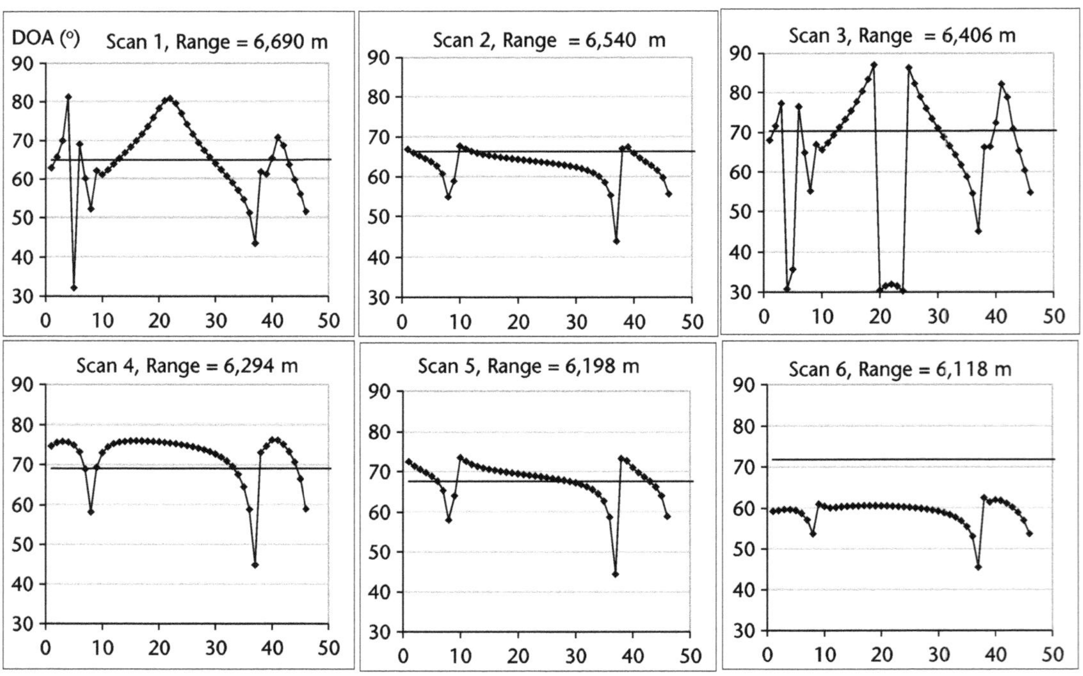

Figure 16.21 Consecutive single-scan DOA profiles for a radar at short range with a single reflection at 0.5° to the direct path.

Table 16.8 Phase Differences for Peak Pulses in Six Consecutive Scans at Short Range

	Peak Pulse Phase Differences			*Correct DOA*	*Calculated DOA*	*DOA Error*
	Antenna 1	*Antenna 2*	*Antenna 3*			
Scan 1	268.6	287.2	331.4	65.0	68.0	3.0
Scan 2	115.2	132.1	175.8	66.3	51.6	–14.7
Scan 3	337.8	353.6	36.5	67.7	67.7	0.0
Scan 4	219.8	233.9	276.1	69.0	76.0	7.0
Scan 5	122.8	135.2	176.5	70.4	54.2	–16.2
Scan 6	49.0	59.7	99.8	71.8	69.1	–2.7

The short range used to generate the scan profiles in Figures 16.20 and 16.21 means that it is effectively a random chance that there are similar profiles within a consecutive set of scans. The difference in phase for each antenna depends on many variables such as the geometry of the reflection and the ESM antenna separation.

The second conclusion on the effect of multipath on DOA at short range is that there is no trend from scan to scan for DOA errors.

Figures 16.22 and 16.23 show a set of six consecutive scans for a radar at long range (~80 km) from the aircraft. A perfect reflection at an angle of 1° to the direct path with a maximum path difference between the direct and reflected path of 10m is used to generate these plots.

Table 16.9 shows the phase difference for the three antennas for the pulse with the greatest DOA error in each of the scans. There is a trend in phase difference for

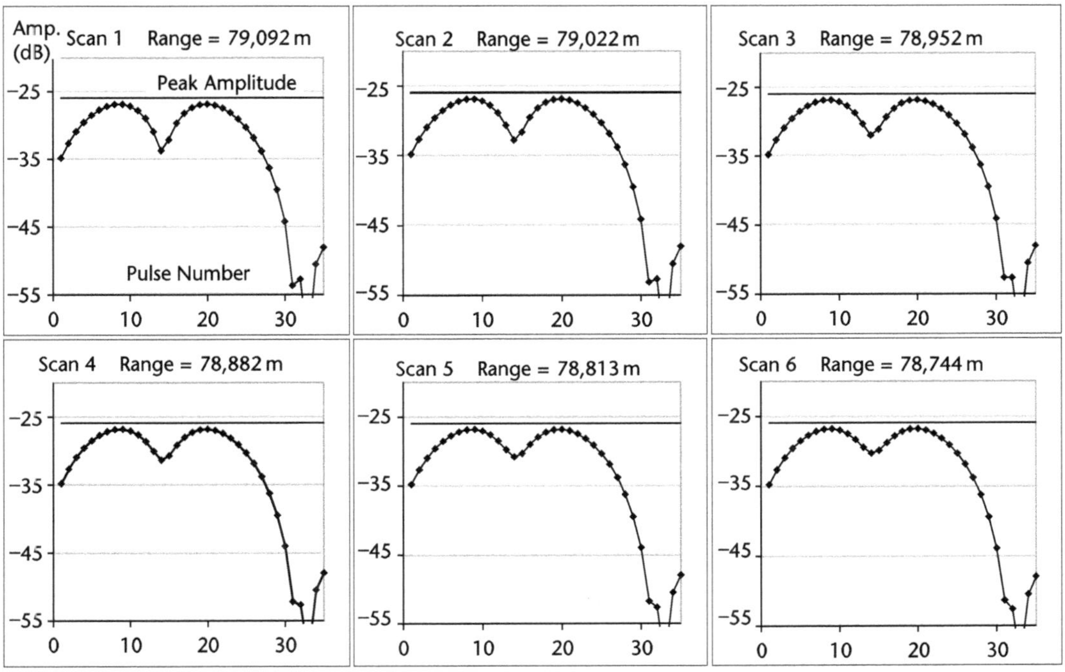

Figure 16.22 Consecutive single scan amplitude profiles for a radar at long range with a single reflection at 1° to the direct path.

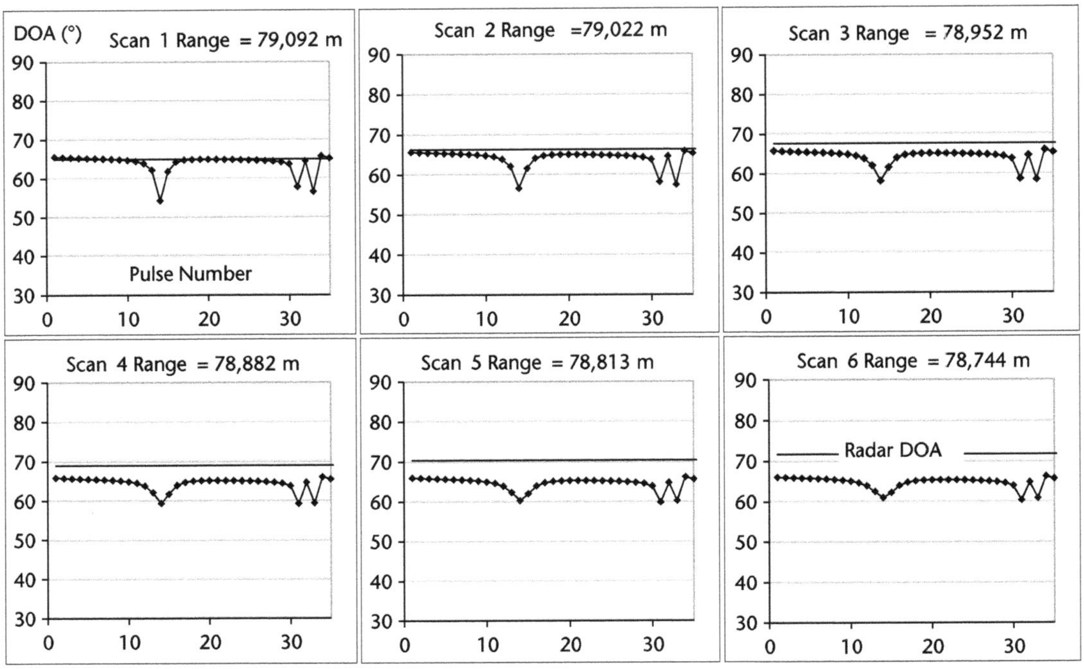

Figure 16.23 Consecutive single scan DOA profiles for a radar at long range with a single reflection at 1° to the direct path.

Table 16.9 Phase Differences for Peak Pulses in Six Consecutive Scans at Long Range

	Peak Pulse Phase Differences			*Correct DOA*	*Calculated DOA*	*DOA Error*
	Antenna 1	*Antenna 2*	*Antenna 3*			
Scan 1	162.3	170.7	179.4	65.0	54.1	–10.9
Scan 2	159.1	167.3	176.2	65.1	56.5	–8.6
Scan 3	155.7	164.1	173.0	65.2	57.8	–7.4
Scan 4	152.5	160.9	169.6	65.3	59.3	–6.0
Scan 5	149.3	157.7	166.5	65.4	60.1	–5.3
Scan 6	145.9	154.6	163.4	65.6	61.1	–4.5

each antenna from scan to scan, hence the similar DOA profiles of the individual scans. This scan-to-scan trend in phase difference was not seen at short range.

The third conclusion on the effect of range on multipath interference is that at long range there will be a trend in DOA error seen from scan to scan. This has the potential to lead to multitracking with updates at the wrong DOA persisting for some time and causing false target situations for the operator of the ESM.

16.5 Reflection Coefficient

The results presented so far have been generated assuming that there is a perfect reflection in each case. In the real world, all reflected signals are subject to attenuation at the reflection surface. The reflection coefficient depends on the grazing angle (i.e., the complement of the angle of incidence), the polarization of the signal and the admittance of the reflecting medium [1]. The admittance is the ratio of the complex dielectric constant (depending on the wavelength and conductivity of the medium) relative to the magnetic permeability.

It should be noted that all the signals considered here are horizontally polarized and the radar and receiver are matched in polarization. Figure 16.24 shows a plot of the reflection coefficients for 3-GHz and 9-GHz horizontally polarized signals for rough ground, smooth sea, and aircraft-based reflections.

For ground reflections, the radar frequency is important for small grazing angles, whereas for reflections from smooth sea, the reflection coefficient is effectively independent of radar RF at frequencies between 2 and 18 GHz and decreases only

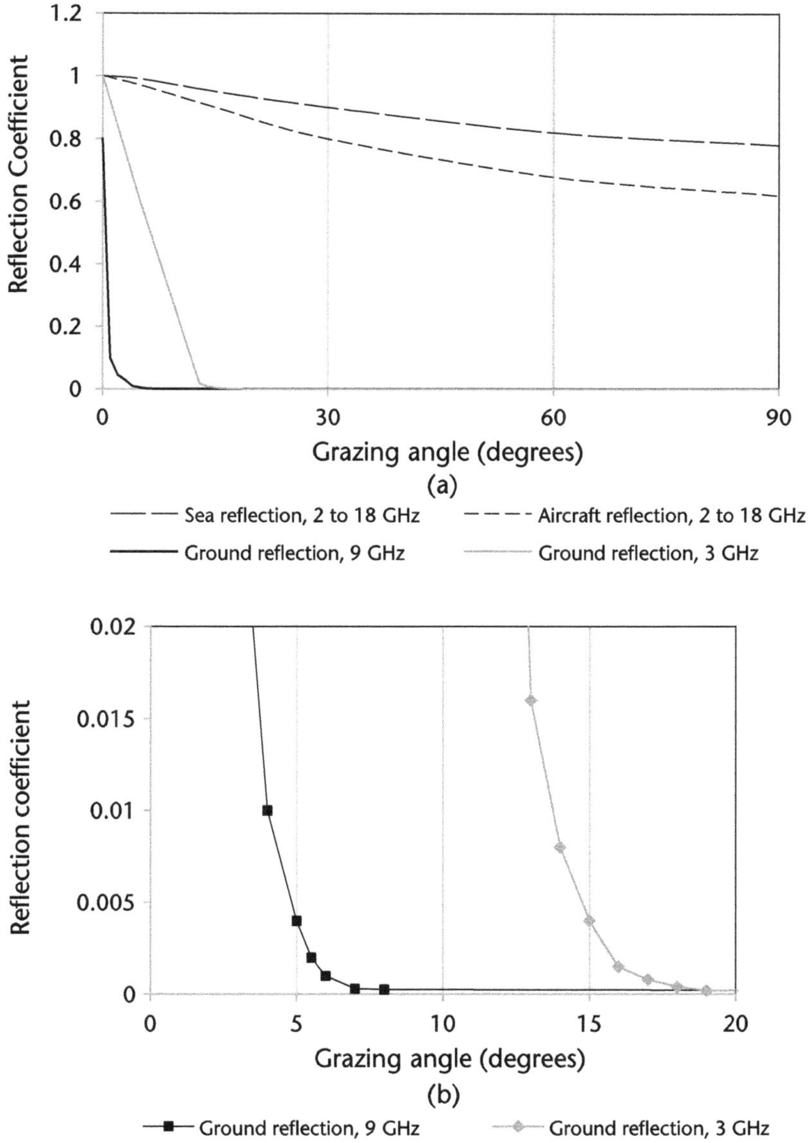

Figure 16.24 Reflection coefficients for rough ground, smooth sea, and aircraft surfaces.

slowly with increase in grazing angle. At an 80° grazing angle, the reflection coefficient from smooth sea has only reduced to 77%. The details of the calculation of reflection coefficients can be found in Appendix B.

Figures 16.25 and 16.26 show the effect of various values of reflection coefficients on the amplitude and the DOA of a scanning radar of a beamwidth of 1.8°.

A reflection coefficient of 1.0, a perfect reflection, produces a significant change in the DOA profile with a maximum error of 20°. Similar profiles are seen as the reflection coefficient is reduced. At a 50% reflection coefficient, the DOA profile is shifted with a DOA error of 5° for the peak pulse of the scan. The only effects seen at 10% and 1% reflection coefficients are small shifts in the DOA profiles such that the DOA errors seen for the peak pulses are 0.5° and 0.2°, respectively.

Figure 16.24 shows that it is only for small grazing angles that the reflection coefficient for ground-based reflections is above 1%, but there are many circumstances where such reflections may occur. However, for reflections from the smooth sea, there may be a significant effect due to multipath as the reflection coefficient is above 80% for all likely grazing angles.

The other possibility for reflection is from the surfaces of the aircraft itself. In this case, it is likely that the reflection coefficient may be as high as 80% and also that a single antenna would be affected. This may have serious consequences on the DOA calculation as seen in Figures 16.27 and 16.28, which are the amplitude and DOA profiles of a reflection of a scanning radar with each of the three relevant antennas being affected in turn for reflection coefficients of 80% and 50%.

Similar DOA profiles are seen for each antenna for the two reflection coefficients considered. The simulations are set up so that each affected antenna had a

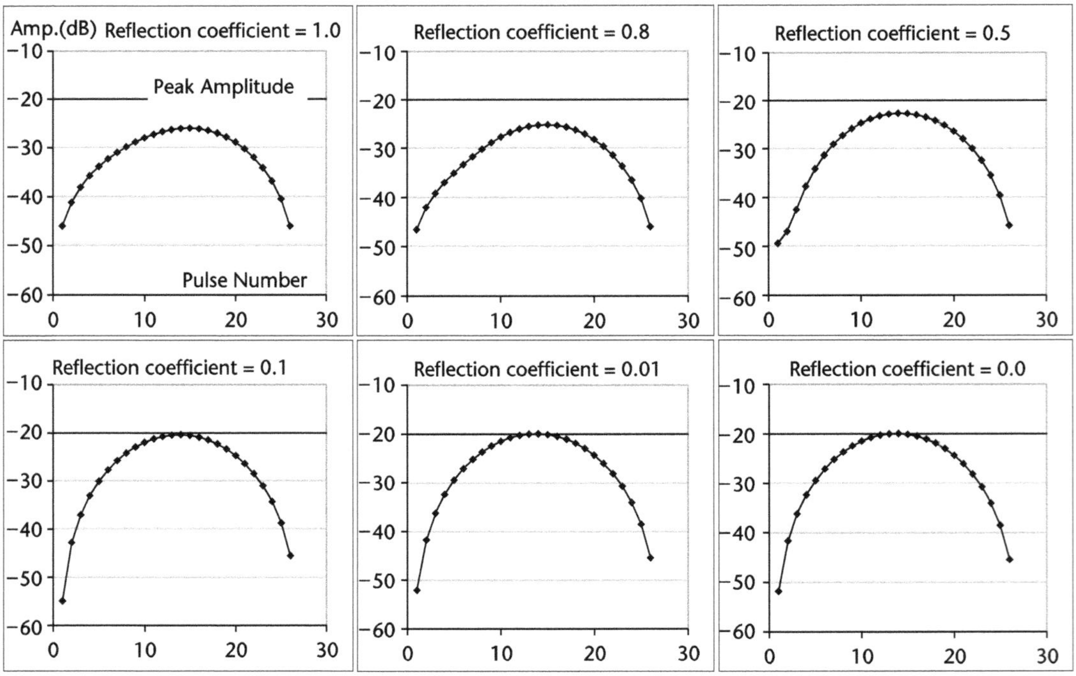

Figure 16.25 Amplitude profiles at a range of 15 km for various reflection coefficients with all antennas affected by multipath for various reflection coefficients.

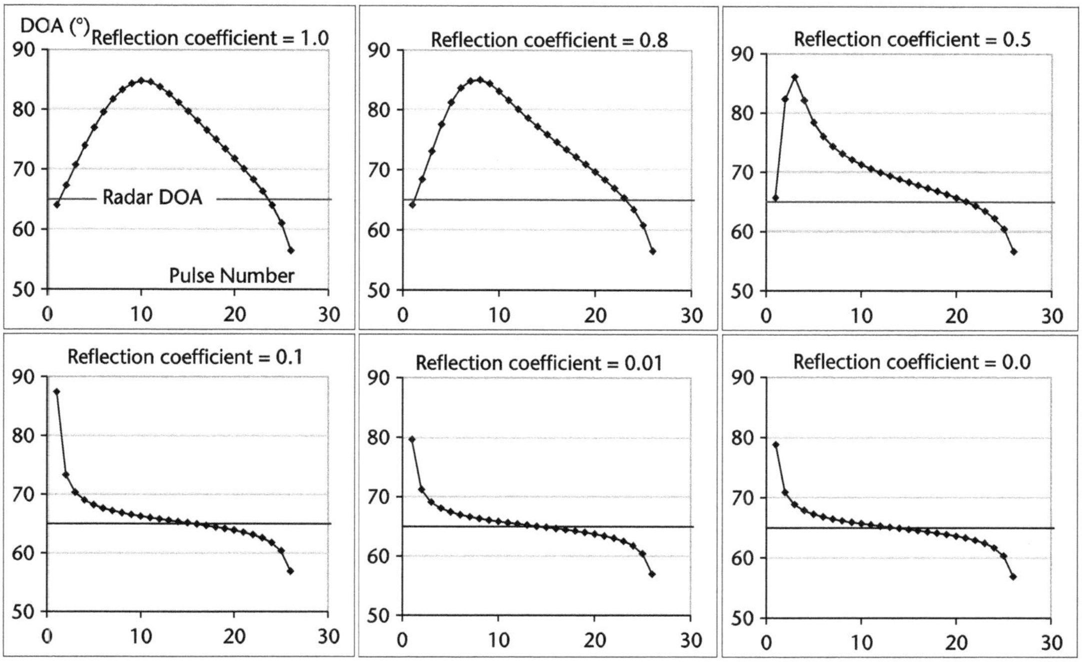

Figure 16.26 DOA profiles of main beam of a scanning radar at a range of 15 km with all antennas affected by multipath for various reflection coefficients.

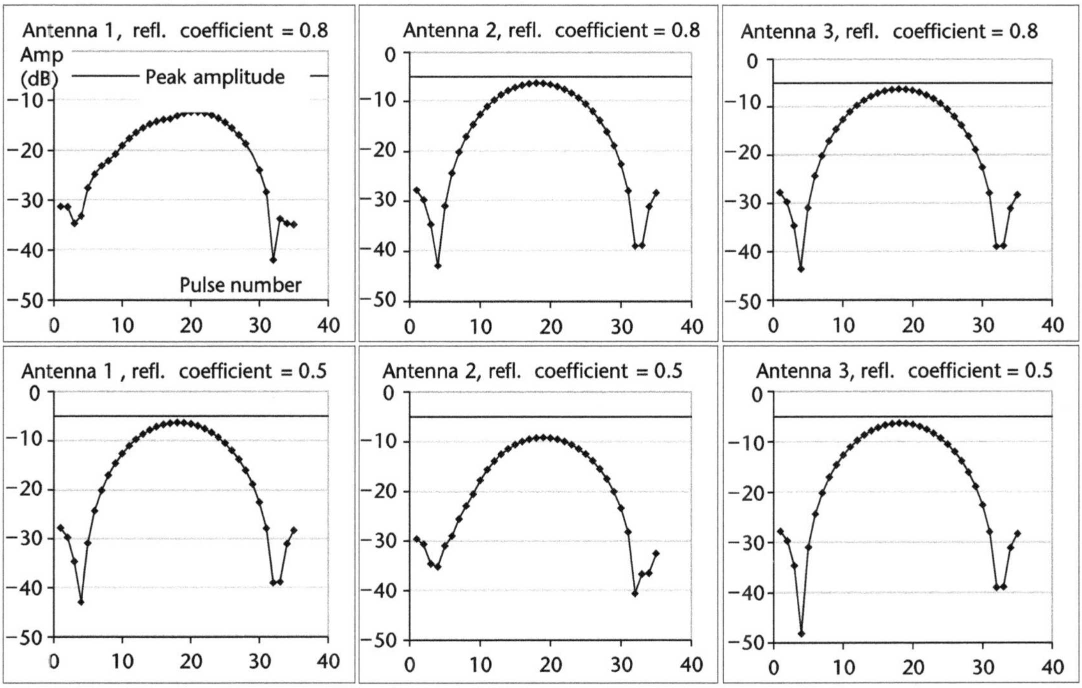

Figure 16.27 Amplitude profiles of the main bean of a radar with individual antennas affected by multipath for reflection coefficients of 0.5 and 0.8.

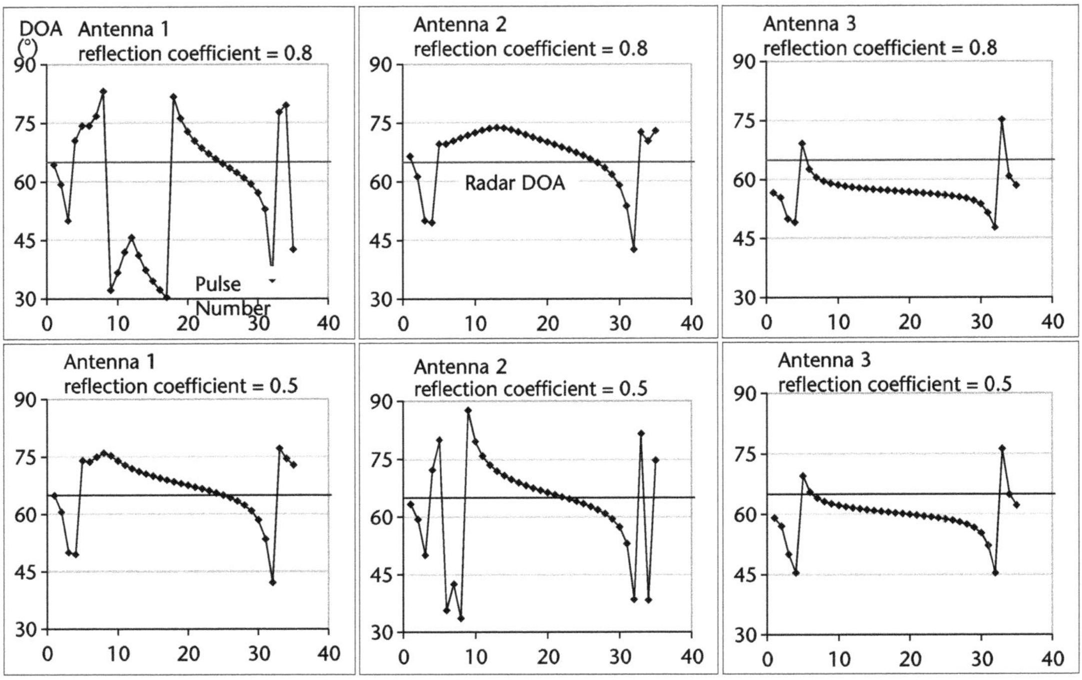

Figure 16.28 DOA profiles of main beam for a scanning radar at a range of 8 km with individual antennas affected by multipath for refection coefficients of 0.5 and 0.8.

similar phase difference between direct and reflected paths so that they were subject to some destructive interference.

Table 16.10 shows the effects of the two reflection coefficients on the amplitude of each of the three antennas and the DOA error, which results for the peak pulse of the scan in each case.

The largest effects are seen when antenna 2, the center antenna, is affected by multipath as the DOA of the radar in these simulations is set to be near the boresight of this antenna. The DOA error changed from –30° to +7° for the peak pulse in this scan for the two reflection coefficients considered here. When using a three-antenna amplitude comparison algorithm for the calculation of the DOA, the algorithm becomes unstable when the antenna that should see the strongest signal has a lower activation level than the adjacent antennas. In this case, antennas 1 and 3 both have higher amplitudes than the center antenna, resulting in the instability in DOA calculation.

The DOA error is positive when antenna 1, the front antenna, is affected and the DOA is increased due to the relatively large amplitudes seen by the other two antennas. When antenna 3, the rear antenna, is affected, the reverse effect is seen and the DOA is decreased.

If the three antennas had been the subjects of constructive interference, the DOA errors would have been reversed in sign. However, as shown in Figure 15.5, the change in amplitude is not linear with phase difference and the effects of a near total constructive interference are not as extreme as those of near total destructive interference.

Table 16.10 Effect of Reflection Coefficient on the Amplitude of Each Antenna for a Single Reflection of a 3-GHz Signal

Antenna 1 affected	*80% Refl. Coefficient*			*50% Refl. Coefficient*		
	Antenna 1	*Antenna 2*	*Antenna 3*	*Antenna 1*	*Antenna 2*	*Antenna 3*
Path diff. (m)	1.5	0.0	0.0	1.5	0.0	0.0
Time diff. (ns)	170.9	0.0	0.0	170.9	0.0	0.0
Phase diff. (°)	5.2	0.0	0.0	5.2	0.0	0.0
Amplitude in (dB)	−23.3	6.3	−13.6	−19.9	−6.4	−13.7
DOA error (°)	+8			+5		
Antenna 2 affected	*80% Refl. Coefficient*			*50% Refl. Coefficient*		
	Antenna 1	*Antenna 2*	*Antenna 3*	*Antenna 1*	*Antenna 2*	*Antenna 3*
Path diff. (m)	0.0	1.5	0.0	0.0	1.5	0.0
Time diff. (ns)	0.0	178.2	0.0	0.0	178.2	0.0
Phase diff. (°)	0.0	5.2	0.0	0.0	5.2	0.0
Amplitude in (dB)	−16.8	18.5	−13.7	−16.8	−9.5	−13.7
DOA error (°)		−30			+7	
Antenna 3 affected	*80% Refl. Coefficient*			*50% Refl. Coefficient*		
	Antenna 1	*Antenna 2*	*Antenna 3*	*Antenna 1*	*Antenna 2*	*Antenna 3*
Path diff. (m)	0.0	0.0	1.6	0.0	0.0	1.6
Time diff. (ns)	0.0	0.0	194.5	0.0	0.0	194.5
Phase diff. (°)	0.0	0.0	5.2	0.0	0.0	5.2
Amplitude in (dB)	−16.7	6.4	−18.9	−16.7	−6.4	−16.4
DOA error (°)			−7			−4

Reference

[1] Beckmann, P., and A. Spizzichino, *The Scattering of Electromagnetic Waves from Rough Surfaces*, Oxford, UK: Pergamon Press, 1963.

CHAPTER 17

Extent of the Multipath Problem and Possible Solutions

The effects of multipath interference are different for ground-based reflections and for reflections from the platform on which the ESM system is mounted. In the first case, the reflection coefficient for each reflected path is dependent on both the grazing angle and the frequency of the radar. However, many of the reflected paths for ground-based reflections are of such low amplitude that they have little impact on the amplitude of the composite pulse. For reflection from ESM platform surfaces, the reflections are all of similar amplitudes and each reflected path contributes significantly to the composite pulse amplitude.

The number of reflected paths is different in each case, with all pulses subject to many ground-based reflections, but fewer reflected paths seen for platform-based reflections. Although pulses subject to platform reflections also suffer from multipath effects due to ground reflections, the two effects are considered separately in this chapter to quantify the extent of the multipath problem.

17.1 Platform-Based Reflections

One of the important factors in determining the effect of multipath is the number of reflections contributing to a typical pulse amplitude. Figure 17.1 shows distributions of the numbers of pulses having amplitudes from 0 to 40 dB when subject to different numbers of reflections with random phase difference between the direct and reflected paths. One thousand pulses are generated for each plot. The individual reflections and direct path are all of the same nominal amplitude of 30 dB with a reflection coefficient of 0.8 applied to all reflected paths.

When only one reflection is present, the maximum possible pulse amplitude is 33 dB, but destructive interference means that the pulse amplitude can be signficantly reduced.

The distribution is broader at the low amplitude end for the case with one reflection than for the distributions with higher numbers of reflections. However, over 70% of the cases with only one reflection have a change in amplitude of less than ±3 dB due to multipath. This percentage reduces as the number of reflections increases, and when there are five or more reflection paths, the same shape distribution is seen for all numbers of reflections. As the number of reflection paths

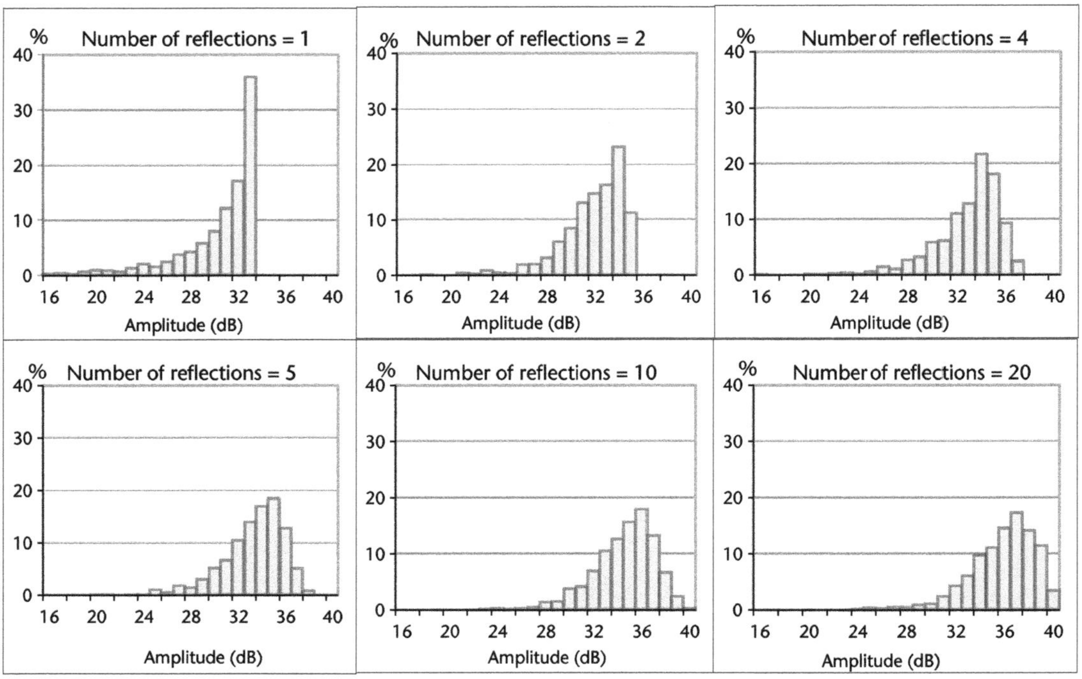

Figure 17.1 Percentage distributions of amplitudes for pulses subject to multipath interference with random phase differences from direct path.

increases, the distribution shifts along the amplitude scale as multiple constructive interference reflections allow for higher resultant pulse amplitudes.

Simple conclusions drawn from Figure 17.1 are that:

- More reflections give a higher percentage of signals with constructive interference.
- Fewer reflections give a higher percentage of signals with destructive interference.
- Once there are as many as 10 reflections of equal strength present, there is little change in the effect of multipath with an increase in the number of reflection paths.

Amplitude comparison systems using two antennas for the calculation of DOA typically show an error of 3° for every decibel difference in the incident amplitude at the two antennas. To see the effect of amplitude errors on the calculation of the DOA, Figure 17.2 shows the percentage of signals having DOA errors from +60° to –60° due to multipath reflections.

To generate Figure 17.2, two cases are considered. The first has a single reflected path, and in the second case the pulses are the subject of 10 reflections. In both cases, the phase differences between the direct and reflected paths are random for each antenna. A constant amplitude direct path pulse of 30 dB is assumed and no amplitude difference due to antenna separation or elevation is included in the

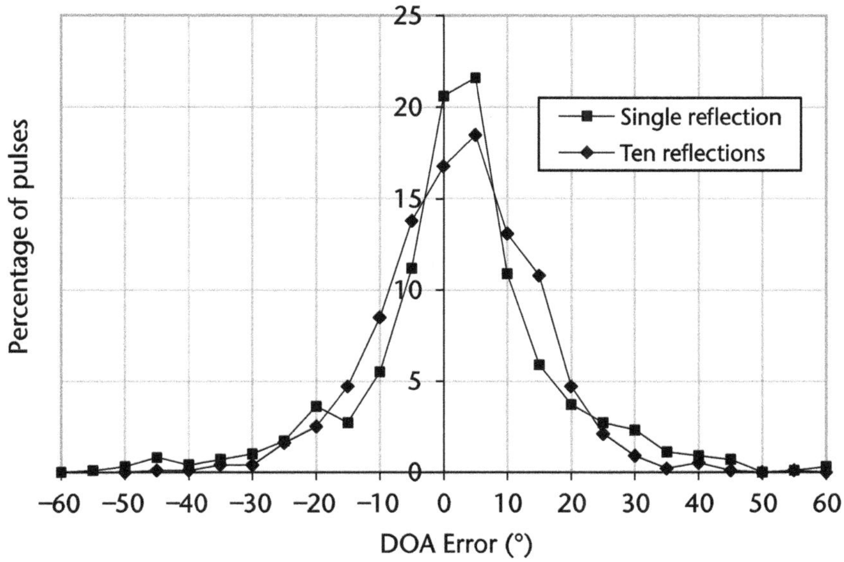

Figure 17.2 Distributions of the DOA error for aircraft-based reflections for a simple two-antenna amplitude comparison algorithm.

calculations. The reflection coefficient for all reflected paths is 0.8, and in both cases 1,000 entries are generated for the plot.

There is little difference in the case for a single reflection or for 10 reflections in Figure 17.2. For the single reflection case, the tail of the distribution is more populated than for the 10-reflection case as expected from the previous results where amplitude differences of more than 20 dB were possible due to both constructive and destructive interference.

For both these cases, DOA errors of more than 10° occur in about 40% of the pulses and errors of more than 20° are likely in about 15% of pulses.

17.2 Ground-Based Reflections

At radar frequencies, the reflection coefficient for ESM platform reflections is not dependent on the frequency of the pulse. The platform surfaces can be considered to be smooth and a scattering coefficient dependent on radar wavelength does not apply to the calculation of the reflection coefficient (see Chapter 16).

However, for ground-based reflections, the effects of multipath are frequency-dependent and two RF values, 3 GHz and 9 GHz, are chosen for the analysis as these represent the two RF bands in which many of the radars of concern to current ESM systems lie.

The grazing angle for each reflection is chosen at random assuming that the entire environment is evenly populated with reflection surfaces so there is an even distribution of all possible grazing angles between 0° and 90°. Once the grazing angle is selected, the reflection coefficient is found using the curves in Figure 16.24.

For the simulation at 3 GHz, only 17% of the reflections have a reflection coefficient above 1% with just 5% having a reflection coefficient above 0.5 (50%).

For the 9-GHz simulations, these numbers reduce to only 5% having a reflection coefficient above 1% with 2% having a reflection coefficient above 0.5.

The plots in Figure 17.3 show cases of 10, 20, and 50 reflection paths with random phase difference between the direct and reflection paths and reflection coefficients randomly chosen to fit the percentages described above. The top three plots in Figure 17.3 are for 3-GHz pulses and the lower three plots are generated for 9-GHz pulses. As expected, the DOA error distributions are narrower in every case for the RF of 9 GHz, meaning that ground-based multipath is less of a problem for 9-GHz pulses than for those of lower RF.

At 3 GHz, when there are 10 reflections present, only 10% of pulses have DOA errors of > 5° due to ground reflections. For 20 reflections, there are 15% of pulses with DOA error > 5°. When the number of discrete reflections is as high as 50 the percentage of pulses with DOA errors > 5° is 40% for 9 GHz pulses and has increased to 60% for pulses at 3 GHz.

For aircraft reflections it could be assumed that either only one antenna is affected, or that the differences in phase difference between the antennas is random. However, for ground-based reflections, where all the antennas are affected by multipath from the same sources, the phase differences between the antennas are not always random, except at short range. A modification must be made to the estimate of the level of DOA corruption to take this into account.

Results of simulations of the phase differences between antennas seen for various reflection geometries for ranges of 5 km to 100 km from the ESM platform where there is a 10-m spacing between all the antennas are shown in Table 17.1. All the examples considered have small azimuth angles to the direct path, as it is for these reflections that the effects of multipath on the relatively high amplitude main beam pulses are significant.

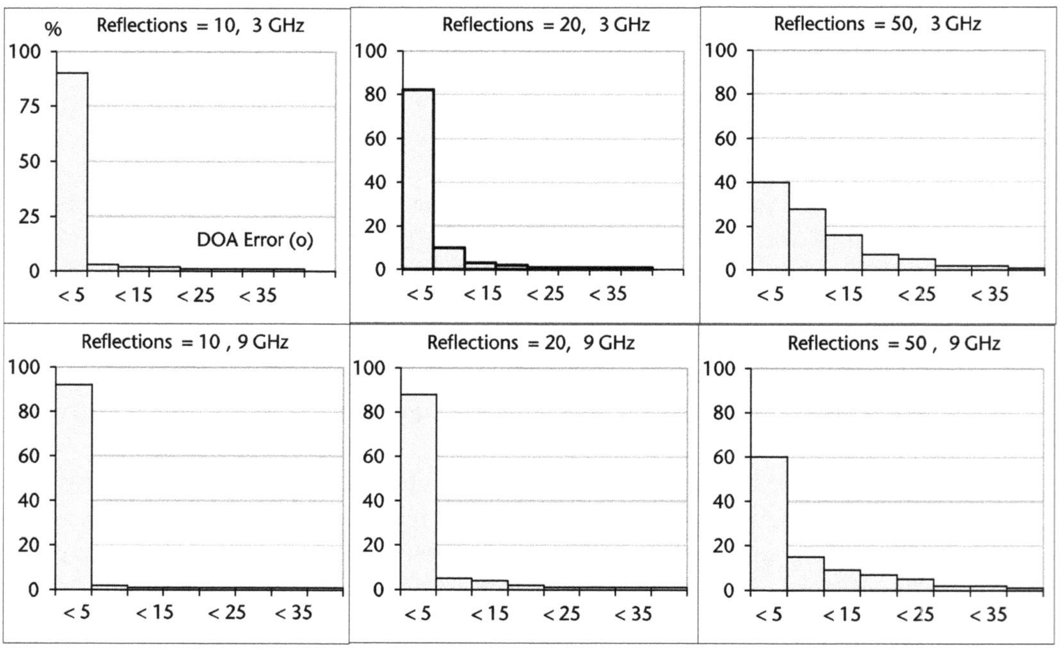

Figure 17.3 Percentage distributions of signals with DOA errors due to multiple reflections.

In all cases, the differences between the phase differences at the three antennas decreases with increasing range. Although only a limited number of examples are shown in Table 17.1, many more cases have been considered in drawing the following conclusions:

- At short range (e.g., 5 km and 8 km), there is effectively a random phase difference between the antennas in 80% and 60% of the cases, respectively.
- At medium range (e.g., 15 km and 25 km), there is a phase difference of > 30° between the receiving antenna phases in 40% and 25% of cases.
- At long range (e.g., 50 km and 100 km), in only 5% and 1% of cases is a phase difference of > 30° seen.

Combining the results of Figure 17.3 and of Table 17.1 leads to estimates of the magnitudes of DOA errors in main beam pulses due to multipath for ground-based reflections. Tables 17.2 and 17.3 show percentages of the DOA errors expected for 3-GHz and for 9-GHz radars.

It should be noted that no account has been taken of situations where the path difference for some of the reflections falls outside the pulse amplitude measurement time for some of the antennas, but not for the others. In these cases different numbers of reflections will contribute to the final amplitude of the pulse at each antenna and unpredictable DOA errors may result at any range.

The estimations of DOA error given in Tables 17.2 and 17.3 refer only to the specific ESM system with 10-m ESM antenna separation. However, they can be

Table 17.1 Phase Differences for Three Antennas for Various Reflection Geometries

Range (km)	*Antenna (1 – 2)*	*Antenna (2 – 3)*	*Grazing Angle (°)*	*Reflection Range (m)*	*Geometry Height (m)*	*Angle from Direct Path (°)*
5	26.2	24.6	3.1	500	50	0.25
8	6.0	6.2	0.7			
15	5.1	6.7	0.9			
25	3.2	4.1	1.7			
50	1.7	2.0	2.3			
100	1.0	1.0	2.5			
5	37.8	111.3	5.2	1,000	50	0.25
8	25.3	4.9	2.4			
15	8.6	9.2	0.5			
25	5.9	6.7	0.3			
50	3.3	3.6	0.8			
100	1.9	2.0	1.1			
5	77.8	32.7	9.8	2,500	50	0.25
8	4.4	66.2	4.3			
15	20.4	17.2	1.1			
25	14.7	17.3	0.6			
50	8.3	9.8	0.1			
100	4.9	5.1	0.3			

Table 17.2 Percentages of DOA Errors for Ground-Based Reflections of a 3-GHz Radar Pulse

	Range (km)					
DOA Error	*5*	*8*	*15*	*25*	*50*	*100*
0° to 5°	53%	65%	77%	85%	97%	99%
5° to 10°	22%	16%	11%	7%	2%	0.3%
10° to 15°	13%	10%	6%	4%	0.8%	0.1%
> 15°	11%	8%	4%	2%	0.6%	0.1%

Table 17.3 Percentages of DOA Errors for Ground-Based Reflections of a 9-GHz Radar Pulse

	Range (km)					
DOA Error	*5*	*8*	*15*	*25*	*50*	*100*
0° to 5°	68%	76%	84%	90%	98%	99%
5° to 10°	16%	12%	8%	5%	1%	0.2%
10° to 15°	9%	7%	4%	3%	0.5%	0.1%
>15°	7%	5%	3%	2%	0.4%	0.1%

used as an indication of the percentage of main beam pulses likely to exhibit gross errors in DOA due to multipath from ground-based reflections.

An estimation of the extent of the multipath problem leads to the conclusions in Table 17.4 for small-angle, ground-based reflections.

The figures quoted in Table 17.4 are for an ESM installation with an antenna separation of 10m and a radar of a frequency of 3 GHz with reflections affecting all the antennas.

Reflections from ESM platform surfaces that usually affect only a single antenna can cause gross DOA errors at any range and for any ESM antenna separation. For example, if there are 10 platform-based reflection paths contributing to the amplitude of a pulse and affecting an individual antenna, it is expected that DOA errors > 10° will occur in about 40% of pulses, with errors > 20° likely in 15% of pulses.

Table 17.4 Percentage of Pulses with DOA Errors >10° and >20° Due to Random Phase Differences Between Direct and Reflected Paths at Various Ranges

	Range (km)	*Percentage of Pulses with Random Δ Phase Difference Between Antennas*	*Percentage of Pulses with DOA Errors >10°*	*Percentage of Pulses with DOA Errors >20°*
Short Range	5	80	24	5
	8	60	18	4
Medium Range	15	40	10	2
	25	25	6	1
Long Range	50	5	1	< 0.1
	100	1	0.2	< 0.1

17.3 Common DOA Profiles for Scanning Radars

Nine common types of a DOA profile are typically seen in pulse data for scanning radars. The classes of DOA profile are no multipath, offset DOA, large DOA changes, opposite slope DOA, wishbone, diamond, pseudo-random DOA, dual RF separation, and extra main beam. Examples of these categories are shown in Figure 17.4.

It is likely that the classification of DOA scan profiles will be used in future ESM systems to improve the DOA performance of these systems.

17.4 Possible Solutions to the Multipath Problem for All Types of ESM

While there are many references to reduce multipath effects in navigation systems such as GPS and GNSS [1–5], there is little information available in the public

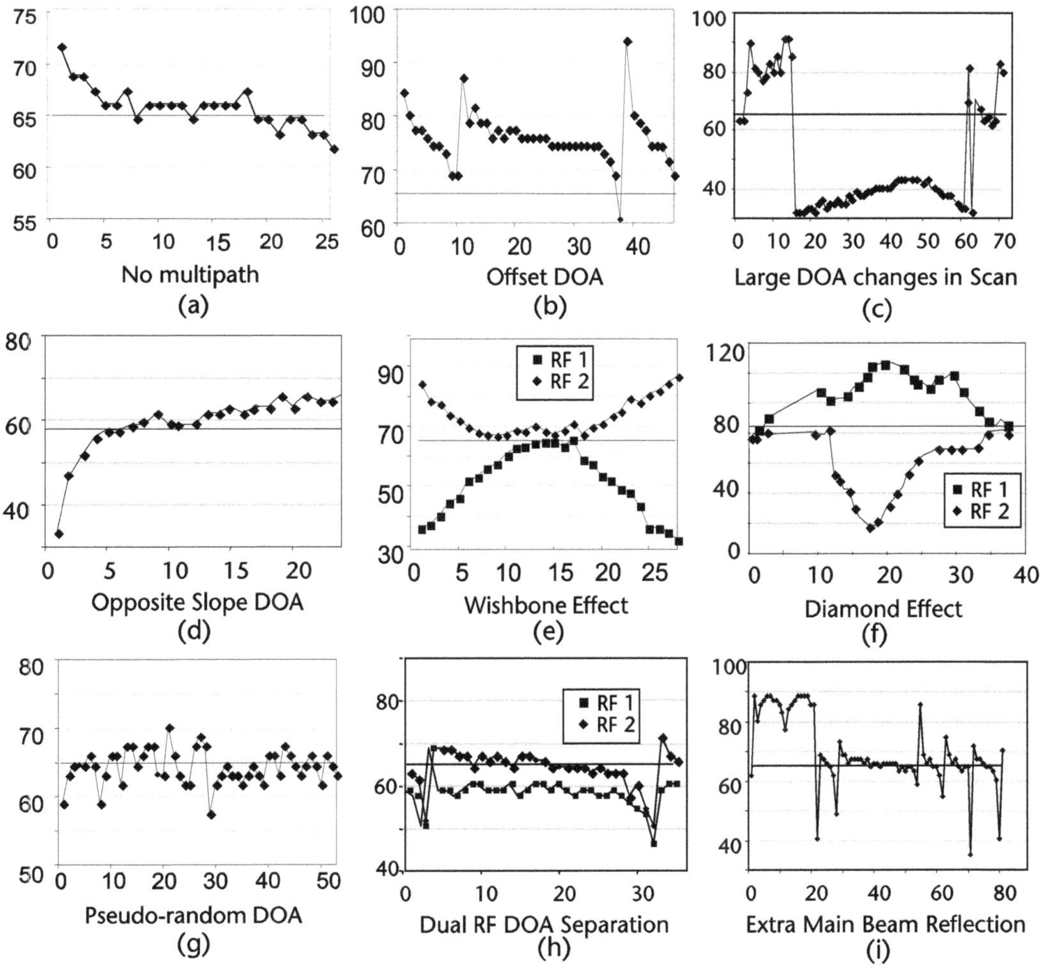

Figure 17.4 Categories of DOA profile for scans of pulses affected by multipath.

domain on the measures that can be taken to mitigate against multipath in ESM systems.

There are two approaches to fixing the multipath problem. The first is to make equipment changes to the system, and the second is to implement processing solutions. The possibilities for fixing the multipath problem are:

- Colocate the ESM antennas.
- Select the pulses for use in radar track creation.
- Reduce the pulse amplitude measurement time.
- Classify the DOA scan type.
- Use intrapulse amplitude profiles.
- For amplitude comparison systems, the DOA may be verified using an adapted amplitude comparison calculation algorithm.
- For TDOA systems, both methods of pulse measurement time (amplitude threshold and amplitude reduction) should be used, possibly with extra baselines and a fine resolution in TOA measurement.
- For phase comparison systems, use the mode of phase difference samples for a particular pulse and use multiple baselines.

17.4.1 Colocate the ESM Antennas

This will significantly reduce the effects of multipath on DOA for ground-based reflections. However, it will not solve the problem of a single antenna being affected by multipath. There are likely to be limitations in the siting of the antennas in an aircraft ESM installation, but doubling up each single antenna may be considered.

17.4.2 Select the Pulses for Use in Track Creation

If there is no multipath present, then it is a good idea to use only pulses from near the peak of the main radar beam for DOA determination and intercept creation. This is due to the antenna separation effect in amplitude comparison and TDOA systems. This does not mean that the threshold of signal detection should be increased as perfectly good main beams may be seen at very low amplitude due to the radar being at long range.

If the performance of the ESM system is unacceptable due to the large number of tracks that are created for each radar at different DOAs, then this measure will probably improve performance for those radars. However, care should be taken not to throw away all the data gathered from pulses that are not near the peak of the main beam. Their parameters, including RF, TOA, and PW, should be allowed to contribute to the calculation of parameters such as PRI sequence and RF and PRI agility.

It should be noted, as discussed in Chapter 16, that there are situations where the DOA error actually increases near the peak of the main beam. For example, when the angle between the direct and reflected paths is small and one antenna suffers destructive interference, a diamond shape is seen in the DOA profile. At long range, the same wrong DOA profile may persist from scan to scan. Using only pulses near the peak of the beam clearly will not help in determining the correct DOA of the radar in these circumstances.

However, it is expected that pulses of high relative amplitude near the peak of the beam will not suffer as much interference from multipath as those of lower relative amplitude, where it is possible for interference from reflections of the pulses in the peak of the beam to affect the pulse amplitude and hence the DOA calculation.

An example of a single scan of pulses from a weather radar shows the DOA performance of an amplitude comparison system for main beam pulses and for those pulses further from the peak of the beam.

Figures 17.5 and 17.6 show a simulation of amplitude and DOA profiles of a single scan from the radar at a range of 40 km. The DOA has a spread of over 30° in Figure 17.6.

The effect of applying amplitude limits of –35 dB, –30 dB, and –23 dB is shown in Figures 17.7 to 17.9.

When all pulses with an amplitude lower than –35 dB are excluded, the DOA spread is slightly improved, but is still about 20°. As Figure 17.8 shows, the DOA spread is reduced when pulses of amplitude <–30 dB are excluded from the data. It is now a maximum of 10°. When all but the main beam is cut out from the pulse data by applying an amplitude limit of >–23 dB, the radar DOA spread is reduced to 5°, as can be seen in Figure 17.9.

A classification of four types of pulses in a single scan is defined according to the position of the pulses on the scan profile

- Main beam peak pulses—within 3 dB of the peak;
- Main beam side pulses—all pulses in the main beam which are more than 3 dB from the peak;
- First sidelobe pulses;
- Far sidelobe pulses.

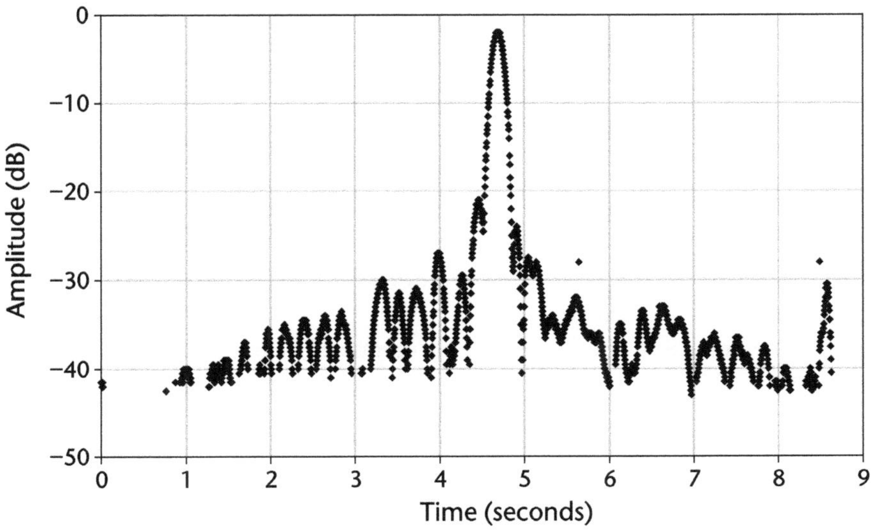

Figure 17.5 Single-scan amplitude profile for simple weather radar.

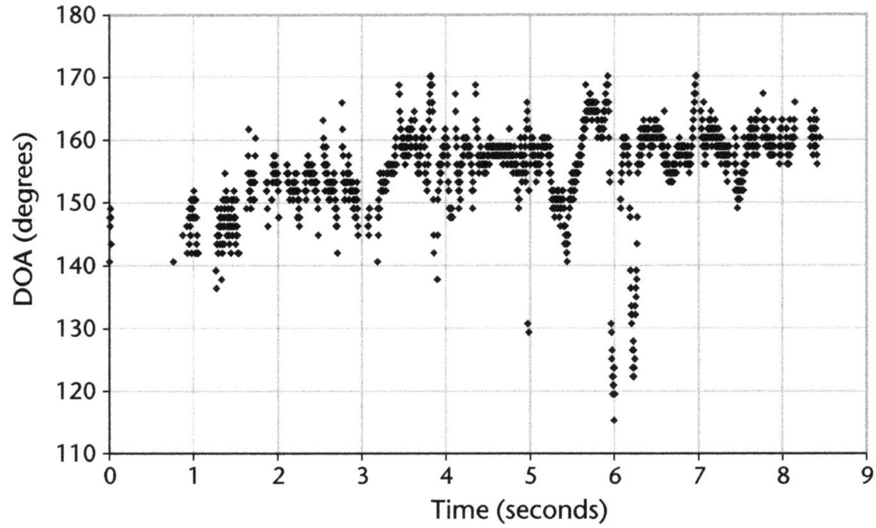

Figure 17.6 Single-scan DOA profile for simple weather radar.

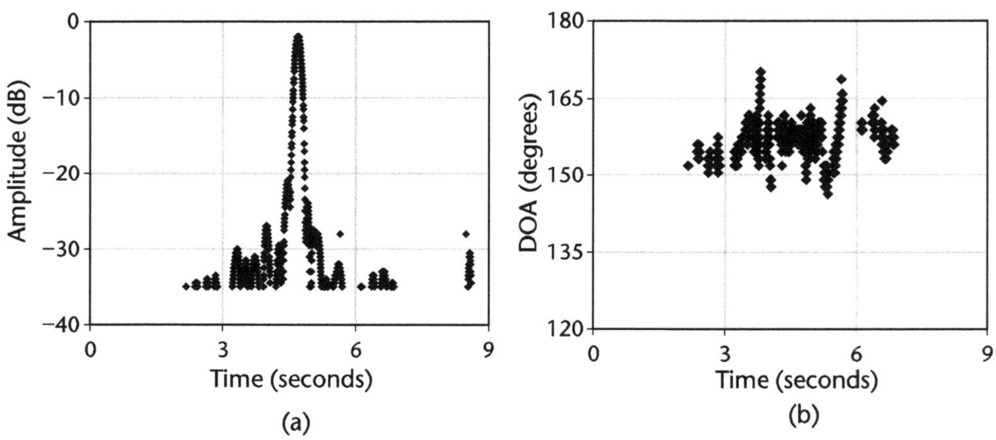

Figure 17.7 Single-scan amplitude (>–35 dB) and DOA profiles for weather radar.

Figure 17.10 shows an example of an ideal beam profile with all four types of pulses.

A simulation of a set of pulse data from a single radar in a 4-minute time period shows the DOA accuracy and spread of each of the four classes of pulses. Figure 17.11 shows all the pulses, and Figure 17.12 shows individual plots for each of the four categories of pulse.

The spread in the DOA is less for pulses near the peak of the scan with an increase in DOA spread from ±6° for the peak pulses to ±20° for far sidelobe pulses. Two further sets of pulses have been simulated and the results for each class of pulse are shown in Table 17.5.

Although the spread in DOA for each category of pulses is less for main beam pulses in all cases, the average DOA of each category of pulses is usually similar.

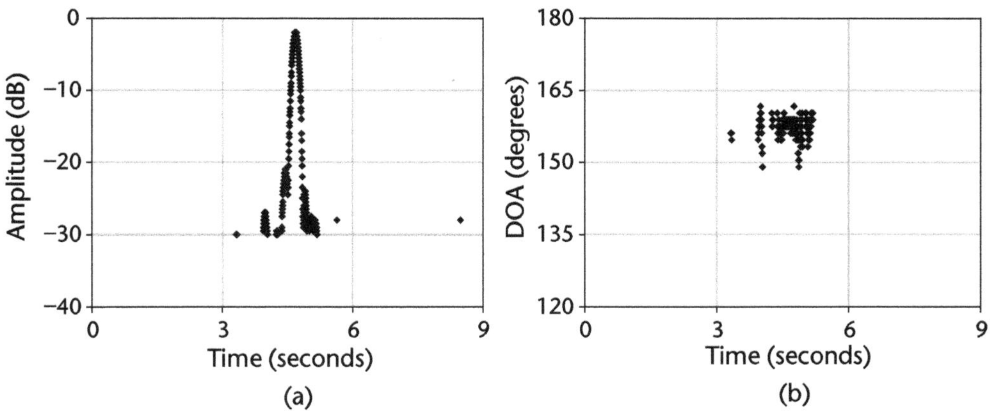

Figure 17.8 Single-scan amplitude (>–30 dB) and DOA profiles for weather radar.

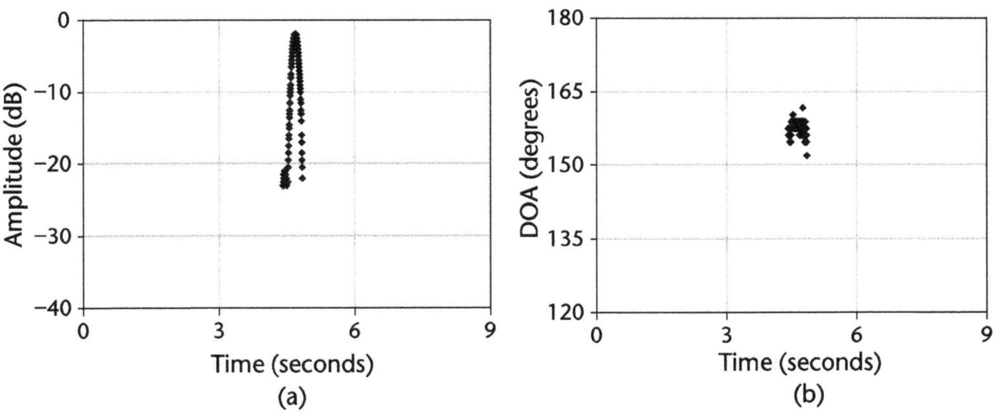

Figure 17.9 Single-scan amplitude (>–23 dB) and DOA profiles for weather radar.

If the pulses in a scan are offset from the real DOA, they are all likely to be offset, including those from the scan peak. However, the advantage of limiting the spread in DOA of pulses used for track DOA estimation is that multitracking for single radars will be reduced.

17.4.3 Use of Several Scan Peaks

It appears from the data presented in the previous section that it may be an advantage to use data from several scan peaks to calculate the DOA of an radar. In many cases, this will be a correct course of action. For example, averaging the DOA of the peak pulses from the first 34 scans of the radar shown in Figure 17.13 results in a correct DOA value.

Scan 35 in Figure 17.13 is offset from correct DOA and its standalone DOA average is 10° below the correct DOA. In many ESM systems, this offset would be enough to generate an additional track. However, including its pulses in an average DOA calculation should restrict the possibility of extra track creation in this case.

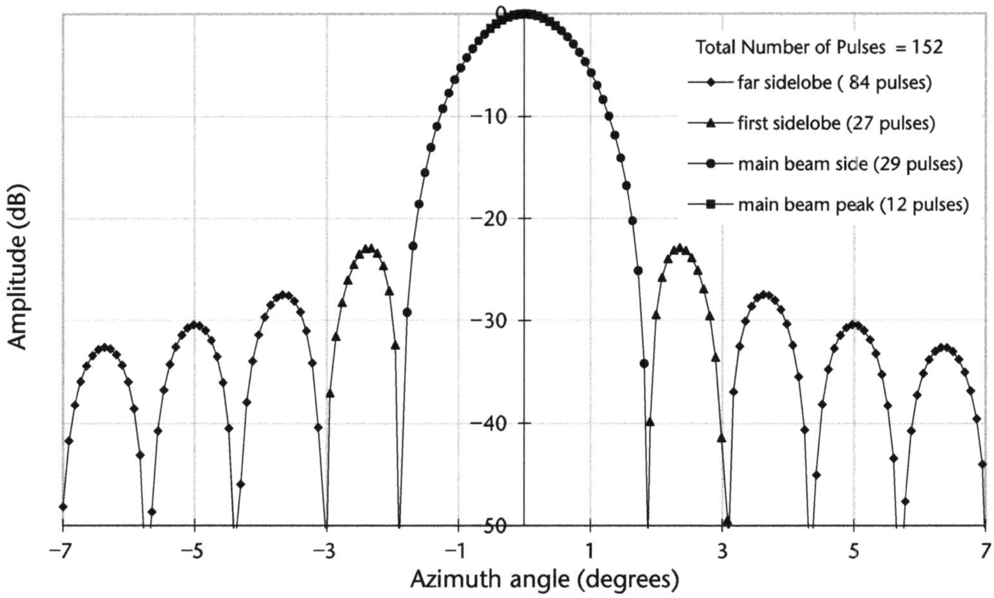

Figure 17.10 Example of an ideal beam profile showing the difference categories of pulse.

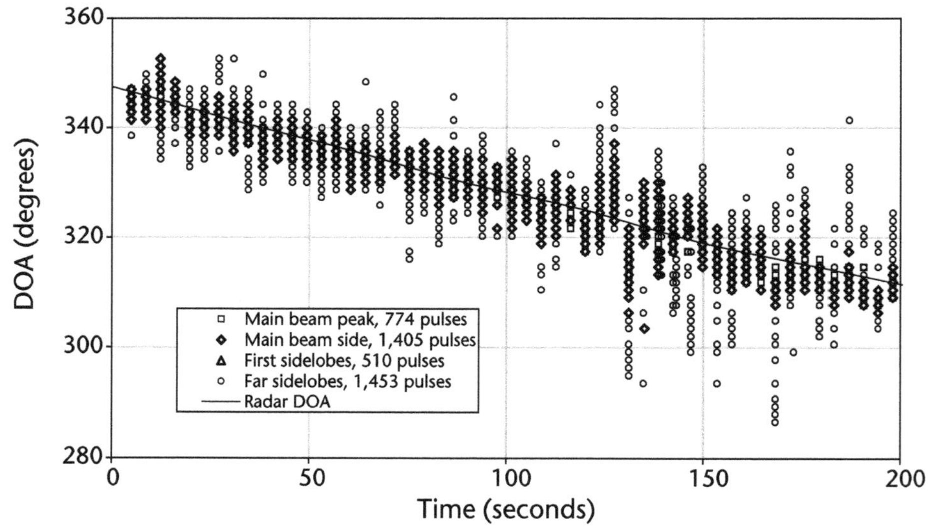

Figure 17.11 DOA of all pulses from a radar to show the DOA of the four categories of pulse.

Analysis of trials data usually shows that the use of peak pulses from several consecutive scans to determine DOA leads to an improvement in the accuracy of the DOA and reduces multitracking for single radars.

At short range (<20 km), the use of peak pulses from several scans to estimate the DOA of the radar is almost certain to show an improvement in DOA performance. This is because the multipath effects cause very different DOA offsets for consecutive scans and DOA errors tend to cancel out as a result.

At medium range (20 to 50 km), the intercept DOA error may be improved by the use of peak pulses from consecutive scans and multitracking is reduced.

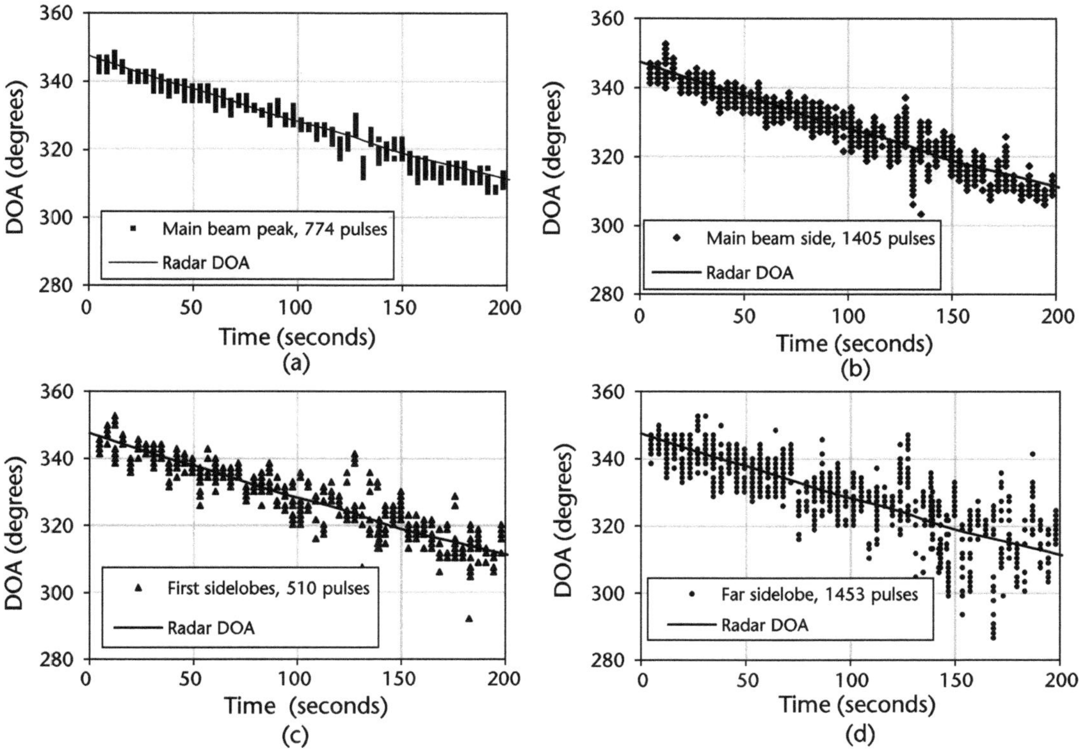

Figure 17.12 DOA of pulse data classified according to pulse type.

Table 17.5 Spread in DOA for Each Category of Pulse Data

	Pulse Data Set 1	*Pulse Data Set 2*	*Pulse Data Set 3*
	Range 50 km	*Range 50 km*	*Range 15 km*
Pulse category		Spread in DOA (°)	
Main beam peak	6	6	8
Main beam side	14	16	30
First sidelobe	16	20	45
Far sidelobe	30	35	60

At long range (>50 km), this solution has a much smaller benefit as multipath effects are similar from scan to scan at long range. Often the same large DOA offsets are seen for many consecutive scans. Although the possibility of extra track creation at long range is reduced by using pulses from several scans, the DOA error due to multipath, which may be as much as 40°, is not corrected using this method.

17.4.4 Reduce the Pulse Amplitude Measurement Time

The reflections causing the most severe problems are those at very small path differences to the direct path. If the pulse amplitude could be measured a very short time after the detection of the pulse, then it may be possible to detect only the direct path pulse.

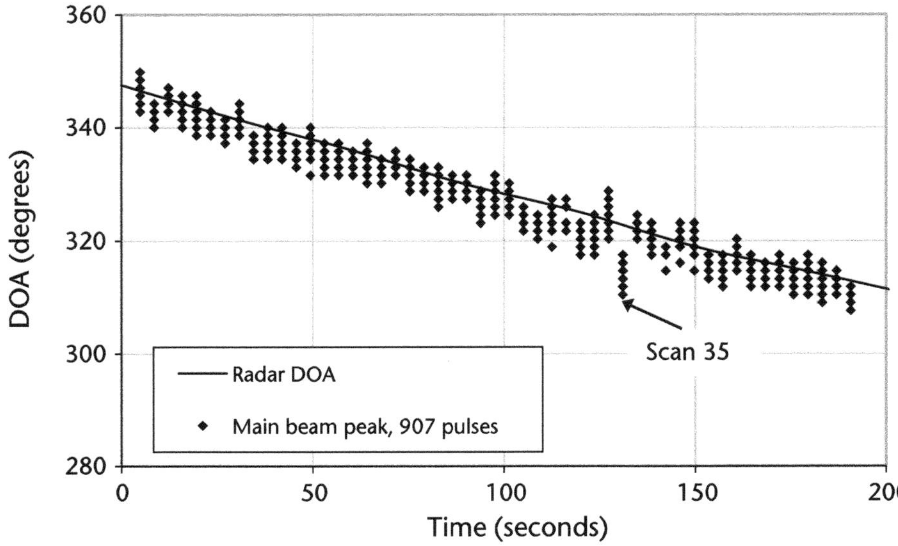

Figure 17.13 DOA of peak pulses from individual radar scans.

Current systems use pulse measurement times ~50 ns leaving them vulnerable to path differences of up to 15m likely to be caused by reflections off surfaces very close to either the radar or to the ESM antenna on the platform. Reducing the amplitude measurement time to only 5 ns makes the ESM system robust to reflections greater than 1.5-m path difference. However, practical limitations are foreseen in that the rise time of many current radars exceeds this value and problems would arise in the timing and amplitude calculations if the amplitude measurement time was reduced significantly.

17.4.5 Classify the DOA Scan Type

A rule-based system could be used to decide if a whole scan of pulses has a reasonable DOA profile, or whether it is likely to be in error, by matching the DOA profile to one of the nine types described in Section 17.4.

For example, if the trend in DOA decreases across the main beam of a scan then the DOA profile falls into category 1 (no multipath). It would be reasonable to assume that multipath has not greatly affected DOA and the pulses from near the peak should be used for the creation of an intercept. Other information such as the DOA profile of a second RF for a dual RF radar and the slope of the DOA profile could be used to improve the DOA estimate in this case.

However, if the DOA profile is of opposite slope to that expected or the slope of the profile alters throughout the scan, then the scan may be classified as category 4 (opposite slope) or category 5 (wishbone), category 6 (diamond), and so forth. Once the DOA profile has been categorized an assessment can be made of which pulses in the scan are likely to be the most useful in terms of estimating the correct radar DOA.

This method would require the ESM system to make a qualitative assessment of the DOA pattern of each scan and it is possible that a neural network would prove suitable for this task.

17.4.6 Use Intrapulse Amplitude Profiles

The pulse amplitude level is the key to knowing whether there is multipath present, as even if the antennas are colocated, the pulse amplitude is altered by multipath interference.

Figure 17.14 shows a typical simulated intrapulse amplitude profile. In this pulse, of nominal width 1 μs, with no intentional modulation, there are more than 10 reflection components.

When multipath affects a pulse with intentional modulation (i.e., one with a variation in frequency across its width (chirp)), it is likely that a rippled intrapulse amplitude profile will result. Figure 17.15 shows a simulation of such an intrapulse amplitude profile.

It may prove difficult to perform an intrapulse amplitude profile classification for pulses with no intentional modulation as all that happens for pulses of this type is that the amplitude level changes every time a new reflection is received.

However, it may be possible to correlate the intrapulse amplitude profiles from each ESM antenna, but this would require a level of processing and data storage not usually available in current ESM equipment.

17.5 Solutions for Multipath in Amplitude Comparison Systems

A correction method for DOA errors in amplitude comparison systems is discussed in Chapter 5. The correction algorithm takes account of the fact that the pulse amplitudes incident at each antenna are not equal due to the shape of the radar beam, the separation of the ESM antennas, and multipath effects.

A way of solving the problem of the different incident powers at each antenna is to adjust the relative gains of the three ESM antennas to compensate for the different incident powers. To obtain an accurate DOA result from the amplitude

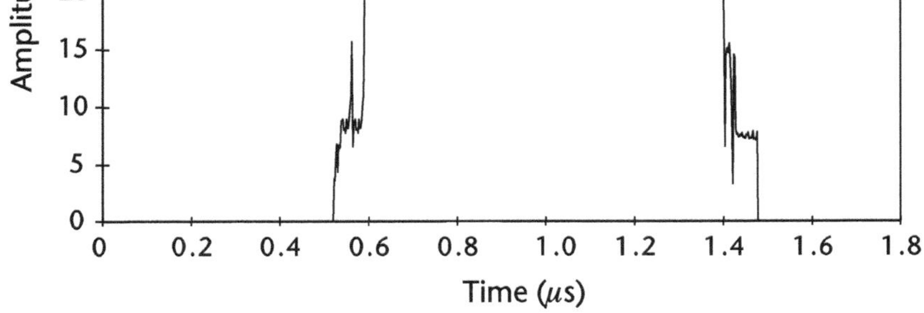

Figure 17.14 Simulation of an amplitude profile of a pulse with multipath interference.

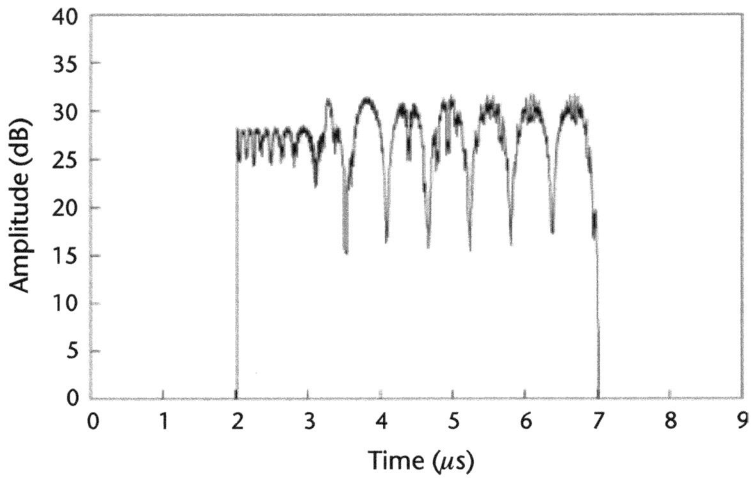

Figure 17.15 Simulation of intrapulse amplitude profile showing interference of two chirped pulses of equal amplitude separated in time by 1 μs.

comparison algorithm, the relative values of the incident powers for each ESM antenna are needed. Unfortunately, this information is not available. All that is known are the measured signal strengths of the individual ESM antennas and the calculated DOA.

The DOA correction process starts by deciding whether the calculated DOA is in error or not. This decision begins by estimating the incident powers to each antenna (in this case a three-antenna system) by reversing the calculation of the amplitude comparison algorithm.

For antenna 1, an estimate of the incident power is made using the equation:

$$A_{1in} = k\left(\theta_c - \theta_0\right)^2 + A_{1out} \tag{17.1}$$

where

$k = 12/\left(\theta_0\right)^2, \theta_b = ESM\,antenna\,3dB\,beamwidth$ θ_c = calculated value of DOA, θ_o = antenna boresight, and A_{1out} = ESM antenna activation level.

Similar calculations lead to estimates of the incident power for antennas 2 and 3, respectively. These are not the correct values, as an incorrect DOA value was used in their calculation. The magnitude of the difference between the estimates of incident powers for the two pairs of antennas, $|A_{2in} - A_{1in}| + |A_{2in} - A_{3in}|$, gives a measure of the DOA error.

To see the relationship between the magnitudes of the differences of the estimates of incident power for the two pairs of antennas and the DOA error, a set of reverse input calculations is carried out with the input amplitudes of antennas 2 and 3 set to 0 dB and the input amplitude of antenna 1 varied from 0 to 9 dB in 1-dB steps.

Figure 17.16 shows a plot of the sum of the differences between the reverse inputs (A_{1in}, A_{2in}, A_{3in}) calculated for the two antenna pairs, that is, $|A_{2in} - A_{1in}|$ +

$|A_{2in} - A_{3in}|$ against the DOA error. In this simple case, there is an easily identifiable correlation between the two parameters and the DOA correction can be calculated without ambiguity using this method.

However, the situation in Figure 17.16 is a special case with two antennas having equal input amplitudes. In reality, the DOA error is not so easily calculated as there are multiple possibilities for the value of the DOA error.

17.6 Solutions for Multipath in TDOA Systems

There are two methods of measuring the TOA of pulses in TDOA ESM systems, described in detail in Chapter 6. The first method uses an amplitude threshold to detect the pulse, and the second method uses an amplitude reduction of 3 dB from the peak amplitude of the pulse to find the pulse TOA.

For the amplitude threshold method, the separation of ESM antennas causes DOA errors because of the difference in TOA for pulses seen at the various ESM antennas, due to the rise time of the pulse (see Section 6.5). Pulses are only seen at equal amplitude by all ESM antennas at the peak of the radar beam for scanning radars. The effect of antenna separation in the absence of multipath is a slope in DOA across the main beam with larger DOA errors in either direction at the edges of the beam and for the sidelobes.

The effects of antenna separation are more extreme for pulses that are near the sensitivity threshold of the receiver, for amplitude threshold systems, whereas pulses of low power that are near the ESM sensitivity level are missed altogether by TDOA systems that use amplitude reduction TOA measurement.

The use of the amplitude reduction method of measuring TOA removes the DOA error except in the presence of multipath, where the effects are worse for this method than for the amplitude threshold method. This is because more time has elapsed before the TOA measurement is made, leading to a greater possibility that the pulse has suffered from multipath before the time measurement is made.

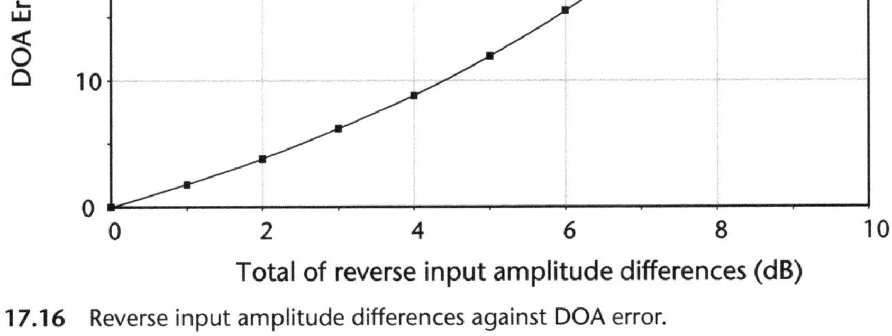

Figure 17.16 Reverse input amplitude differences against DOA error.

Recommendations for improving the DOA in TDOA systems include using more antennas and hence more baselines to improve DOA resolution and the use of both methods of time measurement (amplitude threshold and amplitude reduction) for the recognition and possible correction of TOA measurement errors.

17.7 Solution for Multipath in Phase Comparison Systems

Using phase differences of several pairs of antennas helps to improve the DOA measurement in phase comparison systems. For example, consider a pulse arriving at three antennas arranged in a linear configuration. When there is no multipath present, the correct phase differences are measured between the antenna pairs (i.e., antennas 1 and 2, antennas 1 and 3, and antennas 2 and 3). Figure 17.17 shows a plot of the DOA against the phase difference for the three antenna separations. The measured phase differences all lie on the same horizontal line in this plot, showing the correct DOA of 30°.

Antennas 1 and 2 (separated by 5 cm) have a phase difference of 60°, leading to the correct DOA value. Antennas 1 and 3 (separated by 15 cm) have a phase difference of 270°, also leading to a correct DOA of 30°. Antennas 2 and 3 (separated by 10 cm) have a phase difference of 180°, again giving the correct DOA.

If there is a change in phase difference due to one antenna being affected by multipath, a DOA error will result for all the phase differences calculated using that antenna for that particular phase difference sample. Figure 17.18 shows an example where one of the antennas has been affected by multipath and the other two antennas are unaffected. Only one of the phase differences is correct and that is for the case where the antenna separation is 10 cm.

If any of the antennas suffers from multipath and the others do not, there will be a phase difference error seen for the antenna pairs involving the affected antenna. The plot of the DOA versus the phase difference does not show a horizontal line, thus confirming the DOA error. At this stage, it should be possible to know that there is likely to be a DOA error for the pulse, but not to know the magnitude

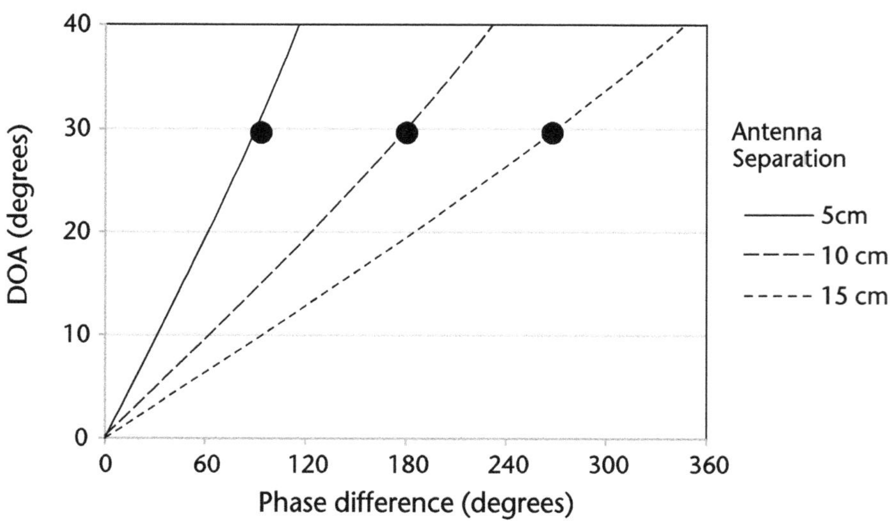

Figure 17.17 DOA versus phase difference for various antenna separations.

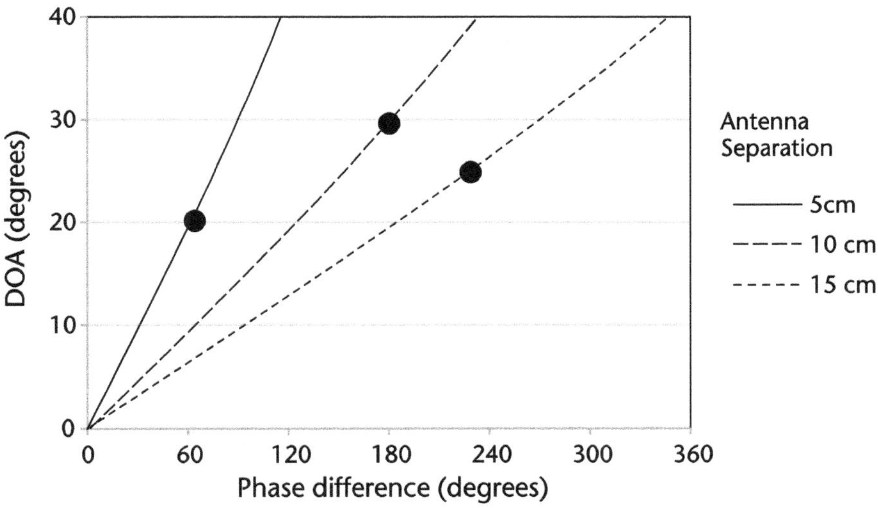

Figure 17.18 DOA versus phase difference for various antenna separations showing an incorrect DOA estimation.

of the DOA error. In the example shown here, it is not possbile to determine the correct DOA using a single sample and with a configuration of only three antennas. However, the results from this phase difference sample do indicate that there is a DOA error and that the sample should not be used as a reliable estimate of the pulse DOA.

Using the mode of many phase difference samples taken throughout the lifetime of a pulse should eliminate spurious phase differences that occur when one antenna has been affected by multipath and the other antenna(s) are unaffected at the time of taking the sample.

17.8 Conclusions on Multipath

There are two sources of multipath reflections causing errors in the calculation of the DOA in the ESM systems. Particular types of ground-based reflections are the first category and reflections from platform surfaces are the second. The types of reflections that cause significant effects due to ground-based multipath are those at a small angle relative to the direct path pulses and with small grazing angles. These reflections are important for the following reasons:

- The path difference between direct and reflected paths is small enough for the amplitude of the pulse to be the subject of interference before it is measured.
- A small angle reflection results in a reflected pulse of similar amplitude to that of the direct path pulse, depending on the reflection coefficient.

The reflection coefficient of the ground is particularly dependent on grazing angle, with reflections of significant amplitude (i.e., 50%) at grazing angles of 5° for 3 GHz and 2° for 9-GHz signals.

Reflections from a smooth sea or from ESM platform surfaces are significant at much larger grazing angles, with a reflection coefficient of over 80% for grazing angles as high as 20° for both types of reflection.

An important contributor to DOA error due to ground-based reflections is the separation of the ESM antennas causing large DOA errors for pulses at the edges of the main beam and the sidelobes of scanning radars, even when there is no multipath present. At a range of 5 km, a typical ESM system with separated antennas sees differences in the DOA of 40° from the first pulse to the last pulse in the main beam scan of a typical surveillance radar of a beamwidth of 0.8°.

It is the difference in phase difference between the direct and reflected paths for each receiver antenna (Δ phase difference) that causes the DOA error due to multipath in an ESM system with separated antennas. If all the ESM antennas are colocated there is no error due to multipath (when all the antennas are affected) even though the pulse amplitudes are affected by interference due to reflections.

At long range, the Δ phase differences for the individual antennas are highly correlated for the small angle reflections considered in this book. This leads to a decrease in the effect of ground-based multipath on the calculation of the pulse DOA for long-range radars. For wider angle reflections, the Δ phase difference may be random even at long range, but the path differences for these reflections are likely to be in excess of the pulse amplitude measurement time, so these reflections have no effect on the pulse amplitude and the subsequent DOA measurement.

The change in the path difference from pulse to pulse across a single scan is small due to the insignificant change in geometry during the PRI. There is therefore a trend in the phase difference between direct and reflected paths from pulse to pulse for each antenna, which means that a trend in the DOA error is seen. This results in a whole set of pulses being detected at an incorrect DOA and the possibility that a new track will be created for them. At long range, this trend in the phase difference persists from scan to scan so that similar incorrect DOA profiles are seen, resulting in the update of extra tracks.

The most usual effect of multipath is to shift the DOA of the main beam pulses by a few degrees and to reduce the impact of antenna separation in causing the DOA error of the sidelobe pulses. There are also some extreme effects seen in sets of consecutive scans and nine classes of DOA profile have been identified as described in Section 17.3.

There are steps that can be taken in the design of ESM systems or the processing of ESM recorded data that will mitigate the effects of multipath. However, it should be noted that multipath has always been the one of the most significant reasons for the poor performance of ESM systems and an understanding of how it affects the DOA measurement helps to manage expectations of system performance by the ESM operator and data analysts.

References

[1] Vera, J. S., "Efficient Multipath Mitigation in Navigation Systems," Ph.D. thesis, University of Catalunya, 2004.

[2] Mao, W. -L., " An Adaptive Multipath Mitigation Archtecture for GPS System," *Proc. of Conference on Recent Advances in Signals and Systems*, Budapest, 2009.

[3] Bhuiyan, M. Z., and E. S. Lohan, "Advanced Multipath Mitigation Techniques for Satellite-Based Positioning Applications," Tampere University of Technology, 2010.

[4] Veytsel, A., "Multipath Mitigation with New Signals and Services," *Meeting of the International Committee on Global Navigation Satellite Systems (ICG-10)*, Boulder, CO, 2015.

[5] Miller, S., et al., "Multipath Effect in GPS Receivers," *Synthesis Lectures on Communications*, Vol. 11, 2015.

CHAPTER 18

The Future for ESM Systems

It is not surprising that there is little information in the public domain about the requirements and plans for future ESM systems. However, there is no doubt that the RF spectrum is going to become more congested and that the processing requirements of ESM systems will be more challenging.

18.1 The Future of the RF Environment

The traditional RF environment contains radars that each use a small number of fixed frequencies with relatively simple PRI and PW. There are already radar developments that cause issues with the way that ESM systems traditionally operate. Broadband radars in the maritime domain and AESA radars based on land, mounted on aircraft and on ships, mean that ESM systems of the future will require new techniques to detect, deinterleave, and identify these radars. Monopulse and software-defined radars are under development so new fingerprinting techniques will be needed to correctly identify these radars.

18.1.1 AESA Radars

Traditional LPI radars use techniques such as RF agility, long complicated PRI sequences, power modulation, and multiple transmit beams to make it difficult for the ESM system to deinterleave a pulse stream from the radar and to make sense of the pulse stream that it is receiving from them. AESA radars take this one step further, using an antenna array of TRMs, each capable of generating and radiating its own independent signal allowing the AESA to produce radar pulses on different frequencies with interleaved pulse streams carrying out several functions simultaneously.

AESA radars are now essential sensors for warships and fighter aircraft [1]. The first AESA radar, the OPS-24 fire control radar, was mounted on a Japanese destroyer launched in 1988. One of the first AESA radars to be deployed on a European warship was the active phased array radar (APAR), a multifunction, 3-D radar. It has four fixed sensor arrays mounted on a pyramidal structure, each array consisting of 3,424 TRMs. APAR is typical of AESA radars as it can track over 200 air targets to a range of 150 km and over 150 surface targets to a range of 32 km. It can perform a horizontal search out to 75 km and a volume search out to

150 km. Thirty-two semiactive radar homing missiles can be guided simultaneously, including 16 in the terminal guidance phase.

Airborne AESAs are typically mounted in the nose cone of a fighter aircraft. An example of such a radar is the AN/APG-77 AESA radar installed on the F-22 RAPTOR [2]. With an antenna composed of 1,956 TRMs, it can perform near-instantaneous beam steering. The radar provides a 120° field of view in azimuth and elevation, the highest possible value for a flat phased array antenna. It is specified to achieve an 86% probability of intercept against a 1 m^2 target at a detection range of 240 km. Figure 18.1 shows some of the modes in which such a radar can operate, for example, volume search, cued search, target tracking, and missile guidance.

It is not only fixed-wing aircraft that take advantage of AESA technology. In 2016, the Osprey AESA radar was introduced. This has multiple fixed antennas distributed around the aircraft, each antenna panel containing 256 TRMs and providing 120° coverage. On rotary-wing aircraft, mechanical antennas normally have to be mounted on the underside of the fuselage, limiting the size of the antenna due to ground clearance restrictions. A major advantage of Osprey is that its antenna arrays can be mounted higher on an aircraft fuselage than traditional mechanically scanned antennas.

Figure 18.2 shows a typical RF and amplitude plot for an AESA radar. Traditional ESM systems have considerable difficulty in deinterleaving pulses for a single AESA radar due to the wide range of frequencies used and the seemingly random pulse amplitude profile for the radar.

In reality, the ESM is only seeing a fraction of the pulses that are transmitted as only those beams that are pointing in the general direction of the ESM will be received. The fact that there are no seemingly repeatable patterns in either of the plots in Figure 18.2 is likely to cause serious difficulty for the ESM system.

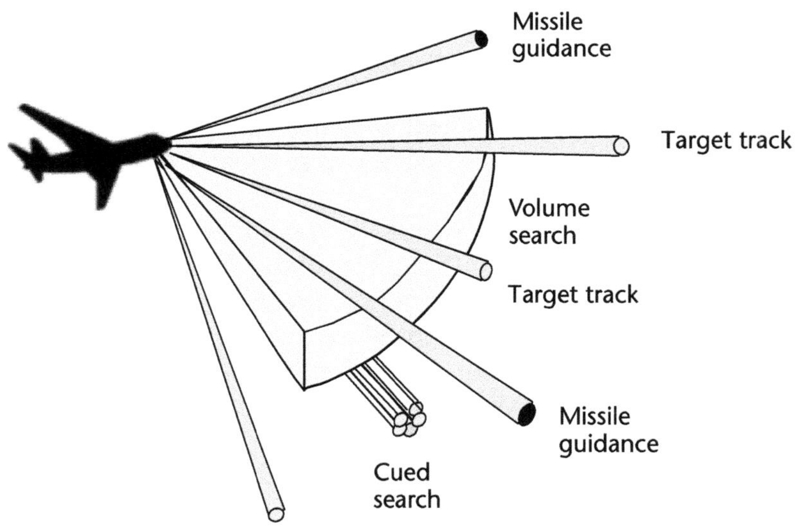

Figure 18.1 Schematic of typical airborne AESA functions.

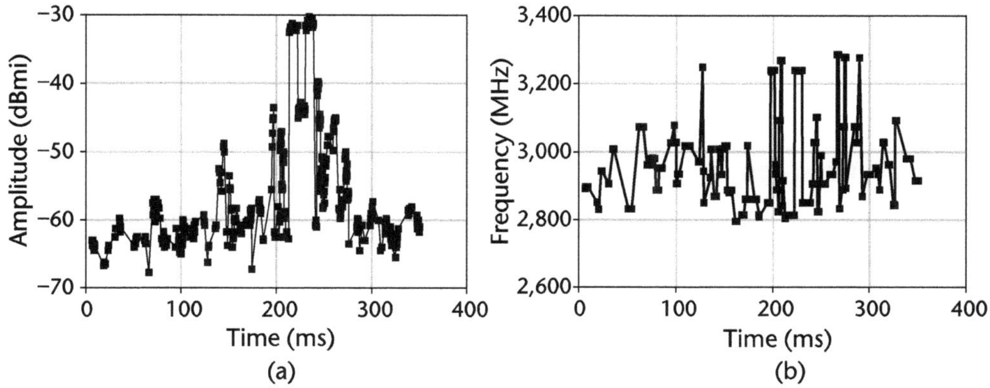

Figure 18.2 Amplitude and frequency profiles for an AESA radar.

18.1.2 Multiple Input Multiple Output Radar

AESA radars use multiple antennas to generate a single pattern with a waveform that does not vary from location to location in the far field. A multiple input multiple output (MIMO) radar is a space-time waveform diversity technique that represents a significant departure from a traditional radar antenna [3]. A MIMO radar produces a spatially diverse transmit waveform so that the radiation pattern, when measured in the far field, varies at different locations in space.

The MIMO radar transmits independent waveforms from a number of independent antennas. Each waveform illuminates the target and is reflected back to the receiver where the coherent sum of the waveforms is received on each antenna. The antennas used to transmit do not have to be the same as those used to receive. MIMO radar systems can be used to improve spatial resolution, and they are very resistant to jamming. They have an improved signal-to-noise ratio over conventional radars so the probability of detection of targets is also increased.

MIMO radars are classified into two categories:

- Those with colocated antennas (monostatic MIMO) where the target is a point target as in traditional radar systems;
- Those with widely separated antennas (distributed or bistatic MIMO) where the target is regarded by each antenna from different angles.

In the colocated antenna case, the transmitting antennas are close enough so that the target radar cross sections (RCS) observed by the transmitting antenna elements are identical. This system is similar to a phased array antenna in which each radiator has its own transceiver module and its own A/D converter. However, in a phased array antenna, each antenna only transmits (possibly time-shifted) a copy of a transmission signal that has been generated in a central waveform generator. In a MIMO radar system, each antenna has its own arbitrary waveform generator so that each antenna uses an individual waveform. This individual waveform is also the basis for an assignment of the echo signals to their source.

In the distributed arrangement of the antennas, the radar data processing is much more complex. In contrast to monostatic MIMO, each radar antenna looks at the target from a different aspect angle. Therefore, the target provides a different RCS for each radar antenna. Figure 18.3 shows the separate transmit and receive antennas in a MIMO radar system.

18.1.3 Monopulse Radars

A monopulse radar is a target tracking radar that can provide accurate range and DOA (azimuth and elevation) information using just a single pulse [4]. This technique is an extension to the lobe-switching method of calculating DOA where the radar antenna points slightly to one side of the target and then rapidly switches to point the other side of the target. The difference in the received signal amplitudes from pulse to pulse in the lobe-switching radar indicates the DOA difference from the boresight of the average of the antenna pointing directions. Limitations of lobe-switching radars occur because separate pulses are needed for the calculation of the DOA. Multipath interference affects all pulses differently, so there is likely to be a variation in the received amplitude from pulse to pulse affecting the DOA calculation. Four pulses are needed for the calculation of both azimuth and elevation, so this increases the scope for multipath effects to occur.

The use of measurements from a single pulse in monopulse radars alleviates this problem. Monopulse radars split the receiver beam into four parts (for the calculation of target azimuth and elevation angles). It measures the amplitude or phase of each separate part of the received pulse simultaneously and produces sum and difference values.

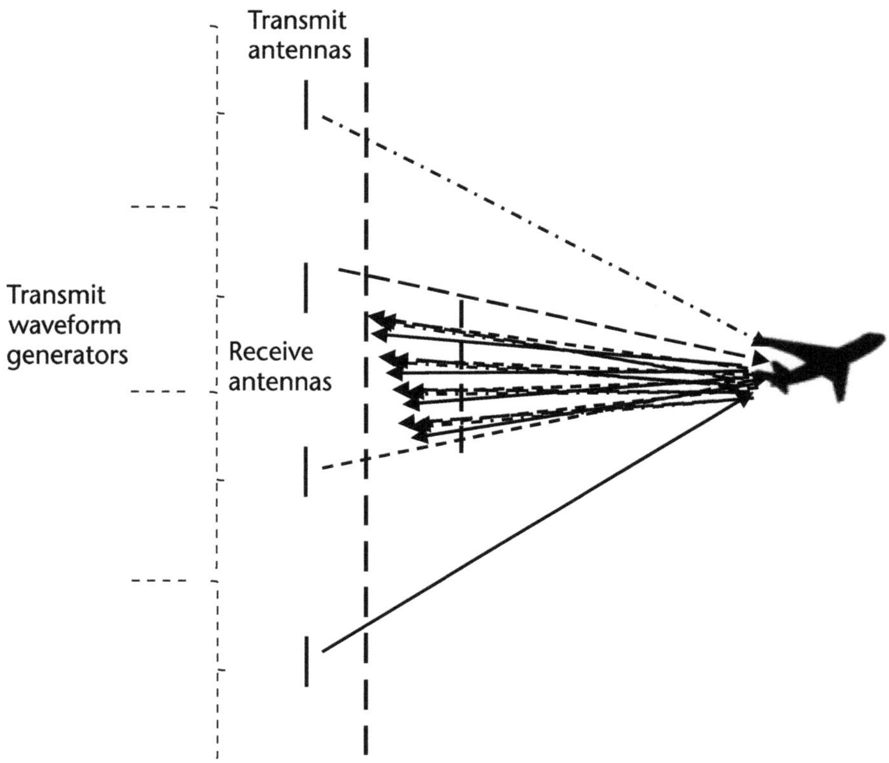

Figure 18.3 The operation of a MIMO radar.

The monopulse receiver antenna is divided up into four quadrants, as shown in Figure 18.4, and the following signals are formed from the received signals in the each quadrant [5]:

$$\textit{Difference azimuth signal},\ \Delta_{Az} = (I + IV) - (II + III) \tag{18.1}$$

$$\textit{Difference elevation signal},\ \Delta_{Ei} = (I + II) - (III + IV) \tag{18.2}$$

The values of the difference signals, Δ_{Az} and Δ_{El}, cause the antenna to move so that it is always pointing towards the target. Since this comparison is carried out during one pulse, typically a few microseconds or less, changes in target position or heading have no effect on the comparison. If the lobes are closely spaced, this method can produce a high degree of pointing accuracy within the beam. Classical conical scan systems generate a DOA accuracy of the order of 0.1°, and monopulse radars can be accurate to 0.01°.

18.1.4 Broadband Radar

This type of radar is of low-power (typically less than 40W) and emits a frequency-modulated continuous wave (FMCW) signal [6]. These radars provide significant improvements over conventional radars for ship navigation and situation awareness. They can operate at more than one range simultaneously and can track

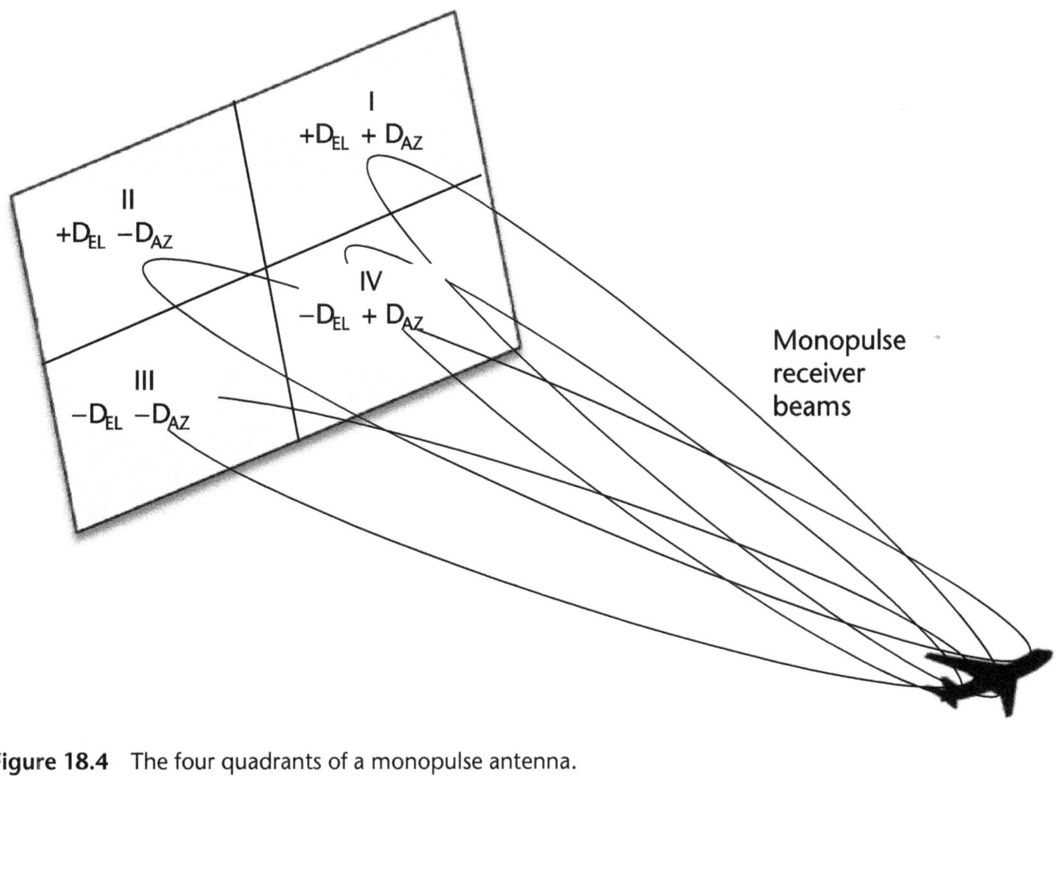

Figure 18.4 The four quadrants of a monopulse antenna.

multiple targets at ranges of up to 36 nm, with great fidelity due to beamshaping techniques.

The basis of operation of the broadband radar is that the difference in frequency between the transmitted signal and the received signal can be used to find the target range.

The FMCW carrier signal frequency f_0 is frequency modulated by frequency $f_m(t)$ such that the transmitted frequency f_t at time t is given by:

$$f_1 = f_0 + f_m(t) \tag{18.3}$$

The received signal frequency, f_r, is given by:

$$f_r = f_0 + f_m(t - t_r) \tag{18.4}$$

where t_r is the time that the signal is received back at the radar.

The frequency difference, f_d, is given by

$$f_d = f_m(t) - f_m(t - t_r) \tag{18.5}$$

The range, R, of the target is given by following equation:

$$R = cf_d / 2a \tag{18.6}$$

where $a = f_d/t_r$ and c is the speed of light.

Figure 18.5 shows the relationship between the transmitted and received frequencies and the signal time measurements.

The direction of travel of the target with respect to the radar can be found by determining whether the frequency change is positive or negative, as shown in Figure 18.6.

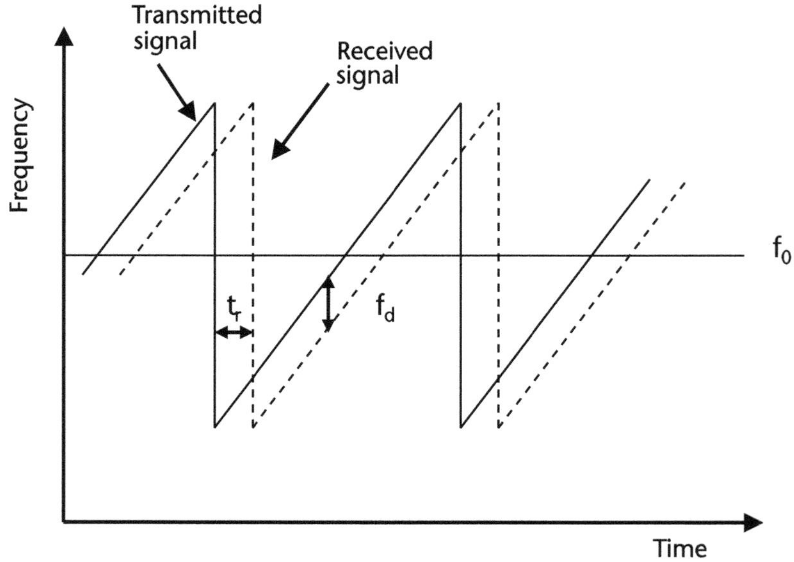

Figure 18.5 Frequency and time measurements for an FMCW signal.

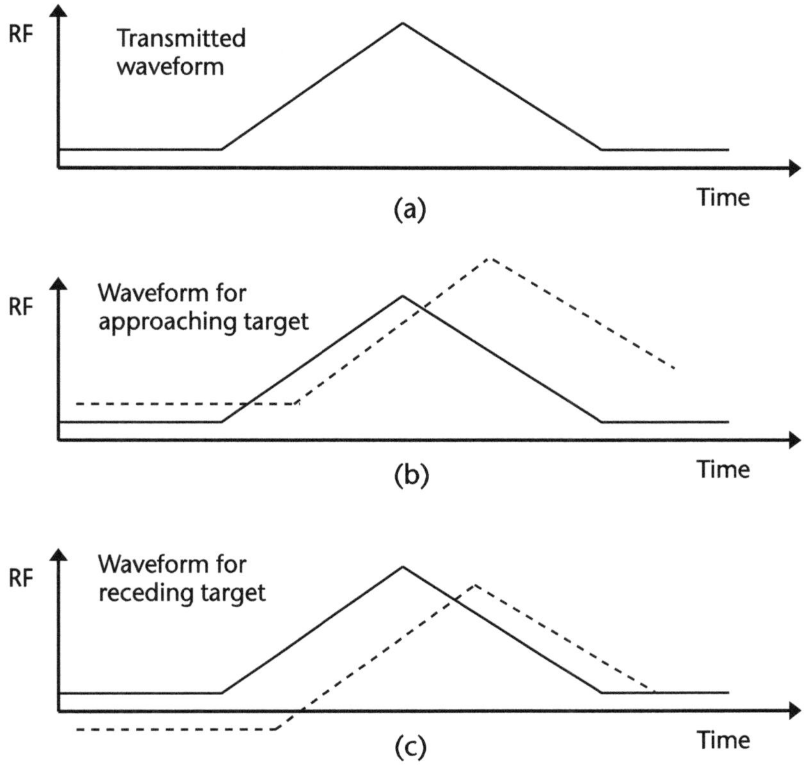

Figure 18.6 FMCW waveforms for approaching and receding targets.

Since broadband radars operate at very low power, for example, 16 dBW, the ESM system cannot detect such radars at a range of more than about a kilometer.

18.2 ESM Processing Considerations

Multipath is the main cause of poor performance of ESM systems. Some of the ideas for limiting the effects of multipath given in Chapter 17 mean that more accurate timing sources will be needed in the future. Pulses should be detected and their characteristics measured very near to the start of the pulse, preferably before multipath interference has occurred. The timing sources, such as atomic clocks, should ideally be accurate to 0.1 ns for improvements in performance to be seen. Techniques for reducing DOA errors and methods for optimizing deinterleavers should also bring performance advantages to ESM systems.

18.2.1 DOA Measurement

Poor measurement of DOA is a major concern for ESM systems. Multitracking almost always occurs due to the DOA errors themselves and due to the effect they have on the correct deinterleaving of pulse sequences. Improvements in DOA measurement are a subject of research, and some of the results are in the public domain.

Reference [7] described a compact antenna system architecture for the DOA estimation in EW environments. The single antenna modules are intended to cover specific frequency ranges This superheterodyne approach allows a very wide bandwidth. The novelty of this approach lies in a spatially shared solution for the intermediate and high-band arrays by means of an integrated architecture. The research performed by the authors of [7] shows how mixed amplitude/phase DOA algorithms allow better angular accuracy than either method used alone.

An experimental study evaluating the performance of a reconfigurable antenna element for the DOA measurement is presented in [8]. A reconfigurable antenna with directive radiation pattern has an electronic control mechanism to continuously scan its beam over space. A multiple signal classification (MUSIC)-type algorithm is used to estimate the DOA using samples collected from different configurations (scanning steps) of the antenna radiation pattern. Experiments were carried out in an anechoic chamber using an actual reconfigurable antenna for different numbers of scanning steps. The results confirm the successful estimation of the DOA even with two scanning steps.

18.2.2 Deinterleavers

Traditional deinterleaving methods are not good at producing accurate intercepts for radars where there are pulses missing from the pulse sequence. Newer deinterleavers such as GT and RCP techniques are more successful in producing the correct chains of pulses for each radar. However, these techniques may not prove to be robust when radars with similar PRI are operating in the same environment.

The RCP process in particular relies on the radar having a crystal-controlled clock that generates a stable PRI. There are many radars, particularly ship-borne ones, that generate a sinusoidal PRI variation using an AC power supply. The presence of these pulses will disrupt the operation of the RCP process.

For pulsed radars that use a crystal-controlled clock to generate stable PRI, on each cycle of the clock, the radar counts down from some initial value known as the DC, and on reaching 0, triggers a pulse. The DC may vary from pulse to pulse (e.g., to generate a staggered PRI). The primary RCP parameter is the CP, derived from the TOA of the pulses. The CP is fundamental to the operation of the RCP technique that does not use any other conventional parameter, such as DOA or RF. The RCP process analyzes an interleaved buffer of pulses to find CPs. It deinterleaves the pulses by sorting them into chains, where each PRI in a chain is an integer multiple of the same CP. The RCP process operates on buffers of pulse data, but does not correlate and track the results from buffer to buffer and does not characterize features such as agility, scan, or parameter levels. The output of the process is a table of pulses for each buffer, sorted into chains.

The CP is a parameter that may help to reduce ambiguities during the radar identification process. It may also lead to specific emitter identification (SEI) if both the radars' crystal is stable in the long term.

Most ESM systems are only able to process pulses from one buffer of data at a time, so the extraction of parameter agility types and levels is difficult. In future ESM systems, the use of pulses from a longer time-frame may be the best way to carry out deinterleaving. Parallel processing techniques that look at each longer buffer of pulses in slightly different ways could be developed. For example, a set of pulses may be filtered with a value of the DOA ± a tolerance in DOA. A second filtering process could offset the initial value of the DOA so that there are no gaps in the pulse sequence due to pulses falling just outside the original DOA range. An intelligent process could then stitch together the pulse sequences and also remove pulses that clutter the original pulse sequences from other radars.

18.3 The Future of ESM Library Matching

The primary task of the ESM system is radar identification [9]. However, to carry out this vital task, an effective radar library must be created and operated effectively, one of the most difficult functions for the ESM system. Current ESM systems have limited capability in terms of the number of radar mode lines available with the capacity of most current ESM systems being only 1,024 mode lines. Significantly increasing the number of mode lines will improve the identification capability of the ESM system. However, as there are thousands of different types of radars in existence, each having several tens of different modes of operation, the generation of the enhanced radar library should very specific to the mission that the ESM platform is to undertake.

There is very little information in the public domain about developments in radar library matching. However, there have been attempts to develop ESM library matching using techniques based on neural networks.

A neural network recognition and tracking system is proposed in [10] for the classification of radar pulses in autonomous ESM systems. Radar-type information is combined with position-specific information from active radars in a scenario. Type-specific parameters of the input pulse stream are fed to a neural network classifier trained on samples of data collected in the field. Meanwhile, a clustering algorithm is used to separate pulses from different radars according to position-specific parameters of the input pulse stream. Classifier responses corresponding to different radars are separated into tracks, one for each active radar, allowing for more accurate identification of radar types based on multiple views of radar data along each radar track. Such a what-and-where fusion strategy is motivated by a similar subdivision of labor in the brain.

A neural network is used to classify streams of pulses according to radar type using their functional parameters. Simulation results obtained with a radar pulse data set indicate that the neural network compares favorably to several other approaches. Modifications improved the classification of data that include missing input pattern components and missing training classes. The neural network was combined with a bank of Kalman filters to group pulses transmitted from different radars based on their position-specific parameters and with a module to accumulate evidence from neural network responses corresponding to the track defined for each emitter. Simulation results demonstrate that the system provides a high level of performance on complex, incomplete, and overlapping radar data.

18.4 Multiplatform ESM

Threat radars may only radiate briefly to search for targets or to track an already detected target. As a consequence, ESM systems might detect these short emissions, but not have enough time to geolocate the radar. ESM platforms need to receive multiple emissions in a time span of several minutes before a geolocation calculation can be started. Integrated air defense systems (IADS) typically radiate for tens of seconds and then are silent for a period of time, meaning that many threat radars cannot be geolocated by a single ESM platform. Combined datasets from multiple ESM platforms on the same target allow for geolocation or an improvement in the accuracy of the coarse geolocation by a single platform [11].

Tactical data links such as Link-11, Link-16, and Link-22 and Improved Data Modem (IDM) over ultrahigh frequency (UHF) or very high frequency (VHF) contain dedicated electronic warfare (EW) words in their message libraries that can be used for communication between ESM platforms, allowing cooperating ESM operations (CESMO). Through these data links, information generated onboard a single platform can be made available to every other participating platform in near real time. An individual ESM aircraft looking for a specific threat radar can be fed instantaneously with measurements from other platforms that are detecting the same radar. The geolocation of the radars becomes a cooperative activity based upon the detection capability of the entire set of ESM sensors in the area. This system of systems approach creates a virtual set of many more LOBs than can be collected by a single platform during the time the radar is active.

Triangulation requires a central function where the LOBs are collected and the position estimate is calculated. This central function could reside in a ground station or can be implemented onboard one or more of the participating ESM platforms. In the future, precision geolocation is likely to become even more important as new radars having extreme parameter agility make radar identification difficult. The location of the emissions will therefore assume great importance in the determination of the picture of the RF environment.

18.5 Autonomous/Intelligent EW Systems

The prolific development of unmanned vehicles in recent years means that ESM operations may be carried out without human intervention. However, the accuracy with which ESM systems can measure DOA and issues with deinterleaving and radar identification mean that truly autonomous ESM systems are unlikely to be viable in the near future. The human operator has always been a crucial part of the ESM system and will remain an important component of these systems for the foreseeable future.

References

[1] Robertson, S. M., "Advantages of AESA Radar Technology," *Naval Forces*, No. V, 2016.

[2] www.f-22raptor.com/af_radar.php.

[3] Bergin, J., and J. R. Guerci, *MIMO Radar: Theory and Application*, Norwood, MA: Artech House, 2018.

[4] Barton, D., and S. Sherman, Monopulse Principles and Techniques, Norwood, MA: Artech House, 2011.

[5] http://www.radartutorial.eu/06.antennas/Monopulse%20Antenna.en.html.

[6] Jankiraman, M., *FMCW Radar Design*, Norwood, MA: Artech House, 2018.

[7] Genova, J., *Electronic Warfare Signal Processing*, Norwood, MA: Artech House, 2018.

[8] Russo, I., P. Baldonero, and A. Manna, "High Resolution ESM/ELINT DOA Estimation with Super-Heterodyne Multi-Octave Antenna System," *Proc. 9th European Conference on Antennas and Propagation (EuCAP)*, 2015.

[9] Vakilian, V., H. V. Nguyen, and S. Abielmona, "Experimental Study of Direction-of-Arrival Estimation Using Reconfigurable Antennas," *Proc. of 27th IEEE Canadian Conference on Electrical and Computer Engineering*, 2014.

[10] Granger, E., et al., *A What-and-Where Fusion Neural Network for Recognition and Tracking of Multiple Radar Emitters*, Defence Research Establishment Canada Technical Report CAS/CNS-TR-2000-029, 2000.

[11] Thaens, R., "NATO Cooperative ESM Operations: Publish or Perish," http://nc3a.info/P/pres/Thaens%20-%20NATO%20Cooperative%20ESM%20Operations.pdf.

APPENDIX A

Radar Beam Pattern Creation

To create a simulation of a radar beamshape, like the one shown in Figure A.1, two parameters are needed: the 3-dB beamwidth of the radar and the first sidelobe level.

A radiation pattern of the general form $(\sin(u)/u)^2$ as shown in Figure A.1 is generated, using the following equation for u:

$$u = \left[C\pi \sin(\theta)\right] / B \tag{A.1}$$

where B is the 3-dB beamwidth in radians and C is a factor dependent on the level of the first sidelobe (–dB) and is found from Table A.1 [1].

For sidelobe levels other than –13.2 dB (which is the default for a $\sin(u)/u$ pattern), the radiation pattern equation must be modified by the factor z (from Table A.1) to give a pattern with the required sidelobe levels. The equation used is:

$$A_\theta = \left(\left[\sin\left(u^2 - z^2\right)^{1/2}\right] / \left[u^2 - z^2\right]^{1/2}\right)^2 \tag{A.2}$$

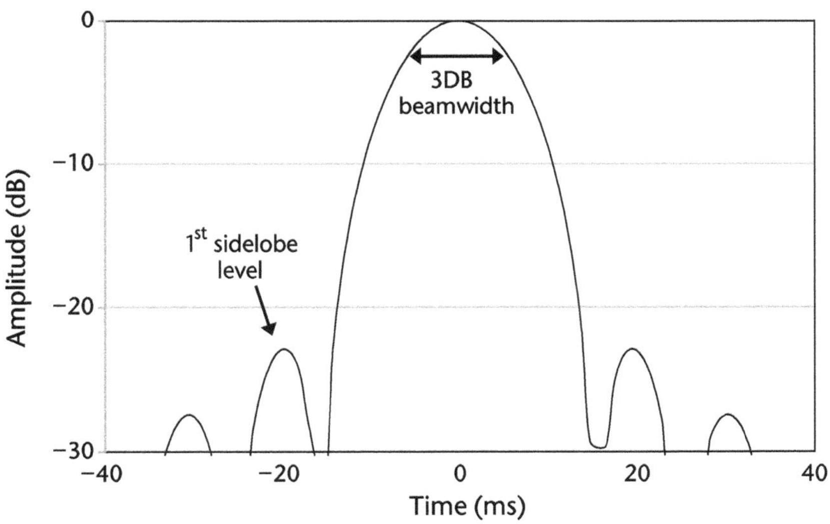

Figure A.1 Radar azimuth beamshape.

Table A.1 Lookup Table for Sidelobe Control

Sidelobe	*z*	C	*Sidelobe*	*z*	C
13.2	0	0.8850	28	3.692	1.1664
15	1.118	0.9230	29	3.852	1.1832
16	1.35	0.9432	30	4.009	1.2000
17	1.59	0.9634	31	4.159	1.2156
18	1.83	0.9836	32	4.309	1.2312
19	2.07	1.0038	33	4.459	1.2468
20	2.32	1.0240	34	4.609	1.2624
21	2.49	1.0424	35	4.755	1.2780
22	2.68	1.0608	36	4.891	1.2926
23	2.85	1.0792	37	5.036	1.3072
24	3.04	1.0976	38	5.181	1.3218
25	3.214	1.1160	39	5.326	1.3364
26	3.373	1.1328	40	5.471	1.3510
27	3.532	1.1496			

Problems arise when $u^2 < z^2$, as the square root of a negative number is needed in the calculation of A_θ.

This is overcome in the following way. Let

$$Let\, u^2 - z^2 = -y^2 \tag{A.3}$$

$$\sqrt{\left(u^2 - z^2\right)} = \sqrt{\left(-y_2\right)} = \sqrt{(-1)y^2} = i\sqrt{y^2} \tag{A.4}$$

Since sin $(i\ y) = i$ sinh (y), it is possible to remove i from the calculation of A_θ:

$$A_\theta = \left[\sin\left(i\sqrt{y^2}\right)^2\right] / \, i\sqrt{y^2} = \left[i\sinh(y)^2\right] / \, iy = \left[\sinh(y)^2\right] / \, y \tag{A.5}$$

Therefore, when $u > z$ use

$$A_\theta = \sin(y)^2 \,/\, y \tag{A.6}$$

when $u < z$ use

$$A_\theta = \sinh(y)^2 \,/\, y \tag{A.7}$$

where

$$y = \sqrt{\left(u^2 - z^2\right)} \tag{A.8}$$

The radiation pattern is altered in two ways by the equation providing the amplitude, A_θ. In addition to the suppression of the sidelobes, the main peak is enhanced in power. The amplitude must be normalized by subtracting the amplitude at the peak of the beam to reduce the effects of the sidelobe generation to a zero gain. This is carried out as follows:

$$A = A_0 - A_\theta \tag{A.9}$$

where A_0 is the calculated power at the peak of the beam (i.e., at $\theta = 0$).

The final stage in the calculation of the radar peak amplitude is the correction for the peak amplitude of the radar at the particular range (r).

The ERP of the radar is given in dBW and is converted into watts as the first stage in this calculation:

$$Erp_{\_W} = \mathrm{a}\log_{10}\left(erp_{\_dbW}\right)/10 = 10^{(erp/10)} \tag{A.10}$$

$$A_{peak\_W} = erp_{\_W} / \left(4\pi r^2\right) \tag{A.11}$$

$$A_{peak} = 10\log_{10}\left(A_{peak\_W}\right) \tag{A.12}$$

The amplitude, A_θ, seen at range (r) from a pulse at azimuth angle θ from the peak of the beam is:

$$A_\theta = A_{peak} - A \tag{A.13}$$

The azimuth positions of the pulses on the amplitude profile of the radar beam, as shown in Figure A.2 depend on the pulse repetition interval (PRI) and the scan period (S).

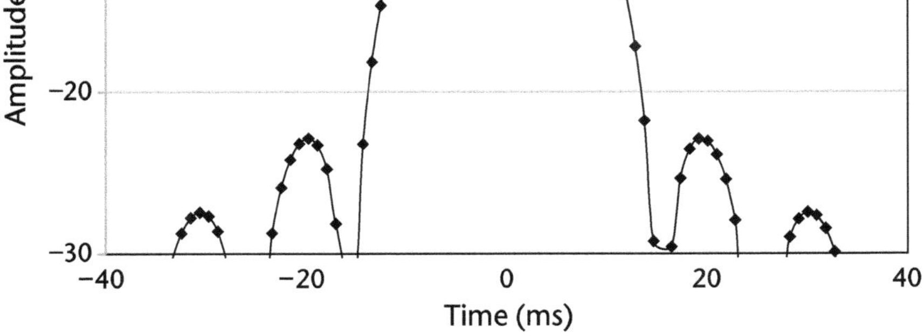

Figure A.2 Radar beamshape with pulse positions.

Equation (A.14) gives the azimuth difference in degrees (θ_n) from the peak of the beam for pulse number n where $n = S \times PRI$ for a single scan so

$$\theta_n = (360 \times n) / (S \times PRI) \tag{A.14}$$

For example, a radar with a fixed PRI of 1 ms and a scan period of 2 seconds emits 2,000 pulses per scan. The number of degrees difference for each pulse is (360/2,000) = 0.18°.

A pulse at position 5 from the peak of the beam is at azimuth (θ_5) = (360 × 5)/(2 × 1,000) = 0.9° from the peak of the beam.

The amplitude (A_n) for pulse (n) at θ_n degrees can then be calculated by following the procedure given above.

Reference

[1] Skolnik, M. I., *Radar Handbook*, First Edition, New York: McGraw-Hill, 1970.

APPENDIX B

Reflection Coefficients

A field scattered by a rough surface is the sum of two components, due to specular and diffuse reflection mechanisms. Specular reflections are directional and obey the laws of classical optics. They are phase coherent and their fluctuations have relatively small amplitude. Diffuse scattering has little directivity and is phase incoherent. The effect of diffuse scattering is sufficiently small to be neglected in the calculation of multipath effects as it causes small unpredictable rapidly changing fluctuations in amplitude of the composite waveform.

The specular reflection coefficient depends on the nature of the reflecting surface, the grazing angle of the incident wave and the scattering coefficient that takes account of the irregularities of the reflecting surface. In the case of scattering from the ground or the surface of the sea, the curvature of the earth is also be taken into account. The general equation defining the specular reflection coefficient (R_s) of a rough earth is:

$$R_s = \rho_s D R_o \tag{B.1}$$

where R_o is the reflection coefficient of a smooth and plane Earth, D is the divergence coefficient caused by the curvature of the earth, and ρ_s is the scattering coefficient taking account of the irregularities of the Earth's surface.

The nature of the reflecting surface is taken into account in the following equations, giving the reflection coefficient for a smooth surface for vertical (R_o^+) and horizontal (R_o^-) polarizations, respectively:

$$R_o^+ = \left[Y^2 \sin\gamma - \sqrt{\left(Y^2 - \cos^2\gamma\right)}\right] / \left[Y^2 \sin\gamma + \sqrt{\left(Y^2 - \cos^2\gamma\right)}\right] \tag{B.2}$$

$$R_o^- = \left[\sin Y - \sqrt{\left(Y^2 - \cos^2\gamma\right)}\right] / \left[\sin\gamma + \sqrt{\left(Y^2 - \cos^2\gamma\right)}\right] \tag{B.3}$$

where γ is the grazing angle of the incident wave.

Y is the normalized admittance of the medium, such that:

$$Y = \varepsilon_{rc} / \mu_{rc} \tag{B.4}$$

ε_{rc} is the relative dielectric constant which is calculated as follows:

$$\varepsilon_{rc} = (\varepsilon / \varepsilon_0) - 60\lambda\sigma \tag{B.5}$$

where ε is the dielectric constant, ε_0 is the dielectric constant of free space, λ is the wavelength in meters, and σ is the conductivity in mho/m.

μ_{rc} is the relative magnetic permeability of the medium and of free space, but in the calculations performed in a multipath simulation it can be assumed that $\mu_{rc} = 1$.

Results quoted in [1] give the values of $\varepsilon/\varepsilon_o$ for dry ground from about 2 in the case of very dry ground to 24 for very wet ground. For the same two cases σ varies from 10^{-4} to several units of 10^{-3} mho/m. For the surface of the sea the values are: $\varepsilon = 80$ and $\sigma = 4$ mho/m.

Figure B.1 shows the reflection coefficient of a perfectly plane Earth and of a very smooth sea for the values of $\varepsilon/\varepsilon_o$ and σ quoted above for values of λ for radars with RF from 1 to 18 GHz. The maximum and minimum wavelengths of interest are 30 cm and 2 cm, respectively, as shown in Table B.1.

The scattering coefficient $<\rho_s>$ is a function of the standard deviation of the normal distribution of heights of irregularities on the surface Δh, of the grazing angle, γ, and of the wavelength, λ, such that:

$$<|\rho_s|^2> = e^{-(\Delta\Phi)2} \tag{B.6}$$

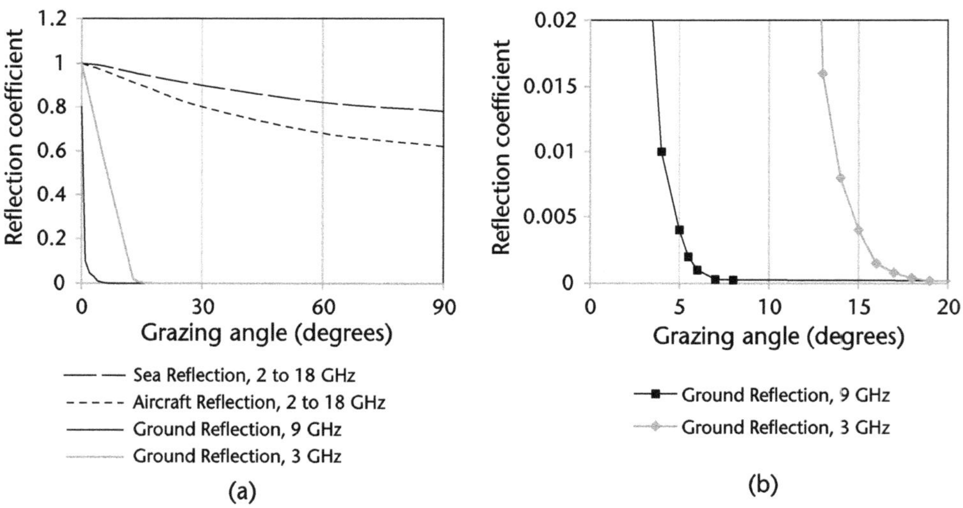

Figure B.1 Reflection coefficients for rough ground, smooth sea, and aircraft surfaces.

Table B.1 Radar Wavelengths in the range 1 to 18 GHz

RF (GHz)	*λ (m)*
1	0.30
3	0.10
9	0.03
18	0.02

where

$$A\Phi = (4\pi\Delta h \sin\gamma) / \lambda \tag{B.7}$$

Figure B.2 shows the values of $<\rho_s>$ expected for various grazing angles. This graph shows that for all except small grazing angles, the effect of the scattering coefficient is to reduce the reflection coefficient of the surface to a small value.

The final part of the equation, the value of D, the Divergence coefficient, taking account of the curvature of the earth is calculated as follows:

$$D = \sqrt{\left[1 + \left(2r_1 r_2 / a(r_1 + r_2)\sin\gamma\right)\right]} \tag{B.8}$$

where $(r_1 + r_2)$ is the reflected path length in meters, γ is the grazing angle, and a is the mean radius of curvature of the Earth = 6,367,445m.

Table B.2 lists the values of $\varepsilon/\varepsilon_o$, σ, and Δ_h for different types of reflection surface. It also indicates which of the three components in the calculation of R_s are used for each surface type. For example, the aircraft surface reflection is not subject to D, the component due to the Earth's curvature. For the perfectly smooth reflector, neither the D nor the ρ_s components are used. Each unused component is set to a value of 1 in the calculation of R_s.

The reflection coefficient for the perfect reflector (surface = 7) is set equal to 1.

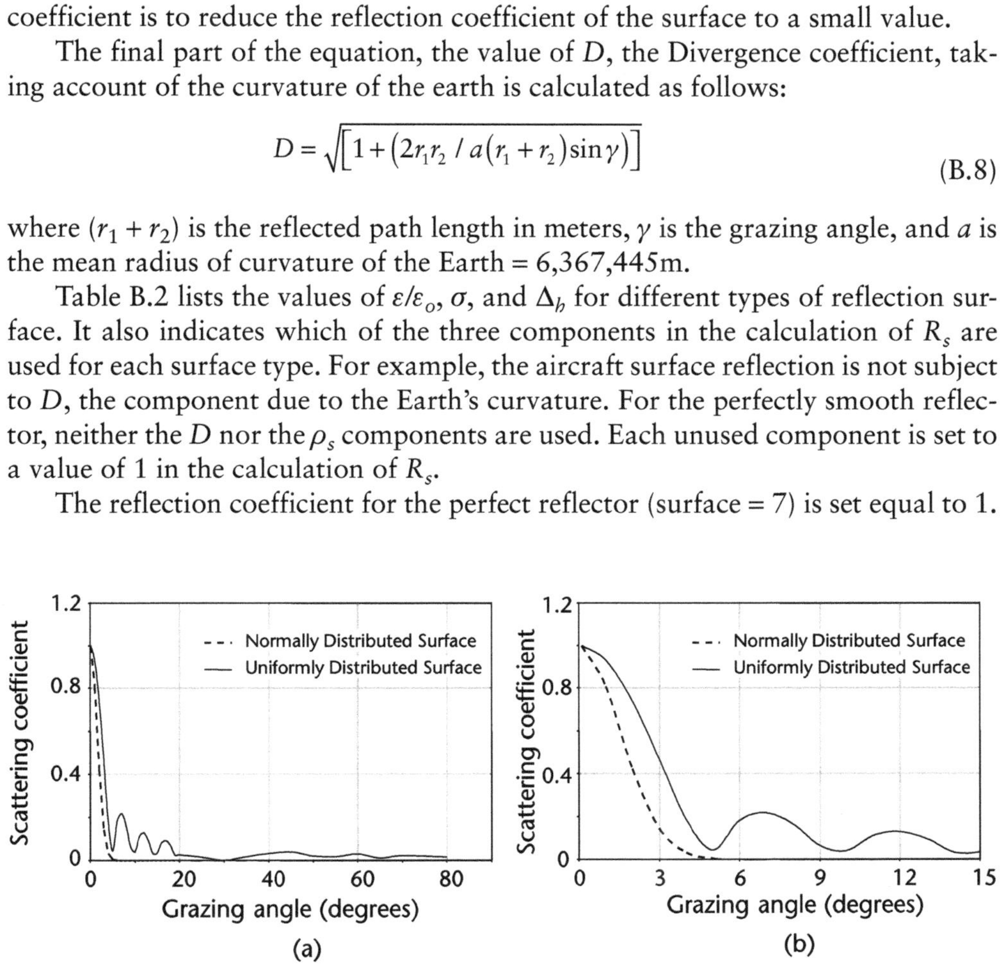

Figure B.2 Scattering coefficient distributions.

Table B.2 Parameters of Reflecting Surfaces Available in the Multipath Model

No.	*Surface Type*	$\varepsilon/\varepsilon_o$	σ *(mho/m)*	Δ_h *(m)*	*Components in Calculation*		
					R_o	D	ρ_s
1	Aircraft	2	0.0005	0.0005	✓		✓
2	Sea, smooth	80	4	0.2	✓		✓
3	Sea, rough	40	4	1	✓	✓	
4	Ground	10	0.005	0.02	✓	✓	✓
5	Chimney stack	4	0.001	0.005	✓	✓	✓
6	Smooth surface	10	0.01	—	✓		
7	Perfect reflector	—	—	—			

Reference

[1] Beckmann, P., and A. Spizzichino, *The Scattering of Electromagnetic Waves from Rough Surfaces*, Oxford, UK: Pergamon Press, 1963.

Acronyms and Abbreviations

2G Second generation mobile network

3G Third generation mobile network

4G Fourth generation mobile network

5G Fifth generation mobile network

A/D Analog to digital

ADRS Advanced dynamic RF simulator

AESA Active electronically scanned array

AI Airborne intercept

AIS Automatic identification system

AOA Angle of arrival

AOC Association of Old Crows

APAR Active phased array radar

ATC Air traffic control

AWACS Airborne warning and control system

BI Beacon interrogator

BIT Built-in test

BN Beacon navigation

CESMO Cooperating ESM operations

CP Clock period

CS Coastal surveillance

CW Continuous wave

dB Decibel

dBmi Decibels milliwatts isotropic

dBW Decibels watts

dBW/m^2 Decibels watts/square metre

DC Down count

DF Direction finding

DOA Direction of arrival

E East

EA Electronic attack

ECM Electronic countermeasures

ELINT Electronic intelligence

EM Electromagnetic

EPL Emitter program library

ERP Effective radiated power

ESM Electronic support measures

ESM Electronic surveillance measures

EW Electronic warfare

FC Fire control

FER Front end receiver

FF Fine frequency

FMCW Frequency modulated continuous wave

FMOP Frequency modulation on pulse

ft Feet

Gbytes Gigabytes

GHz Gigahertz

GNSS Global navigation satellite system

GPS Global Positioning System

GSM Global System for Mobile communications

GT Graph theory

HADF High-accuracy direction finding

HF High frequency

HS Harbor surveillance

Hz Hertz

IADS Integrated air defence system

ICED Incremental clustering for evolving data

IDM Improved data modem

IF Intermediate frequency

IFF Identification friend or foe

IFM Instantaneous frequency measurement

ILA Inter-LOB angle

kbytes Kilobytes

km Kilometer

LO Local oscillator

LOB Line of bearing

LPI Low probability of intercept

LTE Long-term evolution

LWT Link weight

m Meter

Mbytes Megabytes

MG Missile guidance

MH Missile homing

mho/m Milli-ohm/ meter

MHz Megahertz

MIMO Multiple input multiple output

ms Millisecond

μs Microsecond

MUSIC Multiple signal classification

mW Milliwatt

N North

NATO North Atlantic Treaty Organization

nm Nautical mile

NNE North northeast

ns Nanosecond

OT&E Operational test and evaluation

pdf Probability density function

PDW Pulse descriptor word

PFM Preflight message

PGRI Pulse group repetition interval

Pol. Polarization

ppm Pulse position modulation

pps Pulses per second

PRF Pulse repetition frequency

PRI Pulse repetition interval

PRI disp. PRI dispersion

PRI tol. PRI tolerance

PRIMax Maximum pulse repetition interval

PRIMin Minimum pulse repetition interval

PW Pulse width

PWMax Maximum pulse width

PWmin Minimum pulse width

PWT Parameter weight
RCP Radar clock period
RCS Radar cross section
RD Radial distance
RDF Radio direction finding
RF Radar frequency
RF disp. RF dispersion
RF tol. RF tolerance
RFmax Maximum radar frequency
RFmin Minimum radar frequency
rms Root mean square
RPM Revolutions per minute
RWR Radar warning receiver
S South
ScanMax Maximum scan period
ScanMin Minimum scan period
SEI Specific emitter identification
SHipNav Ship navigation radar
SIL Systems integration laboratory
SSW South southwest
superhet Superheterodyne receiver
TA Target acquisition
TCAS Traffic collision avoidance system
TDOA Time difference of arrival
TI Target illuminator
TOA Time of arrival
TOAD Time of arrival difference
TOADH Time of arrival difference histogram
TRM Transmit receive module
tslp Time since last pulse
TT Target tracking
TWS Track while scan
UHF Ultrahigh frequency
UMTS Universal mobile telecommunications system
VHF Very high frequency
W Watt or West

About the Author

Sue Robertson has worked in the field of electronic warfare for more than 25 years and has extensive experience of data analysis and system optimization for airborne and ship ESM systems. She has developed flight test programs and has written hundreds of reports on ESM performance issues. She has developed software for ESM testing and techniques for extracting data from flight trials for the population of ELINT databases. She has made recommendations on system improvements for a variety of airborne platforms backed up by original research into electromagnetic propagation issues.

Dr. Robertson has a Ph.D. in experimental particle physics from University College London and an M.B.A. from City University, London. She is currently working for EW Defence Ltd. in the United Kingdom. In 2016, she was elected to serve on the board of directors of the Association of Old Crows (AOC) as international regional director with responsibility for Europe, Middle East, India, and Africa.

Index

R

S

The Artech House Electronic Warfare Library

Dr. Joseph R. Guerci, Series Editor

Activity-Based Intelligence: Principles and Applications, Patrick Biltgen and Stephen Ryan

Advances in Statistical Multisource-Multitarget Information Fusion, Ronald P. S. Mahler

Antenna Systems and Electronic Warfare Applications, Richard A. Poisel

Electronic Intelligence: The Analysis of Radar Signals, Second Edition, Richard G. Wiley

Electronic Warfare for the Digitized Battlefield, Michael R. Frater and Michael Ryan

Electronic Warfare in the Information Age, D. Curtis Schleher

Electronic Warfare Receivers and Receiving Systems, Richard A. Poisel

Electronic Warfare Signal Processing, James Genova

Electronic Warfare Target Location Methods, Richard A. Poisel

EW 101: A First Course in Electronic Warfare, David L. Adamy

EW 103: Tactical Battlefield Communications Electronic Warfare, David L. Adamy

EW 104: EW Against a New Generation of Threats, David L. Adamy

Foundations of Communications Electronic Warfare, Richard A. Poisel

High-Level Data Fusion, Subrata Das

Human-Centered Information Fusion, David L. Hall and John M. Jordan

Information Warfare and Organizational Decision-Making, Alexander Kott, editor

Information Warfare and Electronic Warfare Systems, Richard A. Poisel

Information Warfare Principles and Operations, Edward Waltz

Introduction to Communication Electronic Warfare Systems, Richard A. Poisel

Introduction to Electronic Defense Systems, 3rd Edition, Filippo Neri

Introduction to Modern EW Systems, 2nd Edition, Andrea De Martino

Knowledge Management in the Intelligence Enterprise, Edward Waltz

Mathematical Techniques in Multisensor Data Fusion, Second Edition, David L. Hall and Sonya A. H. McMullen

Military Communications in the Future Battlefield, Marko Suojanen

www.ingramcontent.com/pod-product-compliance
Ingram Content Group UK Ltd.
Pitfield, Milton Keynes, MK11 3LW, UK
UKHW050947290926
15470UKWH00005B/1699

9 781630 815288